Kinematics and Robot Design VI, KaRD2023

Kinematics and Robot Design VI, KaRD2023

Editor

Raffaele Di Gregorio

MDPI

Basel • Beijing • Wuhan • Barcelona • Belgrade • Novi Sad • Cluj • Manchester

Editor
Raffaele Di Gregorio
University of Ferrara
Ferrara
Italy

Editorial Office
MDPI
St. Alban-Anlage 66
4052 Basel, Switzerland

This is a reprint of articles from the Special Issue published online in the open access journal *Robotics* (ISSN 2218-6581) (available at: https://www.mdpi.com/journal/robotics/special_issues/3E916685J9).

For citation purposes, cite each article independently as indicated on the article page online and as indicated below:

Lastname, A.A.; Lastname, B.B. Article Title. *Journal Name* **Year**, *Volume Number*, Page Range.

ISBN 978-3-7258-1207-3 (Hbk)
ISBN 978-3-7258-1208-0 (PDF)
doi.org/10.3390/books978-3-7258-1208-0

Contents

About the Editor

Raffaele Di Gregorio

Raffaele Di Gregorio is currently a Full Professor of Machine Mechanics at the Engineering Dept. of Ferrara University, Italy. He received a M.Sc. degree in nuclear engineering, a M.Sc. degree in mechanical engineering, and a M.Sc. degree in automotive engineering from the Polytechnic University of Turin in 1982, 1985, and 1988, respectively, and a Ph.D. degree in applied mechanics from the University of Bologna in 1992. In 1983, he spent one year at the FIAT Research Center, Orbassano (IT), with an ATA grant. Since 1984 till 1992, he was an officer of the Italian Army's Technical Corp at the Military School of Turin and at STAVECO of Bologna. In 1993, he taught at ITIS O. Belluzzi of Bologna. Since 1994, he has joined the Engineering Dept. of Ferrara University.

His research interests include the kinematics and dynamics of mechanisms and machines, biomechanics, robotics, vibration mechanics, and vehicle mechanics.

He has been an "observer/member" of IFToMM TC "Computational Kinematics" since 2007 and a "member" of IFToMM PC "Standardization of Terminology" since 2009. He is an ASME member and was a "general member" of the ASME Mechanisms and Robotics Committee since 2007 till 2012; since then, he has continued to serve this Committee as a past member.

He is the author/co-author of more than 90 technical papers published in refereed international journals and as many papers published in conference proceedings, and actively collaborates with many international journals in the field.

Preface

Kinematics is central to nearly all of the design aspects of robotic/mechatronic systems.

The series of Special Issues on "Kinematics and Robot Design" (KaRD) offers, to the vast community of scientists of this field, a virtual agora where new ideas and new applications can be presented and discussed.

The KaRD series started in 2018 and, in 2023, reached its sixth edition (KaRD2023). KaRD2023 has collected fifteen submissions and, after a rigorous review process, accepted and published ten papers.

This reprint collects KaRD2023's published papers.

Raffaele Di Gregorio
Editor

robotics

MDPI

Editorial

Special Issue Kinematics and Robot Design VI, KaRD2023

Raffaele Di Gregorio

Laboratory of Mechatronics and Virtual Prototyping (LaMaViP), Department of Engineering, University of Ferrara, 44122 Ferrara, Italy; raffaele.digregorio@unife.it

What would our concept of life be without motion? Motion is what makes life visible in living beings, and motion description/generation (i.e., kinematics) somehow describes living objects. This perspective guides many scientists working in the field of kinematics, and kinematics in robotics and mechatronics best interprets this.

Robots, first appearing in the form of industrial machines, as an evolution of automatic machines, have today entered into the everyday lives of people in the form of service robots, guiding important changes in lifestyle. Such changes are reshaping society in many fields, and a wide-ranging community of scientists is involved in this social evolution.

Kinematics is central to nearly all the design aspects of robotic/mechatronic systems. Topics, such as the analysis and synthesis of mechanisms; robot modeling and simulation; robot control, mobility and singularity analysis; performance measures; accuracy analysis; path planning and obstacle avoidance; collaborative robotics; novel manipulator architectures; metamorphic mechanisms; compliant mechanism analysis and synthesis; micro/nano-manipulator design; origami-based robotics; medical and rehabilitation robotics; and bioinspired robotics, deal with kinematics. All these topics have a deep social impact and help to delineate future perspectives on human welfare, which makes kinematics a thriving research field with theoretical and applicative subjects.

The series of Special Issues on "Kinematics and Robot Design" (KaRD) offers, to the vast community of scientists involved in these topics, a virtual setting, where new ideas and new applications can be presented and discussed by implementing the "open science" principles. Open science is the way the scientific community shows the rest of the world the strength of cooperation among individuals and nations in attaining better conditions for humankind. In addition, through open science practice, the same community stresses that the highest level of cooperation is attainable only in a peaceful world, where individuals and nations respect each other and refuse wars.

The KaRD series began in 2018, and in 2023, it published its sixth edition (KaRD2023). KaRD2023 collected fifteen submissions and, after a rigorous review process, accepted and published ten papers.

The ten papers published in the Special Issue "Kinematics and Robot Design VI, KaRD2023" address both theoretical and applicative aspects of the mechanism synthesis [1,2], geometric error effects [3], and instantaneous kinematics [4] of industrial robots, medical robotics [5,6], the path planning of mobile robots [7] and under-actuated robots [8], and robot control [9,10].

Citation: Di Gregorio, R. Special Issue Kinematics and Robot Design VI, KaRD2023. *Robotics* **2024**, *13*, 70. https://doi.org/10.3390/robotics13050070

Received: 20 April 2024
Accepted: 28 April 2024
Published: 1 May 2024

Acknowledgments: The members of KaRD series' Scientific Committee deserve to be cited and acknowledged. They are Massimo Callegari (Italy), Juan Antonio Carretero (Canada), Yan Chen (China), Daniel Condurache (Romania), Xilun Ding (China), Mary Frecker (USA), Clement Gosselin (Canada), Just Herder (The Netherlands), Larry Howell (USA), Xianwen Kong (UK), Pierre Larochelle (USA), Giovanni Legnani (Italy), Haitao Liu (China), Daniel Martins (Brazil), Andreas Mueller (Austria), Andrew Murray (USA), Leila Notash (Canada), Matteo Palpacelli (Italy), Alba Perez (Spain), Victor Petuya (Spain), José Maria Rico Martinez (Mexico), Nina Robson (USA), Jon M. Selig (UK), Bruno Siciliano (Italy), Tao Sun (China), Yukio Takeda, Tokyo Institute of Technology (Japan), Federico Thomas (Spain), and Volkert Van Der Wijk (The Netherlands).

Conflicts of Interest: The author declares no conflicts of interest.

References

1. Xu, T.; Myszka, D.H.; Murray, A.P. A Bi-Invariant Approach to Approximate Motion Synthesis of Planar Four-Bar Linkage. *Robotics* **2024**, *13*, 13. [CrossRef]
2. Antonov, A.; Fomin, A.; Glazunov, V.; Petelin, D.; Filippov, G. Type Synthesis of 5-DOF Hybrid (Parallel-Serial) Manipulators Designed from Open Kinematic Chains. *Robotics* **2023**, *12*, 98. [CrossRef]
3. Wang, L.; Li, M.; Yu, G. A Novel Error Sensitivity Analysis Method for a Parallel Spindle Head. *Robotics* **2023**, *12*, 129. [CrossRef]
4. Simas, H.; Di Gregorio, R.; Simoni, R. Instantaneous Kinematics and Free-from-Singularity Workspace of 3-XXRRU Parallel Manipulators. *Robotics* **2023**, *12*, 138. [CrossRef]
5. Vargas-Riaño, J.; Agudelo-Varela, O.; Valera, A. Applying Screw Theory to Design the Turmell-Bot: A Cable-Driven, Reconfigurable Ankle Rehabilitation Parallel Robot. *Robotics* **2023**, *12*, 154. [CrossRef]
6. Bajrami, A.; Palpacelli, M.-C.; Carbonari, L.; Costa, D. Posture Optimization of the TIAGo Highly-Redundant Robot for Grasping Operation. *Robotics* **2024**, *13*, 56. [CrossRef]
7. Badia Torres, J.; Perez Gracia, A.; Domenech-Mestres, C. Driving Strategies for Omnidirectional Mobile Robots with Offset Differential Wheels. *Robotics* **2024**, *13*, 19. [CrossRef]
8. Tonan, M.; Bottin, M.; Doria, A.; Rosati, G. Motion Planning of Differentially Flat Planar Underactuated Robots. *Robotics* **2024**, *13*, 57. [CrossRef]
9. Khan, H.; Khan, S.A.; Lee, M.C.; Ghafoor, U.; Gillani, F.; Shah, U.H. DDPG-Based Adaptive Sliding Mode Control with Extended State Observer for Multibody Robot Systems. *Robotics* **2023**, *12*, 161. [CrossRef]
10. Righettini, P.; Strada, R.; Cortinovis, F.; Tabaldi, F.; Santinelli, J.; Ginammi, A. An Experimental Investigation of the Dynamic Performances of a High Speed 4-DOF 5R Parallel Robot Using Inverse Dynamics Control. *Robotics* **2024**, *13*, 54. [CrossRef]

robotics

MDPI

Article

A Bi-Invariant Approach to Approximate Motion Synthesis of Planar Four-Bar Linkage

Tianze Xu, David H. Myszka and Andrew P. Murray *

Department of Mechanical and Aerospace Engineering, University of Dayton, 300 College Park, Dayton, OH 45469, USA; xut004@udayton.edu (T.X.); dmyszka@udayton.edu (D.H.M.)
* Correspondence: murray@udayton.edu

Abstract: This paper presents a planar four-bar approximate motion synthesis technique that uses only pole locations. Synthesis for rigid-body guidance determines the linkage dimensions that guide a body in a desired manner. The desired motion is specified with task positions including a location and orientation angle. Approximation motion synthesis is necessary when an exact match to the task positions cannot be obtained. A linkage that achieves the task positions as closely as possible becomes desired. Structural error refers to the deviations between the task positions and the linkage's generated positions. A challenge in approximate motion synthesis is that structural error involves metrics that include location and orientation. A best-fit solution is not evident because the structural error is based on an objective function that combines the location and orientation. Such solutions lack bi-invariance because a change in reference for the motion changes the values of the metric. This work uses only displacement poles, described solely by their coordinates, as they sufficiently characterize the relative task positions. The optimization seeks to minimize the distance between the poles of the task positions and the poles of the generated positions. The use of poles results in a bi-invariant statement of the problem.

Keywords: linkage synthesis; approximate motion; bi-invariance

Citation: Xu, T.; Myszka, D.H.; Murray, A.P. A Bi-Invariant Approach to Approximate Motion Synthesis of Planar Four-Bar Linkage. *Robotics* **2024**, *13*, 13. https://doi.org/10.3390/robotics13010013

Academic Editor: Raffaele Di Gregorio

Received: 8 December 2023
Revised: 5 January 2024
Accepted: 9 January 2024
Published: 10 January 2024

1. Introduction

Kinematic synthesis of linkages is a process to determine the arrangement of links, and their corresponding dimensions, to achieve a desired motion. Rigid-body guidance is a category of kinematic synthesis that is concerned with placing a link in a specified set of task positions. The task positions include the location and orientation of the link at various positions of the desired linkage movement. Generated positions are the actual location and orientation that a link achieves as the linkage moves through its range of motion. Linkage motion that exactly generates the desired task positions is only possible for a finite number of positions. A four-bar linkage is capable of generating positions that exactly match a maximum of five task positions [1]. When greater than five task positions are specified, numerical optimization techniques must be used to determine linkage parameters that best fit the prescribed task positions [2]. The difference between the prescribed task position and the position generated by the linkage is termed structural error.

Levitskii [3] is recognized as the first to address approximate linkage synthesis by using an optimization that involved the synthesis equations and minimized the sum of the squares of the structural error. Sarkisyan et al. [4] improved the optimization by combining the algebraic least-squares method with circles and lines since they represent the geometric constraints associated with linkage joints. Ravani and Roth [5] developed an approach that represents the three parameters of a planar displacement as points on a spatial curve. Using this kinematic mapping, which is essentially a quaternion, the synthesis task becomes a curve-fitting problem. Their initial framework for motion synthesis has been applied to a wide range of planar mechanisms, and several authors have improved upon these techniques since this work was published. One such improvement was by Larochelle [6], introducing a method that uses a parameterized representation of a linkage's workspace, called its constraint manifold. While still maintaining the advantages of the quaternion approach by Ravani and Roth, this optimization is posed to minimize the distance between the linkage constraint manifold and the specified task positions. Ge et al. [7] formulated approximate motion optimization by using a more general statement of the constraint manifolds, such that the coefficients become linear. The method views determining the motion trajectories of four-bar mechanisms as fitting a pencil of quadrics to the given motion data. By doing so, Ge et al. are able to develop a unified framework for motion synthesis that is applied to a wide range of planar mechanisms [8]. Gogate and Matekar [9] addressed approximate motion synthesis using evolutionary methods. Three alternative error functions were used for comparison. One of their alternatives includes the displacement pole position error in addition to the normalized angular displacement error. They noted that the location error of the target and generated position is exposed by the error in the poles.

Invariance generally refers to a property that is unaffected under a particular transformation, whereas bi-invariance implies that the property remains unchanged after two operations. The methods described above involve a metric for measuring structural error that combines a distance between locations and orientations of a link. The solution is inherently dependent on the manner in which these quantities are combined. That stated, many of the techniques above are either left or right invariant. In the context of Lie groups, a bi-invariant metric is unaffected after applying a left or right translation to the elements within the group [10]. Planar motion described in SE(2) is a Lie group, and metrics in SE(2) are not bi-invariant [11]. Elements of SE(2) are also called planar homogeneous transforms. Di Gregorio [12] provides an overview of error metrics on distances between spatial rigid-body positions and their representations by elements of SE(3), the special Euclidean group of spatial displacements, which includes SE(2) as a subset. Liu and Yang [13] present an approximate synthesis optimization using polynomial equations based on geometric conditions of a single dyad that leads to a bi-invariant solution, yet assumes optimality of the linkage after joining dyads. The contribution of this paper is a linkage synthesis method that solely uses the location of the displacement poles. The displacement poles properly describe the relative task positions, with the relative locations in a pole triangle ensuring the relative orientations of positions [1]. The presented technique is advantageous because the relative positions of the poles do not change if the coordinate axes used to specify the problem are changed. As such, the error metric has the potential to result in a more consistent optimization process.

The remainder of the paper is organized as follows. Various concepts that are used in rigid-body guidance are outlined in Section 2. Section 3 describes the optimization details. Details about the claims of bi-invariance are presented in Section 4. Section 5 presents examples and comparisons to several of the alternative methods available in the literature. Conclusions are provided in Section 6.

2. Concepts Used in Rigid-Body Guidance

Figure 1a shows three task positions with their corresponding moving reference frames, M_i. In planar, rigid-body guidance, a task position is defined by a location vector $\mathbf{d}_i = [d_{i_x}, d_{i_y}]^\mathsf{T}$ and an orientation angle θ_i, both measured relative to a fixed frame F.

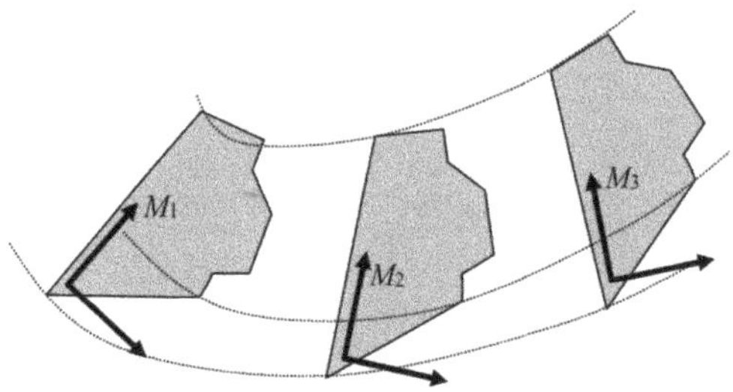

(**a**) Three task positions

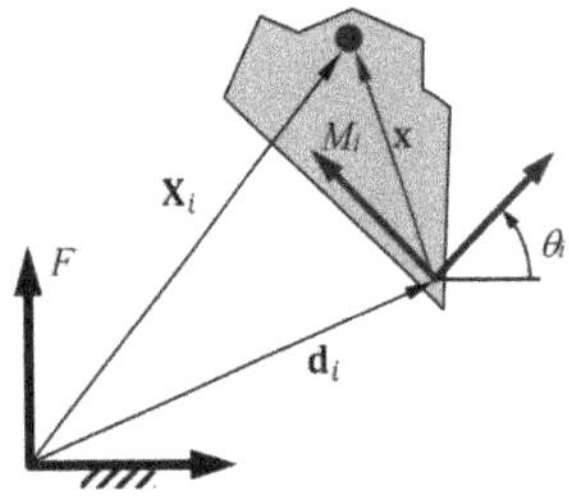

(**b**) Location of a general point on a body

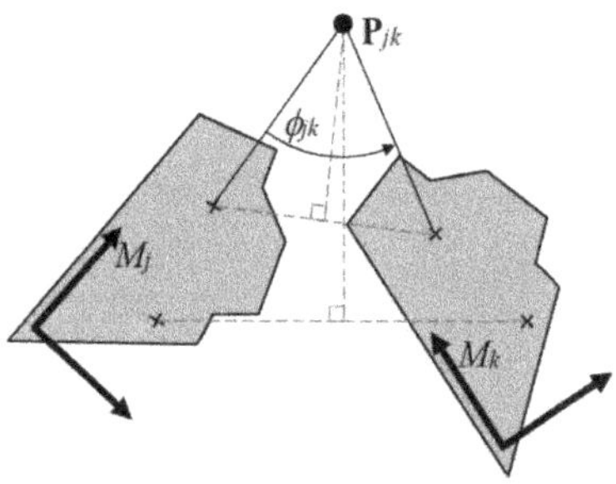

(**c**) Displacement pole

Figure 1. Concepts for rigid-body guidance include identifying target positions, coordinate transformations, and the displacement pole.

A general point in M_i is specified as $\mathbf{x} = [x, y]^\top$ and in F as $\mathbf{X}_i = [X_i, Y_i]^\top$, which are shown in Figure 1b. Relating $\mathbf{x}$ and $\mathbf{X}_i$,

$$\mathbf{X}_i = A_i\mathbf{x} + \mathbf{d}_i, \tag{1}$$

where

$$A_i = \begin{bmatrix} \cos\theta_i & -\sin\theta_i \\ \sin\theta_i & \cos\theta_i \end{bmatrix}. \tag{2}$$

2.1. The Displacement Pole

Any displacement of a rigid body from position j to position k, and vice versa, can be accomplished by a pure rotation about the displacement pole $\mathbf{P}_{jk} = \mathbf{P}_{kj}$, as shown in Figure 1c. The rotation angle about the pole from M_j to M_k is

$$\phi_{jk} = \theta_k - \theta_j. \tag{3}$$

Since the pole is the center of rotation, its location $\mathbf{p}_{jk}$, relative to either M_j or M_k, must be identical. Using Equation (1) for both M_j or M_k,

$$\mathbf{P}_{jk} = A_j\mathbf{p}_{jk} + \mathbf{d}_j = A_k\mathbf{p}_{jk} + \mathbf{d}_k. \tag{4}$$

Solving Equation (4) for $\mathbf{p}_{jk} = [A_j - A_k]^{-1}(\mathbf{d}_k - \mathbf{d}_j)$ and using Equation (1) to determine the coordinates of the displacement pole in the fixed frame gives

$$\begin{aligned} \mathbf{P}_{jk} &= A_j[A_j - A_k]^{-1}(\mathbf{d}_k - \mathbf{d}_j) + \mathbf{d}_j \\ &= A_k[A_k - A_j]^{-1}(\mathbf{d}_j - \mathbf{d}_k) + \mathbf{d}_k = \mathbf{P}_{kj}. \end{aligned} \tag{5}$$

Three successive positions j, k, and l will define the three poles $\mathbf{P}_{jk}$, $\mathbf{P}_{jl}$, and $\mathbf{P}_{kl}$. Note the relationship that $\phi_{jl} = \phi_{jk} + \phi_{kl}$. The three poles form a pole triangle as shown in Figure 2 [1].

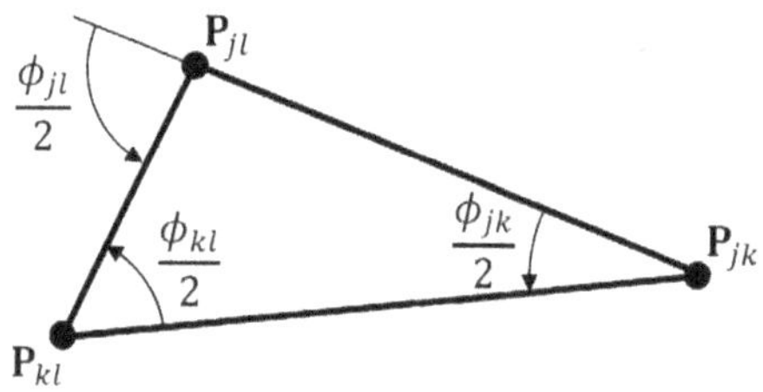

Figure 2. The pole triangle.

2.2. *Using Poles to Determine the Task Positions*

Observe from Figure 2 that with one task position and all pole locations, instances of Equation (3) provide the orientation angles of the remaining task positions. So, given the pole locations in a pole triangle and one position, say θ_j of A_j and $\mathbf{d}_j$, the other angles and their corresponding rotation matrices, A_k and A_l, are readily determined. Equation (5) is then used to solve directly for $\mathbf{d}_k$ and $\mathbf{d}_l$.

2.3. *Counting the Required Poles*

A displacement pole is associated with a pair of specified task positions. As such, the number of displacement poles n_p is related to the number of specified task positions n_t through a pairwise combination [1]:

$$n_p = n_t(n_t - 1)/2. \tag{6}$$

Given one of the positions, however, having the full set of poles is not required (for more than three positions) to find the remaining positions. For example, $n_t = 4$ positions have nine unknowns for θ_j and $\mathbf{d}_j$, $j = 2, 3, 4$. Four positions used in Equation (5) produce 6 vector equations or 12 equations when expanded. The result is 12 equations in 9 unknowns, and the system is observed to be redundant.

For an arbitrary number of positions, and without loss of generality, consider from Section 2.2 that Positions 2 and 3 may be generated, given Position 1, from poles $\mathbf{P}_{12}$, $\mathbf{P}_{13}$ and $\mathbf{P}_{23}$. To generate Position 4, only 2 of the poles from $\mathbf{P}_{14}$, $\mathbf{P}_{24}$, or $\mathbf{P}_{34}$ are needed to complete a pole triangle. Now, to generate Position 5, only 2 of the poles from $\mathbf{P}_{15}$, $\mathbf{P}_{25}$, $\mathbf{P}_{35}$, or $\mathbf{P}_{45}$ are needed to complete the triangle. Observing that the pattern continues, only two poles are needed per position following the recognition of an initial pole triangle. The following general statement can now be made. Given one task position, the number of poles required to determine the location and orientation of the other $n_t - 1$ task positions is

$$n_p^* = 2(n_t - 3) + 3. \tag{7}$$

The preceding rationale introduces the condition that each of the n_t task positions must be represented in at least two of the n_p^* poles to complete a pole triangle. Consider an example with $n_t = 6$ task positions. Equation (6) yields $n_p = 15$ total poles, and Equation (7) specifies that only $n_p^* = 9$ are required to determine θ_j and $\mathbf{d}_j$, $j = 2, \ldots, 6$ given the information for $j = 1$. However, the following set of nine poles $\mathbf{P}_{13}$, $\mathbf{P}_{14}$, $\mathbf{P}_{15}$, $\mathbf{P}_{26}$, $\mathbf{P}_{34}$, $\mathbf{P}_{35}$, $\mathbf{P}_{36}$, $\mathbf{P}_{45}$, and $\mathbf{P}_{56}$ will not generate independent equations because Position 2 is only represented once. Alternately, the set of nine poles $\mathbf{P}_{12}$, $\mathbf{P}_{13}$, $\mathbf{P}_{14}$, $\mathbf{P}_{15}$, $\mathbf{P}_{16}$, $\mathbf{P}_{23}$, $\mathbf{P}_{34}$, $\mathbf{P}_{45}$, and $\mathbf{P}_{56}$ has each position represented at least twice and is sufficient to determine the location and orientation of task Positions 2 through 6.

2.4. Dimensional Synthesis

Dimensional synthesis for rigid-body guidance is a classic problem in linkage design. Well-established methods have been developed to generate a revolute–revolute (RR) dyad to exactly reach as many as five arbitrary target positions [1,14,15]. Figure 3 illustrates two RR dyads that join to form a four-bar linkage.

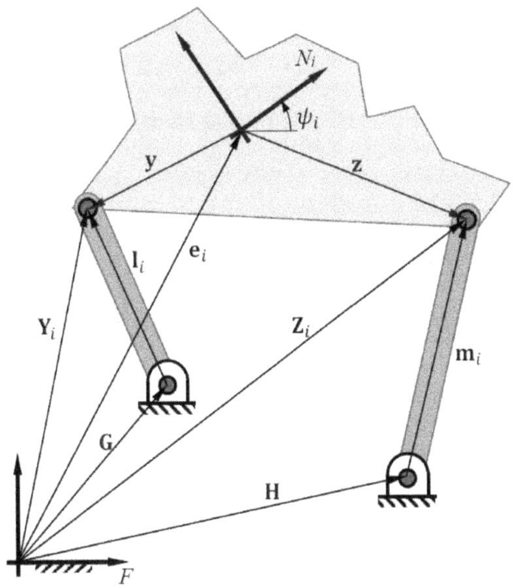

Figure 3. Vector description of a four-bar linkage.

The fixed pivots, also known as the center points, are located in the fixed frame with vectors **G** and **H**. The links attached to the fixed pivot, also known as cranks, have lengths l and m and for the linkage position illustrated have location vectors $\mathbf{l}_i$ and $\mathbf{m}_i$.

The joined link of the two RR dyads is called the coupler and contains a generated reference frame N_i, as illustrated in Figure 3. This reference frame is referred to as "generated" because its location is determined as part of the optimization process detailed below and, being approximate motion synthesis, does not match the target task positions. The position of the generated reference frame is described by a location vector $\mathbf{e}_i = [e_{x_i}, e_{y_i}]^{\mathsf{T}}$ and an orientation angle ψ_i, both stated relative to the fixed frame F.

The location vectors of the moving pivots, also known as circle points, relative to the moving frame are **y** and **z**. Using Equation (1), the locations of the moving pivots relative to the fixed frame are

$$\mathbf{Y}_i = B_i\mathbf{y} + \mathbf{e}_i, \quad (8)$$

$$\mathbf{Z}_i = B_i\mathbf{z} + \mathbf{e}_i, \quad (9)$$

where

$$B_i = \begin{bmatrix} \cos\psi_i & -\sin\psi_i \\ \sin\psi_i & \cos\psi_i \end{bmatrix}. \quad (10)$$

Alternately, the location of the moving pivots relative to the fixed frame as constrained by the cranks are

$$\mathbf{Y}_i = \mathbf{G} + \mathbf{l}_i, \quad (11)$$

$$\mathbf{Z}_i = \mathbf{H} + \mathbf{m}_i. \quad (12)$$

Dimensional synthesis of the RR dyads amounts to determining values of **G**, **y** for one crank and **H**, **z** for the other that place the generated reference frames N_i near the desired positions. The fundamental synthesis condition is that the crank lengths, l and m, remain constant throughout the motion. An expression for the resulting crank length l is generated by combining and squaring Equations (8) and (11), giving

$$l^2 = (B_i\mathbf{y} + \mathbf{e}_i - \mathbf{G})^\mathsf{T}(B_i\mathbf{y} + \mathbf{e}_i - \mathbf{G}). \tag{13}$$

Similar manipulation of Equations (9) and (12) gives

$$m^2 = (B_i\mathbf{z} + \mathbf{e}_i\mathbf{H})^\mathsf{T}(B_i\mathbf{z} + \mathbf{e}_i - \mathbf{H}). \tag{14}$$

Considering a synthesized linkage movement from position j to position k, the displacement pole of the generated reference frame N, which is attached to the coupler, is $\mathbf{Q}_{jk}$. An expression for the location vector of the pole is determined in an identical process as that for Equation (5):

$$\begin{aligned} \mathbf{Q}_{jk} &= B_j[B_j - B_k]^{-1}(\mathbf{e}_k - \mathbf{e}_j) + \mathbf{e}_j \\ &= B_k[B_k - B_j]^{-1}(\mathbf{e}_j - \mathbf{e}_k) + \mathbf{e}_k = \mathbf{Q}_{kj}. \end{aligned} \tag{15}$$

3. Dimensional Synthesis Optimization

The approximate motion synthesis begins with a specified set of task positions, from which the pole locations are determined from Equation (5). A synthesized linkage will produce generated positions, from which the pole locations are determined from Equation (15). In the proposed approach, structural error is defined based on the difference between the task and generated pole locations

$$J = \sum_{i=1}^{n_p} \sum_{j=i+1}^{n_p} (\mathbf{P}_{ij} - \mathbf{Q}_{ij})^\mathsf{T}(\mathbf{P}_{ij} - \mathbf{Q}_{ij}). \tag{16}$$

By defining J as in Equation (16) for the motion optimization, both the orientation error and location error are incorporated into a measure of linear distance. To illustrate that J is invariant, left and right translations will be explored. A left translation is a shift of the task positions in the fixed reference frame F, whereas a right translation is a shift relative to the local reference frame M_i. As discussed further in Section 4, a right translation does not alter the values in the equation in any way. A left translation does alter the individual values in the equation, but since it moves the entire set of poles maintaining the relative dimensions, the value of the metric remains unchanged.

Note that Equation (16) is written to use all n_p pole locations, which creates an unnecessarily large optimization. As stated in Section 2.2, n_p^* poles and one task position provide sufficient information to describe the remaining task positions, given that each position is referenced twice. The first position is arbitrarily selected as the reference with $\psi_1 = \theta_1$ and $\mathbf{e}_1 = \mathbf{d}_1$. Observe that, while efficiently solved, an optimization posed with n_p^* poles can create numerical issues when some poles lie at a great distance from the task positions due to $\theta_i \approx \theta_j$. Further, it has been observed that selecting $\tilde{n}_p$ poles, $n_p^* \leq \tilde{n}_p \leq n_p$, creates a more robust optimization.

To select the $\tilde{n}_p$ poles for the optimization, the centroid of all n_p poles is determined. The distance from each pole to the pole centroid δ_i is determined. The n_p^* poles with the lowest δ values are selected, and their indices are inserted into a set $\mathbb{P}$. As stated in Section 2.2, each task position must be represented in two of the n_p^* poles. To that end, poles are progressively added to $\mathbb{P}$ based on the next lowest δ value until each task position is represented in at least two poles. This being the minimum number of poles used, where some increase above this number is observed to lead to more robustness in the ensuing optimization.

Instances of Equations (13) and (14) are written in terms of $i = 1$ and subtracted from $i = 2, ..., n_t$:

$$\begin{aligned} C_{1i} = \quad & (B_i\mathbf{y} + \mathbf{e}_i - \mathbf{G})^\mathsf{T}(B_i\mathbf{y} + \mathbf{e}_i - \mathbf{G}) \\ & -(B_1\mathbf{y} + \mathbf{e}_1 - \mathbf{G})^\mathsf{T}(B_1\mathbf{y} + \mathbf{e}_1 - \mathbf{G}), \end{aligned} \tag{17}$$

$$\begin{aligned} C_{2i} = \quad & (B_i\mathbf{z} + \mathbf{e}_i - \mathbf{H})^\mathsf{T}(B_i\mathbf{z} + \mathbf{e}_i - \mathbf{H}) \\ & -(B_1\mathbf{z} + \mathbf{e}_1 - \mathbf{H})^\mathsf{T}(B_1\mathbf{z} + \mathbf{e}_1 - \mathbf{H}). \end{aligned} \tag{18}$$

Thus, $C_{1i} = C_{2i} = 0$ ensures that the crank lengths remain constant at all positions. Finally, the approximate motion optimization is posed as

$$\begin{aligned} \min_{\mathbf{G},\mathbf{y}',\mathbf{H},\mathbf{z}'} \quad & J, \quad ij \in \mathbb{P} \\ \text{s.t.} \quad & C_{1i} = 0, \\ & C_{2i} = 0. \end{aligned} \tag{19}$$

where $\mathbf{y}'$ and $\mathbf{z}'$ are temporary values of $\mathbf{y}$ and $\mathbf{z}$, which will be shifted as described later.

The optimization of Equation (19) is constructed in MATLAB using the gradient-based `fmincon` function. Within the optimization, the generated positions $\mathbf{e}_i$ and ψ_i, $i = 2, \ldots, n_t$ are determined from $\mathbf{e}_1$, ψ_1, $\mathbf{Q}_{ij}$, where $ij \in \mathbb{P}$. Since local minimums are detected, numerous solutions to Equation (19) are generated using diverse starting values. The interior-point algorithm was used. Optimizations were performed using various settings of maximum iterations, along with convergence and constraint tolerances. The MATLAB defaults were ultimately used because they provided consistent solutions in acceptable computation time (less than 10 seconds per randomly selected starting value).

Recall that Task Position 1 is selected as the reference to determine all other positions. As such, the generated positions that result from the optimization of Equation (19) will be biased toward Task Position 1. A final process in the presented dimensional synthesis methodology is to distribute the structural error among all task positions. The average location and orientation error is defined as

$$\bar{e}_x = \frac{1}{n_t}\sum_{i=1}^{n_t} d_{i_x} - e_{i_x}, \tag{20}$$

$$\bar{e}_y = \frac{1}{n_t}\sum_{i=1}^{n_t} d_{i_y} - e_{i_y}, \tag{21}$$

$$\bar{\psi} = \frac{1}{n_t}\sum_{i=1}^{n_t} \theta_i - \psi_i. \tag{22}$$

All generated positions are subjected to a right translation defined by $\bar{\mathbf{e}} = [\bar{e}_x, \bar{e}_x]^\mathsf{T}$ and $\bar{\psi}$. As stated in Section 4.1, right translations will not alter the location of the poles, and therefore will not interfere with the optimization results. Instances of Equation (25) are applied to provide the coupler dimensions

$$\mathbf{y} = \bar{B}\mathbf{y}' + \bar{\mathbf{e}}, \tag{23}$$

$$\mathbf{z} = \bar{B}\mathbf{z}' + \bar{\mathbf{e}}. \tag{24}$$

4. Bi-Invariance

The approach used in the techniques of this paper focuses solely on the poles, whose relative locations are bi-invariant. This section further explores this fact.

4.1. Right Translation

A right translation is a shift of a position in the local reference frame. Consider a right translation of all task positions defined by $\bar{A}$ and $\bar{\mathbf{d}}$. The resulting locations of the task positions in the fixed frame are

$$\bar{\mathbf{X}}_i = A_i[\bar{A}\mathbf{x} + \bar{\mathbf{d}}] + \mathbf{d}_i, \quad i = 1, \ldots, n_t. \tag{25}$$

To determine the pole of the right-translated task positions j and k, an expression similar to Equation (4) becomes

$$A_j[\bar{A}\mathbf{p}_{jk} + \bar{\mathbf{d}}] + \mathbf{d}_j = A_k[\bar{A}\mathbf{p}_{jk} + \bar{\mathbf{d}}] + \mathbf{d}_k. \tag{26}$$

Equation (26) is solved for $\mathbf{p}_{jk}$:

$$\mathbf{p}_{jk} = (A_j\bar{A} - A_k\bar{A})^{-1}(A_k\bar{\mathbf{d}} - A_j\bar{\mathbf{d}} + \mathbf{d}_k - \mathbf{d}_j). \tag{27}$$

Equation (1) is used to determine the coordinates of the displacement pole in the fixed frame:

$$\mathbf{P}_{jk} = A_j\Big[\bar{A}\bar{A}^{-1}(A_j - A_k)^{-1}(A_k\bar{\mathbf{d}} - A_j\bar{\mathbf{d}} + \mathbf{d}_k - \mathbf{d}_j) + \bar{\mathbf{d}}\Big] + \mathbf{d}_j. \tag{28}$$

After factoring and simplifying Equation (28), the resulting expression is identical to Equation (5). Thus, a right translation of every task position will not change the location of the poles. Extending further, a right translation will not affect the optimization metric specified in Equation (16). As such, the process outlined is right invariant.

4.2. Left Translation

A left translation shifts the position of the fixed reference frame used to specify the set of task positions:

$$\tilde{\mathbf{X}}_i = \tilde{A}[A_i\mathbf{x} + \mathbf{d}_i] + \tilde{\mathbf{d}}, \quad i = 1, \ldots, n_t. \tag{29}$$

A left translation will change the location of the poles. The observation that initiated the approach developed in this paper is that the relative locations of the poles remain unchanged. Using the distance metric used in Equation (16) for positions j and k,

$$J_{jk} = (\mathbf{P}_{jk} - \mathbf{Q}_{jk})^T(\mathbf{P}_{jk} - \mathbf{Q}_{jk}). \tag{30}$$

Consider the left translation of J_{jk} defined by $\tilde{A}$ and $\tilde{\mathbf{d}}$:

$$\begin{aligned} \tilde{J}_{jk} &= \Big[(\tilde{A}\mathbf{P}_{jk} + \tilde{\mathbf{d}}) - (\tilde{A}\mathbf{Q}_{jk} + \tilde{\mathbf{d}})\Big]^T\Big[(\tilde{A}\mathbf{P}_{jk} + \tilde{\mathbf{d}}) - (\tilde{A}\mathbf{Q}_{jk} + \tilde{\mathbf{d}})\Big], \\ &= \Big[\tilde{A}(\mathbf{P}_{jk} - \mathbf{Q}_{jk})\Big]^T\Big[(\tilde{A}(\mathbf{P}_{jk} - \mathbf{Q}_{jk})\Big], \\ &= \tilde{A}^T\tilde{A}(\mathbf{P}_{jk} - \mathbf{Q}_{jk})^T(\mathbf{P}_{jk} - \mathbf{Q}_{jk}), \\ &= (\mathbf{P}_{jk} - \mathbf{Q}_{jk})^T(\mathbf{P}_{jk} - \mathbf{Q}_{jk}). \end{aligned} \tag{31}$$

Note that the final expression in Equation (31) is the same as Equation (30). As such, the optimization used in the approximate motion synthesis of Equation (16) is left invariant.

5. Examples

In this section, three examples are provided to demonstrate the applicability of the presented method in various scenarios to analyze the accuracy and efficiency of the pole method in solving motion trajectory problems that lack an exact solution. The first example compares the synthesis results using quaternion-based theory by previous researchers with the pole method for ten desired locations. The second example utilizes the case presented by Gogate [9] to demonstrate the effectiveness and accuracy of the pole method. The third example evaluates the claims of bi-invariance. Overall, the results demonstrate the accuracy and effectiveness of the pole method in solving complex motion trajectory problems.

5.1. Example 1: Task Positions Taken from Ref. [5]

The first example presents synthesis results using various methods available in the literature along with the methodology developed in this paper. The 10 task positions used by Ravani and Roth [5] are given in Table 1.

Table 1. The ten task positions for Example 1.

Location	Orientation (Deg)
(0.00, 0.00)	40
(4.50, 4.00)	20
(8.50, 8.00)	0
(13.00, 11.50)	−30
(13.00, 12.50)	−35
(9.50, 14.00)	−35
(5.00, 13.50)	−30
(1.00, 10.50)	−15
(−1.00, 6.50)	0
(−1.50, 3.00)	20

With $n_t = 10$, $n_p = 45$ total displacement poles exist. The minimum number of poles that are required to determine the remaining task positions is $n_p^* = 17$. By selecting the 17 poles closest to the geometric center of the poles, each task position was represented twice. To generate a more robust optimization, $\tilde{n}_p = 39$ poles closest to the geometric center are selected and placed into the set $\mathbb{P}$, because of a large change in distance from the geometric center is observed: $\delta_{38} = 50.41$, $\delta_{39} = 56.41$, and $\delta_{40} = 85.58$. The optimization was conducted with 1000 different starting points. The best solution resulted in $J = 37.4$. The resulting linkage parameters from the method presented in the paper are $\mathbf{G} = [6.7378, 0.7731]^T$, $\mathbf{H} = [15.4473, -13.9909]^T$, $\mathbf{y} = [3.1862, -8.2763]^T$, $\mathbf{z} = [10.1174, 4.3177]^T$, $l = 5.4134$, and $m = 25.8726$. The resulting linkage and generated positions are shown in Figure 4.

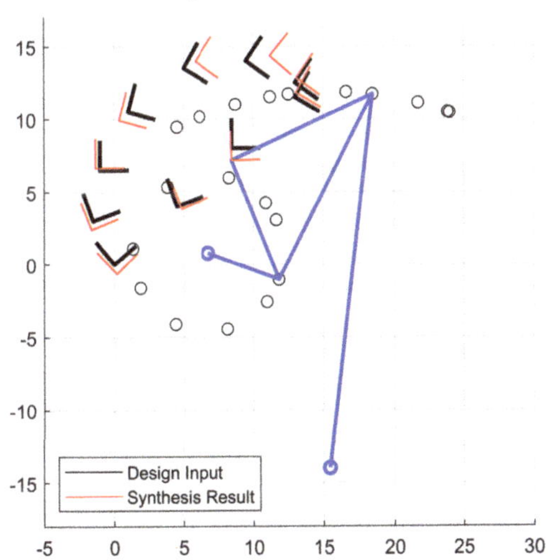

Figure 4. The task frames, generated frames, and resulting linkage from the presented method for Example 1.

The motion curve was generated from the linkage in Figure 4 and shown with the blue curve in Figure 5 (identified as Xu). Using the linkage parameters provided in the literature, the motion curves were generated for Ravani and Roth (shown in red) and Larochelle [6] (shown in green). The method from Ge et al. [7] was reproduced and used with the task positions in Table 1 to generate the motion curve shown in cyan.

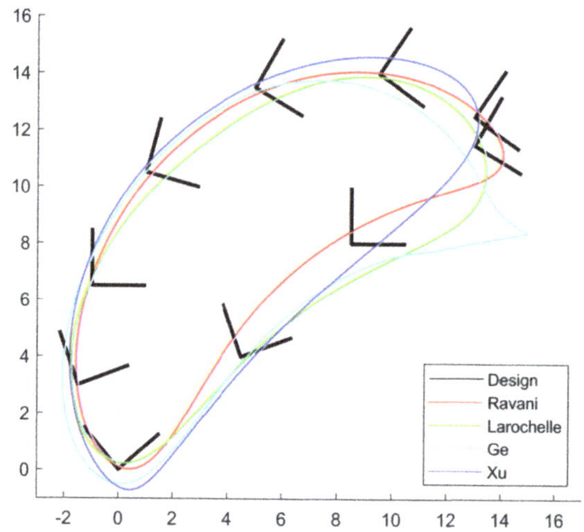

Figure 5. The task frames and generated motion curves from various methods for Example 1.

From the linkage parameters and motion curve, generated frames that approximate the task positions were identified. Those generated frames are shown in Figure 6 for Ravani and Roth (red), Larochelle (green), and Ge et al. (cyan). Using these frames, the error based on the difference of task and generated pole locations was calculated for Ravani and Roth as $J = 120.6$, Larochelle as $J = 132.9$, and Ge et al. as $J = 209.2$. The method described herein uses J as the objective and predictably has a lower value of $J = 37.4$. Figures 5 and 6 show that the alternative methods compare favorably.

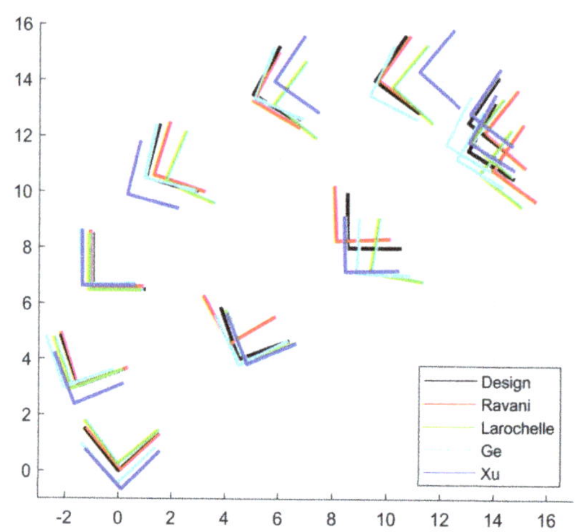

Figure 6. The task frames and generated frames from various methods for Example 2.

5.2. Example 2: Task Positions Taken from Ref. [9]

The second example uses the $n_t = 7$ task positions used by Gogate [9], which are given in Table 2.

Table 2. The seven task positions for Example 2.

Location	Orientation (Deg)
(1.1025, 1.0206)	54.9693
(0.9439, 0.4950)	72.3439
(0.3804, 0.0741)	80.9533
(−0.1340, 0.1014)	72.1922
(−0.3286, 0.5434)	58.6599
(−0.0668, 1.0855)	49.9219
(0.5293, 1.3346)	47.8367

With $n_t = 7$, $n_p = 21$ total displacement poles exist. The minimum number of poles that are required to determine the remaining task positions is $n_p^* = 11$. By selecting the 11 poles closest to the geometric center of the poles, each task position was represented twice. The optimization was conducted using only those 11 poles with 100 different starting points. The best solution resulted in $J = 0.0064$. The resulting linkage parameters are $\mathbf{G} = [-3.6627 \times 10^{-4}, -1.7543 \times 10^{-5}]^T$, $\mathbf{H} = [2.7021, -0.0025]^T$, $\mathbf{y} = [-0.7885, -3.3943 \times 10^{-05}]^T$, $\mathbf{z} = [1.5016, -1.4275]^T$, $l = 0.7477$, and $m = 1.4959$. Figure 7 illustrates the task frames (shown in black), the generated task frames (shown in red), the resulting linkage along with the motion curve (shown in blue), and the motion curve produced by Gogate (shown in green).

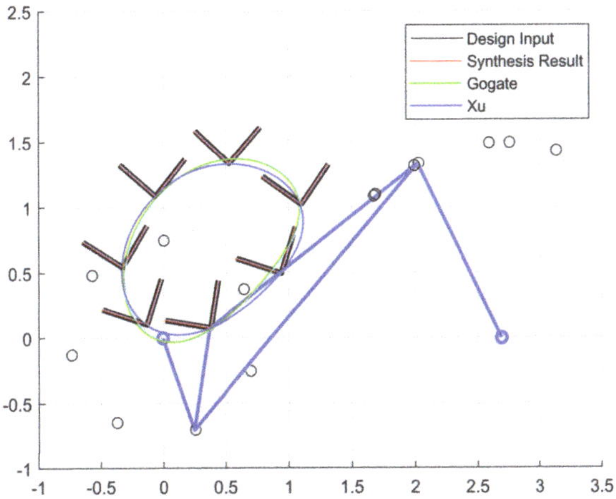

Figure 7. The task frames, generated frames, motion curve, and linkage for Example 2.

5.3. Example 3: Demonstrating Bi-Invariance

The third example uses the developed methodology on twelve arbitrary task positions, which are given in Table 3.

Table 3. The 12 task positions for bi-invariance example.

Location	Orientation (Deg)
(5.4925, −5.5813)	109.2727
(6.5759, −4.0083)	126.6505
(6.3082, −2.3480)	140.3157
(5.0394, −1.1357)	159.4077
(3.4334, −1.8575)	167.6665
(0.8025, −3.7452)	144.6324
(−1.8858, −6.0622)	116.1184
(−1.7438, −7.8746)	103.3444
(−0.8606, −8.8435)	85.9285
(1.0690, −9.5115)	88.2301
(2.2986, −8.8865)	80.0622
(4.5038, −7.8714)	87.0337

Having $n_t = 12$ task positions, the total number of displacement poles is $n_p = 66$. In addition to the first task position, the minimum number of poles that are required to determine the remaining task positions is $n_p^* = 21$. Placing the 21 poles closest to the geometric center into the set $\mathbb{P}$, each task position is not represented twice. Poles are successively added to $\mathbb{P}$, based on the next lowest δ value. When $\tilde{n}_p = 29$, all task positions appear at least twice. For this optimization, $\tilde{n}_p = 53$ poles are included in $\mathbb{P}$ resulting in each task position being represented eight times and having $\delta_{53} = 17.39$. The optimization was conducted with 100 different starting points. The resulting mechanism parameters are given in the first row of Table 4. The task positions (thick black), linkage (solid blue), and the generated frames (thick red) are shown in Figure 8.

Table 4. The generated mechanism result with small and large translations for right and left.

Translation	G	H	y	z	l	m	J
None	(2.2675, −3.0510)	(−4.4649, 2.5947)	(2.5643, −1.2709)	(4.8371, 4.0034)	3.9464	6.9910	72.073
$\bar{A}_S$, $\bar{\mathbf{d}}_S$	(2.2659, −3.0512)	(−4.4641, 2.5767)	(−1.7130, 9.5602)	(1.7188, 4.9564)	3.9409	6.9727	72.073
$\bar{A}_S$, $\bar{\mathbf{d}}_S$	(10.797, −7.8274)	(2.3618, −10.257)	(2.5663, −1.2680)	(4.8394, 4.0052)	3.9421	6.9782	72.073
$\bar{A}_L$, $\bar{\mathbf{d}}_L$	(2.2668, −3.0453)	(−4.4655, 2.6025)	(3.3224, 168.15)	(6.7562, 163.55)	3.9542	6.9982	72.076
$\bar{A}_L$, $\bar{\mathbf{d}}_L$	(95.557, −80.241)	(96.340, −88.973)	(2.5683, −1.2665)	(4.8407, 4.0068)	3.9381	6.9636	72.073

To demonstrate the bi-invariance of the methodology, the twelve task positions were subjected to a small right translation of Equation (25) defined by $\bar{A}_S$ corresponding to a rotation of 120° and $\bar{\mathbf{d}}_S = (10, 5)$. The same number of poles $\tilde{n}_p$ are included in the optimization. The right-translated task positions (thin black), linkage (dashed blue), and the generated frames (thin red) are shown in Figure 8. The resulting mechanism parameters are given in the second row of Table 4. Note that the values of **G**, **H**, l, m, and J are nearly identical to the non-translated results given in the first row. Additionally, a right translation is applied to the results given in the moving frame: $\bar{\mathbf{y}} = \bar{A}_S\,\mathbf{y} + \bar{\mathbf{d}}_S = (2.5771, -1.2636)$, and $\bar{\mathbf{z}} = \bar{A}_S\,\mathbf{z} + \bar{\mathbf{d}}_S = (4.8482, 4.0103)$. These translated values $\bar{\mathbf{y}}$ and $\bar{\mathbf{z}}$ are nearly identical to the non-translated results **y** and **z** given in the first row.

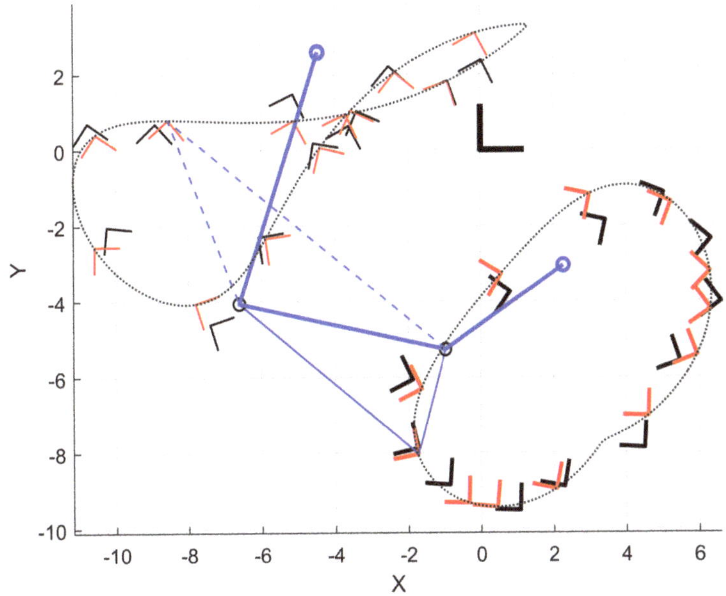

Figure 8. Task positions for Example 3 before (thick black) and after applying a right translation (thin black). Generated positions are shown for task positions before (thick red) and after right translation (thin red). The thick blue lines are the optimized four-bar linkage seen to be identical in the two solutions. The thin blue lines show the coupler to the original positions (solid) and the right translated positions (dashed).

Similarly, the twelve task positions were subjected to a small left translation of Equation (29) defined by $\tilde{A}_S$ corresponding to a rotation of 56° and $\tilde{\mathbf{d}}_S = (7, -8)$. As in the other optimizations, the same number of poles $\tilde{n}_p$ are included. The left-translated task positions (black), linkage, and the generated frames are shown in Figure 9. The resulting mechanism parameters are given in the third row of Table 4. Note that the values of $\mathbf{y}$, $\mathbf{z}$, l, m, and J are nearly identical to the non-translated results given in the first row. Additionally, a left translation is applied to the results given in the fixed frame: $\tilde{\mathbf{G}} = \tilde{A}^{-1}(\mathbf{G} - \tilde{\mathbf{d}}) = (2.2663, -3.0513)$, and $\tilde{\mathbf{H}} = \tilde{A}^{-1}(\mathbf{H} - \tilde{\mathbf{d}}) = (-4.4648, 2.5831)$. These translated values $\tilde{\mathbf{G}}$ and $\tilde{\mathbf{H}}$ are nearly identical to the non-translated results $\mathbf{G}$ and $\mathbf{H}$ given in the first row.

Large right and left translations are also applied, defined by $\bar{A}_L$ of 120°, $\bar{\mathbf{d}}_L = (150, 80)$ and $\tilde{A}_L$ of 135°, $\tilde{\mathbf{d}}_L = (95, -84)$, respectively. Optimizations were conducted with the same parameters as the small translations discussed above. The resulting mechanism parameters are given in the fourth and fifth row of Table 4. For $\bar{A}_L$, $\bar{\mathbf{d}}_L$, the values of $\mathbf{G}$, $\mathbf{H}$, l, m, and J are once again nearly identical to the non-translated results given in the first row. Applying right translation to $\mathbf{y} = \bar{A}_L\bar{\mathbf{y}} + \bar{\mathbf{d}}_L = (2.7166, -1.1977)$, and $\mathbf{z} = \bar{A}_L\bar{\mathbf{z}} + \bar{\mathbf{d}}_L = (4.9834, 4.0760)$ correlates well with $\mathbf{y}$ and $\mathbf{z}$ given in the first row. For $\tilde{A}_L$, $\tilde{\mathbf{d}}_L$, the values of $\mathbf{y}$, $\mathbf{z}$, l, m, and J match the non-translated results given in the first row. Applying the left translation to $\mathbf{G} = \tilde{A}_L^{-1}(\tilde{\mathbf{G}} - \tilde{\mathbf{d}}_L) = (2.2642, -3.0519)$ and $\mathbf{H} = \tilde{A}_L^{-1}(\tilde{\mathbf{H}} - \tilde{\mathbf{d}}_L) = (-4.4640, 2.5689)$ correspond with $\mathbf{G}$ and $\mathbf{H}$ given in the first row.

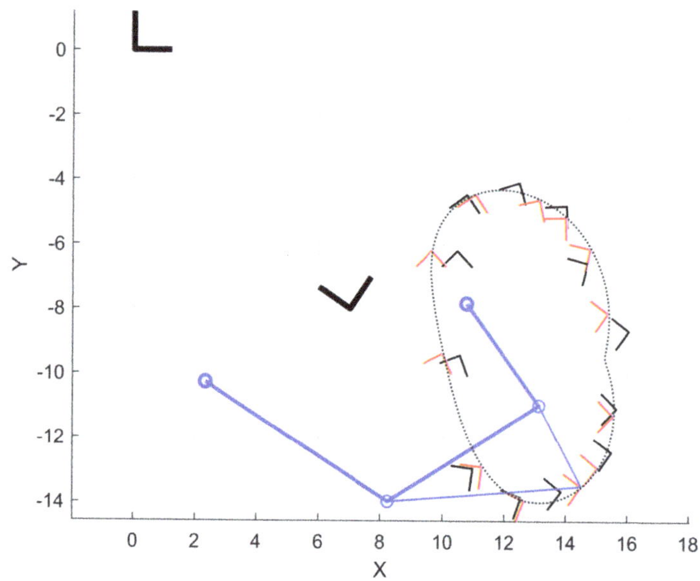

Figure 9. Task positions for Example 3 after applying a small left translation (black) and the associated generated positions (red). The fixed frame after applying a left translation is also shown. The thick blue lines are the optimized four-bar matching the linkage shown in Figure 8. The thin blue lines show the coupler to the left translated positions.

6. Conclusions

This paper presents a technique for the approximate motion synthesis of planar four-bar mechanisms based exclusively on the locations of displacement poles. The benefit is that the displacement poles are described solely by their locations, which includes information about the location and orientation of the task positions. The optimization minimizes the distance between the poles of the task positions and the poles of the positions generated by a synthesized linkage, thereby producing a bi-invariant solution. To generate an efficient optimization process, and to avoid poles very distant or infinitely far from the others, the technique presented uses a subset of the entire set of displacement poles. The technique is illustrated in three examples, where the results compare favorably to alternative methods presented in the literature.

Author Contributions: Conceptualization, A.P.M.; methodology, D.H.M. and A.P.M.; software, T.X. and A.P.M.; validation, T.X.; formal analysis, T.X., D.H.M., and A.P.M.; investigation, T.X. and A.P.M.; resources, T.X.; data curation, T.X.; writing—original draft preparation, T.X., D.H.M., and A.P.M.; writing—review and editing, T.X., D.H.M., and A.P.M.; visualization, T.X., D.H.M., and A.P.M.; supervision, D.H.M. and A.P.M.; project administration, D.H.M. and A.P.M. All authors have read and agreed to the published version of the manuscript.

Funding: This work has been supported in part by the University of Dayton Office for Graduate Academic Affairs through the Graduate Student Summer Fellowship Program.

Informed Consent Statement: Not applicable.

Data Availability Statement: Data are available upon request.

Conflicts of Interest: The authors declare no conflicts of interest.

References

1. McCarthy, J.; Soh, G. *Geometric Design of Linkages*; Springer: New York, NY, USA, 2010.
2. Suh, C.H.; Radcliffe, C.W. *Kinematics and Mechanism Design*; John Wiley and Sons: New York, NY, USA, 1978.
3. Levitskii, N.I. Design of Plane Mechanisms with Lower Pairs. In Proceedings of the AH CCCP Izdatelstvo Akademii Nauk, Leningrad, Moscow, 1950.
4. Sarkisyan, Y.L.; Gupta, K.C.; Roth, B. Kinematic Geometry Associated with the Least-square Approximation of a Given Motion. *ASME J. Eng. Ind.* **1973**, *95*, 503–510. [CrossRef]
5. Ravani, B.; Roth, B. Motion Synthesis using Kinematic Mappings. *ASME J. Mech. Transm. Autom. Des.* **1983**, *105*, 460–467. [CrossRef]
6. Larochelle, P.M. Approximate Motion Synthesis of Open and Closed Chains via Parametric Constraint Manifold Fitting: Preliminary Results. In Proceedings of the ASME Design Automation Conference, Chicago, IL, USA, 2–6 September 2003; pp. 1049–1057.
7. Ge, Q.J.; Zhao, P.; Purwar, A.; Li, X. A Novel Approach to Algebraic Fitting of a Pencil of Quadrics for Planar 4R Motion Synthesis. *J. Comput. Inf. Sci. Eng.* **2012**, *12*, 0041003. [CrossRef]
8. Ge, Q.J.; Purwar, A.; Zhao, P.; Dephpande, S. A Task-Driven Approach to Unified Synthesis of Planar Four-Bar Linkages Using Algebraic Fitting of a Pencil of G-Manifolds. *ASME J. Mech. Robot.* **2017**, *17*, 031011. [CrossRef]
9. Gogate, G.R.; Matekar, S.B. Optimum synthesis of motion generating four-bar mechanisms using alternate error functions. *Mech. Mach. Theory* **2012**, *54*, 41–61. [CrossRef]
10. Milnor, J. Curvatures of left invariant metrics on lie groups. *Adv. Math.* **1976**, *21*, 293–329. [CrossRef]
11. Martinez, J.M.R.; Duffy, J. On the Metrics of Rigid Body Displacements for Infinite and Finite Bodies. *J. Mech. Des.* **1995**, *117*, 41–47. [CrossRef]
12. Di Gregorio, R. Metrics proposed for measuring the distance between two rigid-body poses: Review, comparison, and combination. *Robotica* **2024**, *42*, 302–318. [CrossRef]
13. Liu, A.X.; Yang, T.L. Finding All Solutions to Unconstrained Nonlinear Optimization for Approximate Synthesis of Planar Linkages Using Continuation Method. *J. Mech. Des.* **1999**, *121*, 368–374. [CrossRef]
14. Erdman, A.G.; Sandor, G.N.; Kota, S. *Mechanism Design: Analysis and Synthesis*, 3rd ed.; Prentice Hall: Englewood Cliffs, NJ, USA, 2001; Volume 1.
15. Sandor, G.N.; Erdman, A.G. *Advanced Mechanism Design: Analysis and Synthesis*; Prentice Hall: Englewood Cliffs, NJ, USA, 1984; Volume 2.

robotics

MDPI

Article

Type Synthesis of 5-DOF Hybrid (Parallel-Serial) Manipulators Designed from Open Kinematic Chains

Anton Antonov, Alexey Fomin *, Victor Glazunov, Daniil Petelin and Gleb Filippov

Mechanisms Theory and Machines Structure Laboratory, Mechanical Engineering Research Institute of the Russian Academy of Sciences (IMASH RAN), 101000 Moscow, Russia; antonov.av@imash.ru (A.A.); vaglznv@mail.ru (V.G.); petelin_daniil@inbox.ru (D.P.); filippov.gleb@gmail.com (G.F.)
* Correspondence: alexey-nvkz@mail.ru

Abstract: The article proposes an approach for synthesizing hybrid (parallel-serial) manipulators with five degrees of freedom (5-DOF) using open kinematic chains. The method idea consists in taking an open kinematic chain, selecting a subchain within it, and replacing the subchain with a parallel mechanism. The article considers 5-DOF open chains and 3-DOF subchains, substituted for 3-DOF parallel mechanisms with the same motion pattern as the subchain. Thus, synthesized hybrid manipulators have a 3-DOF parallel part and a 2-DOF serial part. First, we grouped 26 structures of open chains with revolute and prismatic joints into five types and 78 subtypes. Next, for each type, we selected one subtype and presented several hybrid mechanisms that can correspond to it. We considered hybrid manipulators that included 3-DOF parallel mechanisms with planar, spherical, and other commonly used motion types. The suggested synthesis method is intuitive for a designer, and it does not need any mathematical formulations like screw theory or group theory approaches.

Keywords: type synthesis; 5-DOF hybrid (parallel-serial) mechanism; 3-DOF parallel mechanism; serial (open) kinematic chain; 3T2R and 3R2T motion patterns

Citation: Antonov, A.; Fomin, A.; Glazunov, V.; Petelin, D.; Filippov, G. Type Synthesis of 5-DOF Hybrid (Parallel-Serial) Manipulators Designed from Open Kinematic Chains. *Robotics* **2023**, *12*, 98. https://doi.org/10.3390/robotics12040098

Academic Editor: Raffaele Di Gregorio

Received: 29 May 2023
Revised: 3 July 2023
Accepted: 7 July 2023
Published: 9 July 2023

Correction Statement: This article has been republished with a minor change. The change does not affect the scientific content of the article and further details are available within the backmatter of the website version of this article.

1. Introduction

Development of technologies implies creation of novel robotic, technological, medical, research, and other systems, which are usually based on effectively designed mechanisms. The first step in creating any mechanism is to synthesize its structural diagram, which specifies the types of links and joints used in the mechanism and their relative arrangement. This fundamental stage establishes the functional properties and required characteristics of the developed system. In this regard, the major task of mechanism type synthesis is designing its most optimal structural diagram according to the application.

Familiar methods of type synthesis can be classified into three categories [1]. The first category represents motion-based approaches, which focus on possible motions of the mechanism links. This category includes methods based on group theory [2], generalized function sets (G_F sets) [3,4], position and orientation characteristic sets (POC sets) [5], linear transformations [6], finite screws [7], and conformal geometric algebra [8]. The second category represents constraint-based methods, which consider constraints imposed on the mechanism links. This category comprises techniques that rely on instantaneous screws [9–12], virtual kinematic chains [13], Grassmann line geometry [14–17], and motion constraint generators [18,19]. The third category includes other methods, like the ones based on the mobility formulae [20], graph theory [21], and Denavit–Hartenberg parameters [22]. The papers [1,23] discuss most of these methods and provide a comprehensive bibliography with their applications.

Over the last few years, scholars have focused on studying hybrid manipulators. These mechanical systems include kinematic chains with a parallel structure (parallel-parallel mechanisms) or parallel and serial structures (parallel-serial mechanisms). Such manipulators have several advantages compared to conventional parallel or serial manipulators. One of the major advantages over the parallel manipulators is an increased workspace. This is especially useful in fields like machining (for processing large-sized parts) [24], agriculture (for planting and picking operations) [25], and medicine (for large movements of medical equipment relative to the patient) [26]. Compared to serial manipulators, hybrid ones possess a higher rigidity and motion accuracy, which is also important for precise applications [27].

Approaches for type synthesis of hybrid mechanisms are mainly based on the approaches mentioned above. There are few works devoted to systematic type synthesis of these mechanisms, and most studies focus on parallel-parallel structures. One of the first studies in this field was the work by Chakarov and Parushev [28], who used mobility formulae and Assur groups with linear drives. These groups were placed between the links of the primary open-chain mechanism and did not affect its mobility. This approach allowed displacing the drives of the primary mechanism from its joints to the prismatic joints of the structural groups. Campos et al. [29] developed a similar method based on Assur groups. The authors considered symmetrical hybrid manipulators obtained from a primary zero-DOF mechanism by "cutting" one of its links. Mobility formulae and Assur groups were also used by Alizade and Bayram [30], but the authors focused mainly on parallel-parallel manipulators.

In the synthesis approaches discussed above, the authors considered only the number of DOFs, but ignored the motion type of the output link. Other works examined the nature of these DOFs (rotational or translational). For example, Zeng and Fang [31] proposed an approach based on group theory and represented mechanisms as logical matrices. The method relies on performing some logical operations which can automate type synthesis, as the authors notice. The authors advanced their approach in works [32,33] and presented several novel hybrid mechanisms. Another study to be mentioned is by Shen et al. [34], who applied the POC sets to synthesize hybrid manipulators with three to five DOFs. The authors considered various combinations of parallel and serial kinematic chains and obtained many novel mechanisms.

The studies enumerated above suggested systematic approaches for type synthesis of hybrid manipulators with a different number of DOFs and different motion types. Among the variety of hybrid manipulators, mechanical systems with five DOFs take a special place. In most cases, the output link of such manipulators has either three translational and two rotational DOFs (a 3T2R motion pattern) [35] or three rotational and two translational DOFs (a 3R2T motion pattern) [36]. These manipulators are often used for applications that do not need an excessive rotational or translational freedom, or where this freedom is achieved by a separate drive. Such applications include surgical operations [37,38], pick-and-place tasks [39], machining [40,41], polishing [42], and welding [43].

There are several studies that considered the type synthesis of 5-DOF hybrid manipulators, and most works focused mainly on synthesizing parallel mechanisms. For example, Cao et al. [44,45] applied G_F sets to design 3T2R hybrid manipulators, whose parallel parts had three or four DOFs: the authors obtained over three hundred various mechanisms. Later, the authors used that method to synthesize 3R2T hybrid manipulators and also obtained numerous novel mechanisms [46]. The POC sets method, proposed above by Shen et al. [34], was applied by the authors in paper [47]: the authors developed a novel 5-DOF hybrid manipulator with a 2-DOF serial part and a 3-DOF parallel part. Xu et al. [48] applied screw theory techniques to design 5-DOF hybrid manipulators with 3- and 4-DOF parallel parts; later, in the works [49–51], the authors synthesize planar and spatial 3-DOF parallel mechanisms within various 5-DOF hybrid manipulators. Xie et al. [24] performed a research similar to [49] and considered type synthesis of 3-DOF 1T2R parallel mechanisms, which represented a parallel part of other 5-DOF hybrid manipulators. The authors' method relied on Grassmann line geometry.

In the current article, we introduce a novel approach for the type synthesis of 5-DOF hybrid manipulators based on open kinematic chains. The advantage of this approach over the existing ones is that it is more intuitive and does not require any profound mathematical formulations like other methods based on group theory, screw theory, POC sets, or G_F sets.

The rest of the paper has the following organization. Section 2 presents an original synthesis method. Section 3 shows structural diagrams of 5-DOF open kinematic chains, taken as primary mechanisms, and hybrid manipulators, designed using the suggested techniques. Section 4 discusses the obtained results and possible improvements of the presented approach. Section 5 recaps the entire study and mentions directions for future research.

2. Synthesis Method

In the current section, we will look at the proposed type synthesis method. The method idea consists in taking a 5-DOF open kinematic chain and transforming it into a hybrid mechanism. In particular, we select a 3-DOF subchain within the open chain and transform it into a parallel mechanism, so the synthesized hybrid manipulator has a 3-DOF parallel part and a 2-DOF serial part. We consider parallel parts with three DOFs because 3-DOF parallel mechanisms usually have a simple and symmetrical design, and they prevail among other lower-DOF parallel mechanisms [52].

As primary open kinematic chains, we only consider chains with revolute (R) and prismatic (P) joints: we can always decompose any multiple-DOF joint into a combination of R and P joints [13] (p. 28). In addition, open kinematic chains should satisfy the following conditions to avoid any local mobility within the chain [53]:

1. There are no coaxial P joints;
2. The greatest number of (coplanar) P joints is three;
3. There are no coaxial R joints;
4. The greatest number of R joints with parallel axes is three;
5. The greatest number of R joints with intersecting axes is three.

Conditions 1 and 2 guarantee there is no local translational motion. Conditions 3–5 guarantee there is no local rotational motion. We can find number n of different kinematic chains, which meet these conditions, using the following formula [54] (p. 4):

$$n = U(2,5) - \sum_{m=4}^{5} C(5,m) = 2^5 - \frac{5!}{4!(5-4)!} - \frac{5!}{5!(5-5)!} = 26, \quad (1)$$

where $U(2,5)$ is a number of permutations with unrestricted repetitions, equal to a number of 5-DOF chains with P and R joints; $C(5,m)$ is a number of combinations, equal to a number of 5-DOF chains with m P joints. The subtraction of terms $C(5,m)$ in Equation (1) corresponds to Condition 2 and allows us to exclude chains with four and five P joints.

Table 1 presents all 26 structures of open kinematic chains with R and P joints. Note that the same structure can have different joint arrangements. For example, the RRRRR chain can include subchains of planar, spherical, or other types.

Table 1. Possible structures of primary open kinematic chains.

(1) RRRRR	(2) PRRRR	(3) RPRRR	(4) RRPRR	(5) RRRPR	(6) RRRRP
(7) PPRRR	(8) PRPRR	(9) PRRPR	(10) PRRRP	(11) RPPRR	(12) RPRPR
(13) RPRRP	(14) RRPPR	(15) RRPRP	(16) RRRPP	(17) PPPRR	(18) PPRPR
(19) PPRRP	(20) PRPPR	(21) PRPRP	(22) PRRPP	(23) RPPPR	(24) RPRPP
		(25) RPPRP	(26) RRPPP		

Next, we use the enumerated open kinematic chains to synthesize hybrid manipulators as follows. We select a 3-DOF subchain within the 5-DOF primary kinematic chain and replace it with a 3-DOF parallel mechanism, which has the same motion pattern as the substituted subchain (it can also be possible that the parallel mechanism has a motion type different from the subchain—this situation is addressed in Section 4). For each 5-DOF kinematic chain, three possibilities exist to select a 3-DOF subchain, so there are $26 \times 3 = 78$ possible structures of hybrid mechanisms. Table 2 demonstrates all these structures, where the underline signifies the substituted subchain. The structures are classified according to the number of P and R joints in the primary open kinematic chain. There are four distinct types of structures: 5R, 4R1P, 3R2P, and 3P2R. Each type includes several subtypes: the 5R type has three subtypes, the 4R1P type has fifteen subtypes, and both 3R2P and 3P2R types have thirty subtypes.

Table 2. Possible structures of hybrid manipulators designed from open kinematic chains.

5R Type					
(1) RRRRR		(2) RRRRR		(3) RRRRR	
4R1P type					
(4) PRRRR	(5) PRRRR	(6) PRRRR	(7) RPRRR	(8) RPRRR	(9) RPRRR
(10) RRPRR	(11) RRPRR	(12) RRPRR	(13) RRRPR	(14) RRRPR	(15) RRRPR
(16) RRRRP		(17) RRRRP		(18) RRRRP	
3R2P type					
(19) PPRRR	(20) PPRRR	(21) PPRRR	(22) PRPRR	(23) PRPRR	(24) PRPRR
(25) PRRPR	(26) PRRPR	(27) PRRPR	(28) PRRRP	(29) PRRRP	(30) PRRRP
(31) RPPRR	(32) RPPRR	(33) RPPRR	(34) RPRPR	(35) RPRPR	(36) RPRPR
(37) RPRRP	(38) RPRRP	(39) RPRRP	(40) RRPPR	(41) RRPPR	(42) RRPPR
(43) RRPRP	(44) RRPRP	(45) RRPRP	(46) RRRPP	(47) RRRPP	(48) RRRPP
3P2R type					
(49) PPPRR	(50) PPPRR	(51) PPPRR	(52) PPRPR	(53) PPRPR	(54) PPRPR
(55) PPRRP	(56) PPRRP	(57) PPRRP	(58) PRPPR	(59) PRPPR	(60) PRPPR
(61) PRPRP	(62) PRPRP	(63) PRPRP	(64) PRRPP	(65) PRRPP	(66) PRRPP
(67) RPPPR	(68) RPPPR	(69) RPPPR	(70) RPRPP	(71) RPRPP	(72) RPRPP
(73) RPPRP	(74) RPPRP	(75) RPPRP	(76) RRPPP	(77) RRPPP	(78) RRPPP

As mentioned in the paragraph above Table 1, each structure of the primary open kinematic chain can characterize various mechanisms with different joint arrangements. This diversity allows us to synthesize various hybrid manipulators within the same subtype of Table 2. For example, in the RRRRR subtype, the parallel part, indicated by the underline, can represent a planar, spherical, or other parallel mechanism. In the next section, we will exemplify this situation and show several mechanisms corresponding to the same subtype.

3. Results of Method Application

In this section, we will show how to synthesize hybrid manipulators using the proposed approach. We will consider several examples for each of the four types given in Table 2 (5R, 4R1P, 3R2P, and 3P2R). For each type, we will look at one subtype and present various mechanisms that correspond to it. Other subtypes can be analyzed in the same manner. We will focus on open kinematic chains (hence, hybrid manipulators) whose output link has either 3T2R or 3R2T motion pattern: as we saw in Section 1, these are two typical motion patterns of 5-DOF mechanical systems. In this regard, we will consider primary open chains with the following joint arrangement [13] (p. 52):

1. For the 3T2R motion pattern, the axes of all R joints should remain parallel to a common plane;
2. For the 3R2T motion pattern, the axes of all R joints should intersect a common line, orthogonal to the axes of all P joints.

To ensure full-cycle mobility, the conditions above should be satisfied for any configuration of the open chain.

3.1. Type 5R, Subtype RRRRR

The RRRRR subtype (# 1 in Table 2) can correspond to different hybrid manipulators. For example, consider an RRRRR primary open kinematic chain, where the axes of the first three R joints intersect at a common point, whereas the axes of the remaining R joints are parallel (Figure 1a). In the figure, the red curve indicates the replaced subchain, and the blue curve—the remaining serial part. Figure 1b,c illustrate two hybrid manipulators developed from this chain. The manipulators have similar serial parts, but different parallel parts. In Figure 1b, the parallel part is a 3-RRR spherical mechanism [55]; in Figure 1c, the parallel part is a 3-RUS/S spherical mechanism [56] (U and S indicate a universal and a spherical joint, respectively). Both manipulators provide its output link with a 3R2T motion pattern, but different parallel parts affect the manipulator characteristics. The 3-RRR mechanism is overconstrained, so we expect the manipulator in Figure 1b to have higher stiffness and motion accuracy than the one in Figure 1c. On the other hand, the circular rail used in the 3-RUS/S parallel mechanism allows the manipulator in Figure 1c to rotate unlimitedly around the rail axis. Such an unlimited rotation is impossible in the 3-RRR mechanism, so the orientation workspace of the manipulator in Figure 1c can be greater than in Figure 1b.

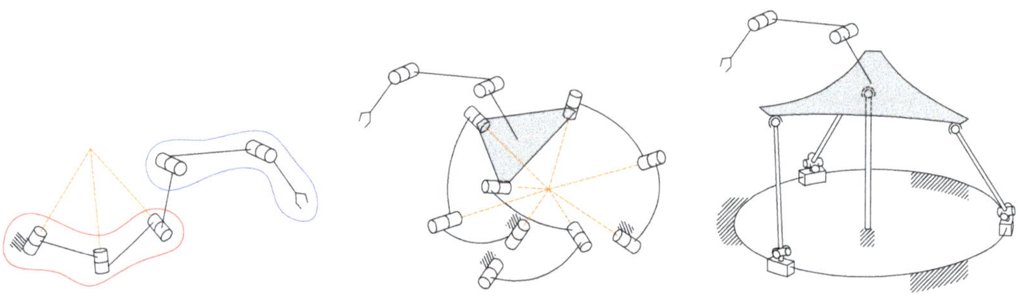

(a) (b) (c)

Figure 1. 5R type, RRRRR subtype (# 1 in Table 2): (**a**) primary open kinematic chain; (**b**,**c**) synthesized 5-DOF 3R2T hybrid manipulators with a 3-DOF spherical parallel mechanism.

An RRRRR primary open kinematic chain can also have a joint arrangement that leads to hybrid manipulators with a 3T2R motion pattern. For example, let the chain comprise two sets of joints with parallel axes (Figure 2a). Figure 2b,c show two hybrid manipulators synthesized from this chain. In Figure 2b, the parallel part is a 3-RRR planar mechanism [57]; in Figure 2c, the parallel part is a 3-RRP planar mechanism [58]. The 3-RRR mechanism includes only the revolute joints and has a simple design. The circular rail and unactuated prismatic joints of the 3-RRP mechanism make its design more cumbersome, but its output link can rotate unlimitedly about the rail axis, like the mechanism in Figure 1c. In addition, the 3-RRP planar mechanism is free of singular configurations [58].

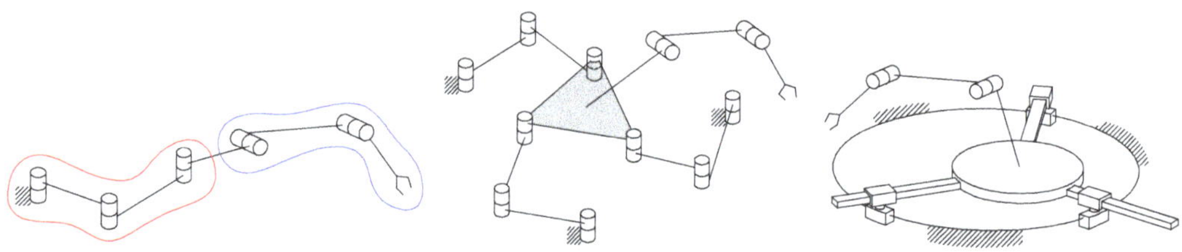

(a) (b) (c)

Figure 2. 5R type, RRRRR subtype (# 1 in Table 2): (**a**) primary open kinematic chain; (**b**,**c**) synthesized 5-DOF 3T2R hybrid manipulators with a 3-DOF planar parallel mechanism.

3.2. *Type 4R1P, Subtype RPRRR*

The next example considers hybrid manipulators that can be designed from the RPRRR subtype (# 7 in Table 2). As an example, we look at an RPRRR primary open chain with the following joint arrangement (Figure 3a). The axes of the first two R joints are orthogonal to each other and the axis of the P joint; the axis of the third R joint is parallel to the axis of the second R joint; the axis of the last R joint is orthogonal to the axis of the third R joint and intersects it as well as the axis of the first R joint. Figure 3b,c illustrate two hybrid manipulators designed from the open chain. Both manipulators include a 3-DOF RPR-equivalent parallel part with a 2R1T motion pattern [2] (ch. 8). The axes of the R joints in the serial part are arranged in such a way that the output link has a 3R2T motion pattern. In Figure 3b, the 3-DOF parallel mechanism has a 2-PUR/PRU structure, where the use of P joints placed on the base increases the mechanism stiffness [59]. On the other hand, the 3-DOF 2-URR/4R parallel mechanism depicted in Figure 3c does not have prismatic joints, and each of its branches includes a 3-RRR planar mechanism [60]. Such a structure makes the mechanism more compact, so it can fold to almost a planar state.

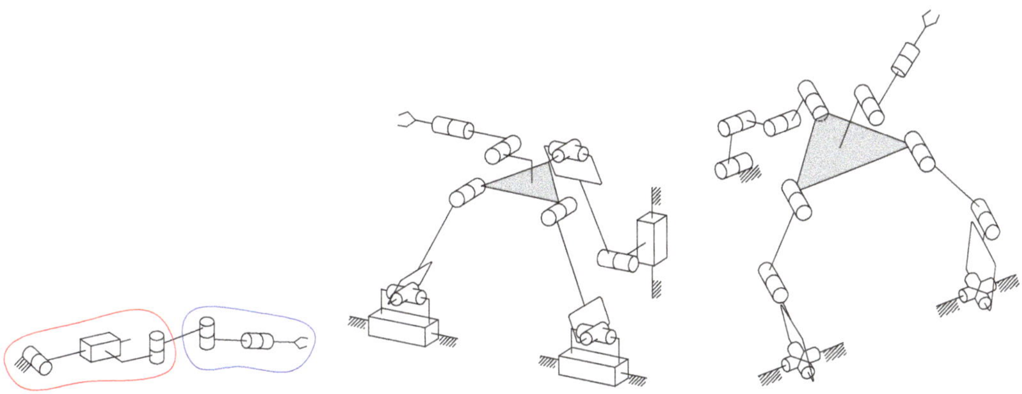

(**a**) (**b**) (**c**)

Figure 3. 4R1P type, RPRRR subtype (# 7 in Table 2): (**a**) primary open kinematic chain; (**b**,**c**) synthesized 5-DOF 3R2T hybrid manipulators with a 3-DOF RPR-equivalent parallel mechanism.

The RPRRR subtype can also give rise to hybrid manipulators with a 3T2R motion pattern. For example, if the RPR subchain corresponds to a planar motion and the axes of the remaining R joints are parallel, we can develop manipulators similar to Figure 2b,c.

3.3. *Type 3R2P, Subtype PPRRR*

This subsection examines hybrid manipulators that we can synthesize from the PPRRR subtype (# 19 in Table 2). As an example, we consider a PPRRR primary open chain, where the axis of the first R joint is parallel to the plane defined by the axes of the P joints, whereas the axes of the two remaining R joints are parallel to each other (Figure 4a). This open chain allows us to synthesize hybrid manipulators with a 3T2R motion pattern. Figure 4b,c show two such manipulators, where the parallel part is a 3-DOF PPR-equivalent parallel mechanism with a 2T1R motion pattern [13] (ch. 8). The parallel mechanism in Figure 4b has a 2-RPU/UPU structure [61], and the parallel mechanism in Figure 4c has a 2-PRU/PRC structure [62], where C indicates a cylindrical joint. If we compare these mechanisms, we see that the second one is overconstrained. This means it will be more challenging to manufacture and assemble this mechanism, but it will have higher stiffness if it is designed properly. All these properties will be inherent to the synthesized 5-DOF hybrid manipulators too.

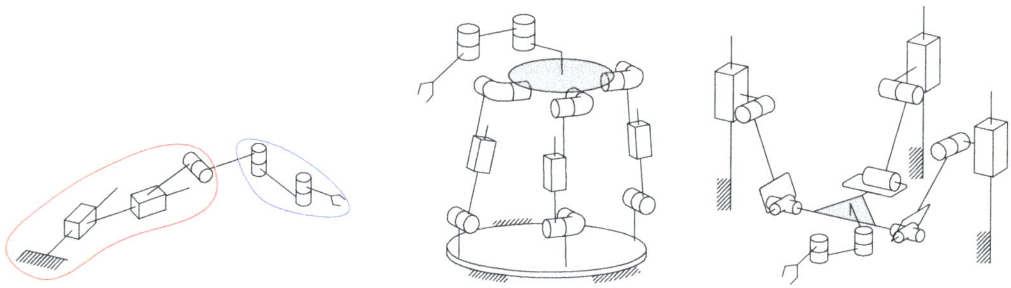

(a) (b) (c)

Figure 4. 3R2P type, PPRRR subtype (# 19 in Table 2): (**a**) primary open kinematic chain; (**b**,**c**) synthesized 5-DOF 3T2R hybrid manipulators with a 3-DOF PPR-equivalent parallel mechanism.

We can also use the PPRRR subtype to develop hybrid manipulators with a 3R2T motion pattern: the axes of the three R joints of the primary open chain should intersect at a common point.

3.4. *Type 3P2R, Subtype PRRPP*

The final example considers hybrid manipulators that correspond to the PRRPP subtype (# 64 in Table 2). Because 3P2R open kinematic chains include only two R joints, the developed hybrid manipulators can have only a 3T2R motion pattern. Figure 5a shows a primary open kinematic chain, where two R joints are equal to a single U joint. Using this chain, we can develop 3T2R hybrid manipulators with a 3-DOF PU-equivalent parallel part, which has a 2R1T motion pattern [2] (ch. 9). Figure 5b presents one such manipulator with a 3-PRS parallel mechanism [24]. The parallel mechanism has a symmetric design with prismatic joints placed on the base; the axes of all these P joints are parallel. We expect this manipulator to have a high stiffness and an elongated workspace along the direction of the P joints. It is known, however, that such a 3-PRS mechanism has parasitic motions: any rotation of the platform causes linear displacements of its center [63]. In contrast, a 3-DOF 2-SPS/PU parallel mechanism shown in Figure 5c does not have such a parasitic motion [64]. Orientation of the output link of the hybrid manipulator is determined directly by the PU branch of the 3-DOF parallel mechanism.

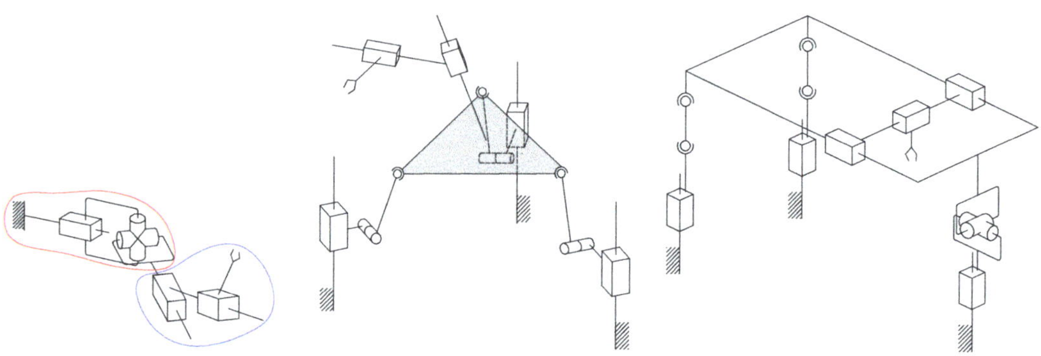

(a) (b) (c)

Figure 5. 3P2R type, PRRPP subtype (# 64 in Table 2): (**a**) primary open kinematic chain; (**b**,**c**) synthesized 5-DOF 3T2R hybrid manipulators with a 3-DOF PU-equivalent parallel mechanism.

4. Discussion

In the previous section, we showed how to apply the proposed method to synthesize 5-DOF hybrid manipulators from open kinematic chains. We focused on subtypes where the parallel part was near the base, because most practical applications mentioned in Section 1 use exactly these manipulators. However, some applications like machining [50] or manipulation [65] can also use hybrid manipulators with a parallel part near the output link. The synthesis method we introduced here is suitable for designing these manipulators too.

Although the suggested approach for type synthesis looks straightforward, it has never been proposed before. Unlike the studies mentioned in Section 1, this approach does not need any mathematical background used in those studies, so it can be more intuitive for a designer. Indeed, open kinematic chains usually include subchains with joints whose axes are parallel, orthogonal, or intersecting. Hence, it can often be possible to select a 3-DOF subchain and substitute it for a parallel mechanism with the same motion pattern.

In Section 3, we considered spherical, planar, RPR-equivalent, PPR-equivalent, and PU-equivalent parallel mechanisms—these are typical 3-DOF parallel mechanisms, which were the subject of numerous studies. There are two more widely used classes of 3-DOF parallel mechanisms we missed here: translational [66] and UP-equivalent [67], which can also be used in the proposed method. As we have seen in the examples, each class of 3-DOF parallel mechanisms can be represented by different mechanical structures with their pros and cons, which will affect the entire 5-DOF hybrid manipulator. Moreover, several possibilities exist to select a subchain within the same open kinematic chain. For example, we can transform the PRRPP open chain in Figure 5a into a hybrid manipulator that includes a PU-equivalent parallel mechanism (subtype PRRPP), a UP-equivalent parallel mechanism (subtype PRRPP), or an (inverted) PPR-equivalent parallel mechanism (subtype PRRPP). A designer should select the substituted subchain and the structure of the 3-DOF parallel mechanism according to the application.

Apart from diverse possibilities of selecting the substituted subchain, there is a non-unique choice of the actuation scheme of the 3-DOF parallel mechanism. A properly chosen actuation scheme should make the manipulator controllable in any general (non-singular) configuration [13] (p. 40). It is also preferable to have the actuated joints located close to the manipulator base and distributed evenly among the kinematic chains. Most of the parallel mechanisms presented in Section 3 meet these conditions: the actuated joints of all these mechanisms, except for Figure 4b, can be placed on the base, and all the mechanisms are controllable. Some 3-DOF parallel mechanisms, however, accept several actuation schemes. For example, the 3-RRR planar mechanism shown in Figure 2b can have actuators in the middle R joints. Such an actuator arrangement affects the singularity loci of this mechanism and its kinematic performance [68], and this actuation scheme can also be suitable for practical applications.

We should note that it is also possible to synthesize a hybrid manipulator whose output link repeats the motion pattern of the primary open kinematic chain, but the parallel part does not replicate the motion pattern of the substituted subchain. To demonstrate this situation, we consider an RRRRR subtype (# 1 in Table 2) and an open kinematic chain presented in Figure 6a. The chain has two sets of R joints with parallel axes, but in contrast to Figure 2c, the selected subchain does not correspond to a planar motion. Furthermore, it does not correspond to any 3-DOF motion type we have examined so far, and it is difficult to find a parallel mechanism that can replace this subchain. We can, however, substitute the subchain with another parallel mechanism. Figure 6b shows a synthesized hybrid manipulator with a 3-RPR planar parallel mechanism [57]. Although the motion pattern of this planar mechanism differs from the RRR subchain in Figure 6a, the output link of the hybrid manipulator has the same 3T2R motion pattern as the primary open kinematic chain. This mismatch between the motion patterns allows us to generate many novel hybrid manipulators that will keep motion patterns of their primary open chains. Developing this approach, however, will require techniques of screw theory or group theory, which are beyond the current paper.

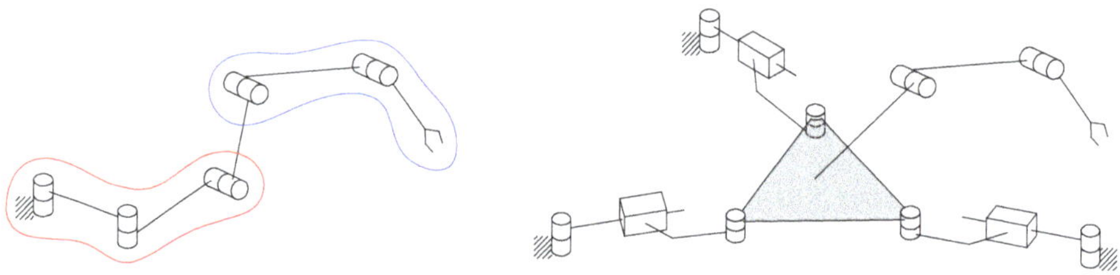

(**a**) (**b**)

Figure 6. 5R type, RRRRR subtype (# 1 in Table 2): (**a**) primary open kinematic chain; (**b**) synthesized 5-DOF 3T2R hybrid manipulator, whose parallel part has a motion pattern different from the replaced subchain.

Following the proposed type synthesis method, we have designed and analyzed several novel hybrid manipulators with five DOFs and developed their virtual prototypes. Figure 7a shows one such manipulator, whose structural diagram corresponds to Figure 5c. We examined the kinematics, workspace, and singularities of this manipulator in works [35,69]. A relatively large and elongated workspace of the manipulator makes it suitable for processing or inspecting long-shaped objects with complex geometry. Figure 7b illustrates another 5-DOF 3T2R hybrid manipulator, which we developed and studied in papers [70,71]. The manipulator structure corresponds to either an RPRPR or an RRRPR subtype from Table 2, with the parallel part being a redundantly actuated 3-DOF 2-RPR/2-RRR planar parallel mechanism. A redundant chain makes the mechanism design symmetrical and suitable for applications like those listed above. Using different kinematic chains is a trade-off: a 4-RRR mechanism is less rigid than a 4-RPR one, while the latter is prone to architecture singularities [72]. In addition, the redundant actuation affects the manipulator performance [73], and it can also be used in other hybrid manipulators we presented in Section 3.

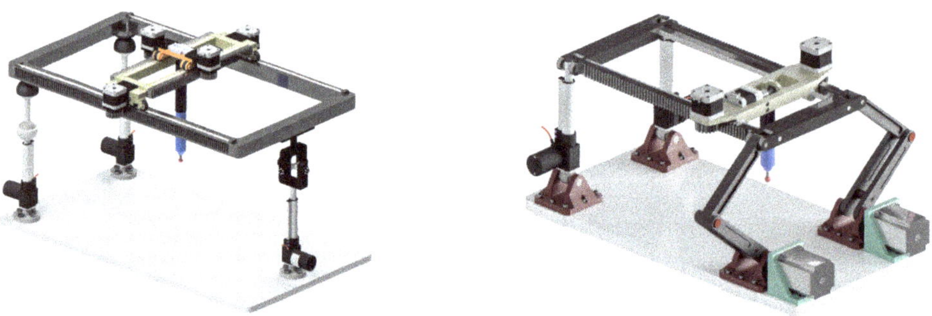

(**a**) (**b**)

Figure 7. Virtual prototypes of two 5-DOF 3T2R hybrid manipulators developed by the proposed method: (**a**) manipulator with a 3-DOF PU-equivalent parallel mechanism and a PP serial chain; (**b**) manipulator with a redundantly actuated 3-DOF planar parallel mechanism and a PR serial chain.

5. Conclusions

The article has proposed a novel approach for synthesizing 5-DOF hybrid manipulators from open kinematic chains. In contrast to other synthesis methods that rely on screw theory, group theory, POC sets, or G_F sets, the suggested approach does not need any extensive mathematical formulations, and it is more intuitive for a designer. The method idea is in taking a 5-DOF open kinematic chain, selecting a 3-DOF subchain, and replacing it with a 3-DOF parallel mechanism that has the motion pattern of this subchain. The choice of the substituted subchain and the parallel mechanism determines the performance of the synthesized hybrid manipulator, and it should be carried out by a designer according to the application.

In this paper, we have considered open kinematic chains with revolute (R) and prismatic (P) joints and found 26 possible structures of these chains, which we have classified into four types (5R, 4R1P, 3R2P, and 3P2R). For each type, there are three possibilities to select a subchain, so we have found 78 subtypes of hybrid manipulators. We have studied mechanisms with 3T2R and 3R2T motion patterns and presented several hybrid manipulators for each subtype. The synthesized hybrid manipulators include spherical, planar, RPR-equivalent, PPR-equivalent, and PU-equivalent parallel mechanisms.

In the future, we will aim to enhance the proposed synthesis method and consider the cases when the substituted subchain and the parallel mechanism have different motion patterns, while the output link of the designed hybrid manipulator keeps the motion pattern of its primary open kinematic chain. A further development of the current work includes its extension for manipulators with another number of DOFs. The proposed approach is certainly not limited to 5-DOF open chains and 3-DOF parallel mechanisms, so we will look at other combinations in the future. Another extension of the performed research is a comprehensive analysis of synthesized manipulators, such as the ones we presented here. The drive location, actuation redundancy, and joint and link design affect the manipulator workspace, singular configurations, and stiffness, so our forthcoming studies will aim to address these issues too.

Author Contributions: Conceptualization, A.A., A.F., V.G. and G.F.; methodology, A.A., A.F., V.G. and G.F.; validation, A.A. and A.F.; formal analysis, A.A. and A.F.; investigation, A.A. and A.F.; resources, A.A. and A.F.; writing—original draft preparation, A.A. and A.F.; writing—review and editing, A.A. and A.F.; visualization, A.A., A.F. and D.P.; supervision, A.A., A.F. and V.G.; project administration, A.A. and A.F.; funding acquisition, A.A. and A.F. All authors have read and agreed to the published version of the manuscript.

Funding: This research was supported by Russian Science Foundation (RSF) under grant No. 22-79-10304, https://rscf.ru/project/22-79-10304/.

Institutional Review Board Statement: Not applicable.

Informed Consent Statement: Not applicable.

Data Availability Statement: The data presented in this study are available on request from the corresponding author.

Conflicts of Interest: The authors declare that they have no conflict of interest.

References

1. Ye, W.; Li, Q. Type synthesis of lower mobility parallel mechanisms: A review. *Chin. J. Mech. Eng.* **2019**, *32*, 38. [CrossRef]
2. Li, Q.; Hervé, J.M.; Ye, W. *Geometric Method for Type Synthesis of Parallel Manipulators*; Springer: Singapore, 2020. [CrossRef]
3. Gao, F.; Yang, J.; Ge, Q.J. Type synthesis of parallel mechanisms having the second class G_F sets and two dimensional rotations. *J. Mech. Robot.* **2011**, *3*, 011003. [CrossRef]
4. Zhang, J.; Jin, Z.; Feng, H. Type synthesis of a 3-mixed-DOF protectable leg mechanism of a firefighting multi-legged robot based on G_F set theory. *Mech. Mach. Theory* **2018**, *130*, 567–584. [CrossRef]
5. Yang, T.L.; Liu, A.X.; Shen, H.P.; Hang, L.B.; Luo, Y.F.; Jin, Q. *Topology Design of Robot Mechanisms*; Springer: Singapore, 2018. [CrossRef]
6. Gogu, G. *Structural Synthesis of Parallel Robots: Part 1—Methodology*; Springer: Dordrecht, The Netherlands, 2008. [CrossRef]
7. Sun, T.; Yang, S.; Lian, B. *Finite and Instantaneous Screw Theory in Robotic Mechanism*; Springer: Singapore, 2020. [CrossRef]
8. Song, Y.; Han, P.; Wang, P. Type synthesis of 1T2R and 2R1T parallel mechanisms employing conformal geometric algebra. *Mech. Mach. Theory* **2018**, *121*, 475–486. [CrossRef]
9. Huang, Z.; Li, Q. Type synthesis of symmetrical lower-mobility parallel mechanisms using the constraint-synthesis method. *Int. J. Robot. Res.* **2003**, *22*, 59–79. [CrossRef]
10. Huang, Z.; Li, Q.; Ding, H. *Theory of Parallel Mechanisms*; Springer: Dordrecht, The Netherlands, 2013. [CrossRef]
11. Borges Dos Santos, J.V.; Simoni, R.; Carboni, A.P.; Martins, D. A new method for type synthesis of parallel mechanisms using screw theory and features of genetic algorithms. *J. Braz. Soc. Mech. Sci. Eng.* **2020**, *42*, 615. [CrossRef]

12. Hu, B.; Bai, P. Type synthesis of serial kinematic chains with screw type terminal constraints based on an adding joint method. *Mech. Mach. Theory* **2023**, *184*, 105277. [CrossRef]
13. Kong, X.; Gosselin, C. *Type Synthesis of Parallel Mechanisms*; Springer: Berlin-Heidelberg, Germany, 2007. [CrossRef]
14. Hopkins, J.B.; Culpepper, M.L. Synthesis of multi-degree of freedom, parallel flexure system concepts via freedom and constraint topology (FACT)—Part I: Principles. *Precis. Eng.* **2010**, *34*, 259–270. [CrossRef]
15. Hopkins, J.B.; Culpepper, M.L. Synthesis of multi-degree of freedom, parallel flexure system concepts via freedom and constraint topology (FACT). Part II: Practice. *Precis. Eng.* **2010**, *34*, 271–278. [CrossRef]
16. Xie, F.; Li, T.; Liu, X. Type synthesis of 4-DOF parallel kinematic mechanisms based on Grassmann line geometry and atlas method. *Chin. J. Mech. Eng.* **2013**, *26*, 1073–1081. [CrossRef]
17. Zhang, Y.; Huang, H.; Mei, T.; Li, B. Type synthesis of single-loop deployable mechanisms based on improved atlas method for single-DOF grasping manipulators. *Mech. Mach. Theory* **2022**, *169*, 104656. [CrossRef]
18. Kuo, C.H.; Dai, J.S. Structural synthesis of serial robotic manipulators subject to specific motion constraints. In Proceedings of the ASME 2010 International Design Engineering Technical Conferences and Computers and Information in Engineering Conference, Montreal, QC, Canada, 15–18 August 2010; Volume 2, pp. 907–915. [CrossRef]
19. Kuo, C.H.; Dai, J.S. Task-oriented structure synthesis of a class of parallel manipulators using motion constraint generator. *Mech. Mach. Theory* **2013**, *70*, 394–406. [CrossRef]
20. Tsai, L.W. Systematic enumeration of parallel manipulators. In *Parallel Kinematic Machines*; Boër, C.R., Molinari-Tosatti, L., Smith, K.S., Eds.; Springer: London, UK, 1999; pp. 33–49. [CrossRef]
21. Lu, Y.; Leinonen, T. Type synthesis of unified planar–spatial mechanisms by systematic linkage and topology matrix-graph technique. *Mech. Mach. Theory* **2005**, *40*, 1145–1163. [CrossRef]
22. Ramirez, D. Automatic Generation of Task-Specific Serial Mechanisms Using Combined Structural and Dimensional Synthesis. Ph.D. Thesis, Gottfried Wilhelm Leibniz Universität, Hannover, Germany, 2018. [CrossRef]
23. Meng, X.; Gao, F.; Wu, S.; Ge, Q.J. Type synthesis of parallel robotic mechanisms: Framework and brief review. *Mech. Mach. Theory* **2014**, *78*, 177–186. [CrossRef]
24. Xie, F.; Liu, X.J.; Li, T. Type synthesis and typical application of 1T2R-type parallel robotic mechanisms. *Math. Probl. Eng.* **2013**, *2013*, 206181. [CrossRef]
25. Lin, G.; Huang, P.; Wang, M.; Xu, Y.; Zhang, R.; Zhu, L. An inverse kinematics solution for a series-parallel hybrid banana-harvesting robot based on deep reinforcement learning. *Agronomy* **2022**, *12*, 2157. [CrossRef]
26. Pisla, D.; Gherman, B.; Vaida, C.; Suciu, M.; Plitea, N. An active hybrid parallel robot for minimally invasive surgery. *Robot. Comp. Int. Manuf.* **2013**, *29*, 203–221. [CrossRef]
27. Zhou, M.; Yu, Q.; Huang, K.; Mahov, S.; Eslami, A.; Maier, M.; Lohmann, C.P.; Navab, N.; Zapp, D.; Knoll, A.; et al. Towards robotic-assisted subretinal injection: A hybrid parallel–serial robot system design and preliminary evaluation. *IEEE Trans. Ind. Electron.* **2020**, *67*, 6617–6628. [CrossRef]
28. Chakarov, D.; Parushev, P. Synthesis of parallel manipulators with linear drive modules. *Mech. Mach. Theory* **1994**, *29*, 917–932. [CrossRef]
29. Campos, A.; Budde, C.; Hesselbach, J. A type synthesis method for hybrid robot structures. *Mech. Mach. Theory* **2008**, *43*, 984–995. [CrossRef]
30. Alizade, R.; Bayram, Ç. Structural synthesis of parallel manipulators. *Mech. Mach. Theory* **2004**, *39*, 857–870. [CrossRef]
31. Zeng, Q.; Fang, Y. Structural synthesis of serial-parallel hybrid mechanisms based on representation and operation of logical matrix. *J. Mech. Robot.* **2009**, *1*, 041003. [CrossRef]
32. Zeng, Q.; Fang, Y. Algorithm for topological design of multi-loop hybrid mechanisms via logical proposition. *Robotica* **2012**, *30*, 599–612. [CrossRef]
33. Zeng, Q.; Fang, Y. Structural synthesis and analysis of serial–parallel hybrid mechanisms with spatial multi-loop kinematic chains. *Mech. Mach. Theory* **2012**, *49*, 198–215. [CrossRef]
34. Shen, H.; Zhao, H.; Deng, J.; Meng, Q.; Zhu, W.; Yang, T. Type design method and the application for hybrid robot based on freedom distribution and position and orientation characteristic set. *J. Mech. Eng.* **2011**, *47*, 56–64. [CrossRef]
35. Antonov, A.; Fomin, A.; Glazunov, V.; Kiselev, S.; Carbone, G. Inverse and forward kinematics and workspace analysis of a novel 5-DOF (3T2R) parallel–serial (hybrid) manipulator. *Int. J. Adv. Robot. Syst.* **2021**, *18*. [CrossRef]
36. Wang, C.; Fang, Y.; Guo, S. Design and analysis of 3R2T and 3R3T parallel mechanisms with high rotational capability. *J. Mech. Robot.* **2016**, *8*, 011004. [CrossRef]
37. Kim, S.K.; Shin, W.H.; Ko, S.Y.; Kim, J.; Kwon, D.S. Design of a compact 5-DOF surgical robot of a spherical mechanism: CURES. In Proceedings of the 2008 IEEE/ASME International Conference on Advanced Intelligent Mechatronics, Xi'an, China, 2–5 July 2008; pp. 990–995. [CrossRef]
38. Pisla, D.; Plitea, N.; Gherman, B.G.; Vaida, C.; Pisla, A.; Suciu, M. Kinematics and design of a 5-DOF parallel robot used in minimally invasive surgery. In *Advances in Robot Kinematics: Motion in Man and Machine*; Lenarčič, J., Stanišić, M.M., Eds.; Springer: Dordrecht, The Netherlands, 2010; pp. 99–106. [CrossRef]
39. Tsai, C.Y.; Wong, C.C.; Yu, C.J.; Liu, C.C.; Liu, T.Y. A hybrid switched reactive-based visual servo control of 5-DOF robot manipulators for pick-and-place tasks. *IEEE Syst. J.* **2015**, *9*, 119–130. [CrossRef]

40. Gao, F.; Peng, B.; Zhao, H.; Li, W. A novel 5-DOF fully parallel kinematic machine tool. *Int. J. Adv. Manuf. Tech.* **2006**, *31*, 201–207. [CrossRef]
41. Tian, W.; Mou, M.; Yang, J.; Yin, F. Kinematic calibration of a 5-DOF hybrid kinematic machine tool by considering the ill-posed identification problem using regularisation method. *Robot. Comp. Int. Manuf.* **2019**, *60*, 49–62. [CrossRef]
42. Li, Y.; Tan, D.; Wen, D.; Ji, S.; Cai, D. Parameters optimization of a novel 5-DOF gasbag polishing machine tool. *Chin. J. Mech. Eng.* **2013**, *26*, 680–688. [CrossRef]
43. Sun, T.; Wu, H.; Lian, B.; Qi, Y.; Wang, P.; Song, Y. Stiffness modeling, analysis and evaluation of a 5 degree of freedom hybrid manipulator for friction stir welding. *Proc. Inst. Mech. Eng. C J. Mech. Eng. Sci.* **2017**, *231*, 4441–4456. [CrossRef]
44. Cao, Y.; Qin, Y.; Chen, H.; Ge, S.; Zhou, H. Structural synthesis of 5-DOF hybrid mechanisms based on G_F set. *Trans. Chin. Soc. Agr. Mach.* **2015**, *46*, 392–398. [CrossRef]
45. Cao, Y.; Zhou, R.; Qin, Y.; Ge, S.; Ding, R. Structural synthesis of fully-isotropic five degree-of-freedom hybrid kinematic mechanisms. *J. Mech. Eng.* **2018**, *54*, 29–37. [CrossRef]
46. Zhou, H.; Qin, Y.; Chen, H.; Ge, S.; Cao, Y. Structural synthesis of five-degree-of-freedom hybrid kinematics mechanism. *J. Eng. Des.* **2016**, *27*, 390–412. [CrossRef]
47. Shen, H.; Yin, H.; Li, J.; Deng, J.; Liu, A. Position and orientation characteristic based method and enlightenment for topology characteristic analysis of typical parallel mechanisms and its application. *J. Mech. Eng.* **2015**, *51*, 101–115. [CrossRef]
48. Xu, Y.; Yao, J.; Zhao, Y. Type synthesis of spatial mechanisms for forging manipulators. *Proc. Inst. Mech. Eng. C J. Mech. Eng. Sci.* **2012**, *226*, 2320–2330. [CrossRef]
49. Xu, Y.; Zhao, Y.; Yue, Y.; Xi, F.; Yao, J.; Zhao, Y. Type synthesis of overconstrained 2R1T parallel mechanisms with the fewest kinematic joints based on the ultimate constraint wrenches. *Mech. Mach. Theory* **2020**, *147*, 103766. [CrossRef]
50. Zhang, D.; Zheng, Y.; Wei, L.; Wu, J.; Xu, Y.; Zhao, Y. Type synthesis of 2T1R planar parallel mechanisms and their moduling development applications. *IEEE Access* **2021**, *9*, 72217–72227. [CrossRef]
51. Zhang, D.; Xu, Y.; Yao, J.; Zhao, Y. Design of a novel 5-DOF hybrid serial-parallel manipulator and theoretical analysis of its parallel part. *Robot. Comp. Int. Manuf.* **2018**, *53*, 228–239. [CrossRef]
52. Merlet, J.P. *Parallel Robots*, 2nd ed.; Springer: Dordrecht, The Netherlands, 2006. [CrossRef]
53. Fang, Y.; Tsai, L.W. Enumeration of a class of overconstrained mechanisms using the theory of reciprocal screws. *Mech. Mach. Theory* **2004**, *39*, 1175–1187. [CrossRef]
54. Riordan, J. *Introduction to Combinatorial Analysis*, Dover ed.; Dover Publications: Mineola, NY, USA, 2002. Available online: https://scholar.google.com/scholar?cluster=17160231405262501443 (accessed on 10 July 2023).
55. He, P.; Kantu, N.T.; Xu, B.; Swami, C.P.; Saleem, G.T.; Kang, J. A novel 3-RRR spherical parallel instrument for daily living emulation (SPINDLE) for functional rehabilitation of patients with stroke. *Int. J. Adv. Robot. Syst.* **2021**, *18*. [CrossRef]
56. Khoshnoodi, H.; Rahmani Hanzaki, A.; Talebi, H.A. Kinematics, singularity study and optimization of an innovative spherical parallel manipulator with large workspace. *J. Intell. Robot. Syst.* **2018**, *92*, 309–321. [CrossRef]
57. Si, G.; Chen, F.; Zhang, X. Comparison of the dynamic performance of planar 3-DOF parallel manipulators. *Machines* **2022**, *10*, 233. [CrossRef]
58. Bonev, I.A.; Yu, A.; Zsombor-Murray, P. XY-Theta positioning table with parallel kinematics and unlimited theta rotation. In Proceedings of the 2006 IEEE International Symposium on Industrial Electronics, Montreal, QC, Canada, 9–13 July 2006; Volume 4, pp. 3113–3117. [CrossRef]
59. Li, Q.; Xu, L.; Chen, Q.; Ye, W. New family of RPR-equivalent parallel mechanisms: Design and application. *Chin. J. Mech. Eng.* **2017**, *30*, 217–221. [CrossRef]
60. Li, Q.; Hervé, J.M. Type synthesis of 3-DOF RPR-equivalent parallel mechanisms. *IEEE Trans. Robot.* **2014**, *30*, 1333–1343. [CrossRef]
61. Kong, X.; Gosselin, C.M. Type synthesis of 3-DOF PPR-equivalent parallel manipulators based on screw theory and the concept of virtual chain. *J. Mech. Des.* **2005**, *127*, 1113–1121. [CrossRef]
62. Xie, F.; Liu, X.J.; You, Z.; Wang, J. Type synthesis of 2T1R-type parallel kinematic mechanisms and the application in manufacturing. *Robot. Comp. Int. Manuf.* **2014**, *30*, 1–10. [CrossRef]
63. Yao, Y.; Wu, W.; Li, R.; Zhao, Y. Parasitic motions of 3-PRS parallel mechanisms with two different branch chain arrangements. *Appl. Sci.* **2023**, *13*, 5425. [CrossRef]
64. Hu, B.; Huang, Z. A family of 2R1T parallel manipulators with intersecting rotational axes. In *Advances in Reconfigurable Mechanisms and Robots II*; Ding, X., Kong, X., Dai, J.S., Eds.; Springer: Cham, Switzerland, 2016; pp. 287–295. [CrossRef]
65. Sun, P.; Li, Y.B.; Wang, Z.S.; Chen, K.; Chen, B.; Zeng, X.; Zhao, J.; Yue, Y. Inverse displacement analysis of a novel hybrid humanoid robotic arm. *Mech. Mach. Theory* **2020**, *147*, 103743. [CrossRef]
66. Prause, I.; Charaf Eddine, S.; Corves, B. Comparison of parallel kinematic machines with three translational degrees of freedom and linear actuation. *Chin. J. Mech. Eng.* **2015**, *28*, 841–850. [CrossRef]
67. Ye, W.; Li, Q.; Chai, X.X. New family of 3-DOF UP-equivalent parallel mechanisms with high rotational capability. *Chin. J. Mech. Eng.* **2018**, *31*, 12. [CrossRef]
68. Zhang, Z.; Wang, L.; Shao, Z. Improving the kinematic performance of a planar 3-RRR parallel manipulator through actuation mode conversion. *Mech. Mach. Theory* **2018**, *130*, 86–108. [CrossRef]

69. Laryushkin, P.; Antonov, A.; Fomin, A.; Essomba, T. Velocity and singularity analysis of a 5-DOF (3T2R) parallel-serial (hybrid) manipulator. *Machines* **2022**, *10*, 276. [CrossRef]
70. Antonov, A.V.; Fomin, A.S. Inverse kinematics of a 5-Dof hybrid manipulator. *Autom. Remote Control* **2023**, *84*, 281–293. [CrossRef]
71. Antonov, A.; Fomin, A. Velocity analysis of a 5-DOF hybrid manipulator. In *New Advances in Mechanisms, Transmissions and Applications*; Laribi, M.A., Nelson, C.A., Ceccarelli, M., Zeghloul, S., Eds.; Springer: Cham, Switzerland, 2023; pp. 161–170. [CrossRef]
72. Wu, X.; Bai, S. Analytical determination of shape singularities for three types of parallel manipulators. *Mech. Mach. Theory* **2020**, *149*, 103812. [CrossRef]
73. Gosselin, C.; Schreiber, L.T. Redundancy in parallel mechanisms: A review. *Appl. Mech. Rev.* **2018**, *70*, 010802. [CrossRef]

robotics

MDPI

Article

A Novel Error Sensitivity Analysis Method for a Parallel Spindle Head

Liping Wang [1,2], Mengyu Li [1,2] and Guang Yu [1,2,*]

[1] State Key Laboratory of Tribology, Department of Mechanical Engineering, Tsinghua University, Beijing 100084, China
[2] Beijing Key Laboratory of Precision/Ultra-Precision Manufacturing Equipment and Control, Beijing 100084, China
* Correspondence: gyu@tsinghua.edu.cn

Abstract: Geometric errors are the main factors affecting the output accuracy of the parallel spindle head, and it is necessary to perform a sensitivity analysis to extract the critical geometric errors. The traditional sensitivity analysis method analyzes the output position and orientation errors independently, defining multiple sensitivity indices and making it difficult to determine critical geometric errors. In this paper, we propose sensitivity indices that can comprehensively consider position and orientation errors. First, the configuration of the hybrid machine tool is introduced, and the TCP position error model is derived. Then, the tool radius and the effective cutting length are introduced, and the sensitivity indices are defined. After that, the sensitivity analysis of the 3-DOF parallel spindle head is performed using the proposed sensitivity indices, and six critical geometric errors are extracted. The machining accuracy of the parallel spindle head can be greatly improved by improving the critical geometric errors. The proposed sensitivity analysis method can provide important guidance for machine tool accuracy design.

Keywords: parallel spindle head; sensitivity analysis; sensitivity indices; critical geometric errors

Citation: Wang, L.; Li, M.; Yu, G. A Novel Error Sensitivity Analysis Method for a Parallel Spindle Head. *Robotics* **2023**, *12*, 129. https://doi.org/10.3390/robotics12050129

Academic Editor: Raffaele Di Gregorio

Received: 13 August 2023
Revised: 4 September 2023
Accepted: 10 September 2023
Published: 11 September 2023

1. Introduction

Due to the large demand for complex parts, five-axis machine tools are widely used in production practice [1,2]. Among them, hybrid machine tools combine the advantages of high stiffness, low inertia, and high flexibility of the parallel mechanism and the advantages of a large working space for the series mechanism, which have higher machining efficiency compared with traditional five-axis machines and become ideal configurations for aerospace machining [3–5].

Accuracy performance is the main index that evaluates the effectiveness of machining, fabrication, and deburring [6–8]. The factors affecting the accuracy of machine tools mainly include geometric errors, thermal errors, load-induced structural deformation errors, and servo errors [9,10]. Among them, geometric error is the main error source, accounting for 40–50% of the total error source [11]. For the hybrid machine tool, due to the existence of a large number of limbs and passive joints, the geometric error of the parallel spindle head is the main factor causing the error [12,13]. Therefore, how to improve the geometric accuracy of the parallel spindle head is the key to ensuring the machining accuracy of the hybrid machine tool.

There are two ways to improve the geometric accuracy of parallel spindle heads: precision design and error compensation [14]. Precision design refers to the elimination of possible error sources in the design, manufacturing, and assembly of machine tools by optimizing materials and structures and ensuring assembly quality [15]. The corresponding error compensation technology, mainly through the establishment of an error model and the measurement and identification of each error parameter, compensates the identified error in real time in practical application to improve the accuracy [16,17]. Whether it is a precision design or error compensation, it is important to evaluate the error parameters. Therefore, establishing the relationship between the error parameters and output errors and performing a sensitivity analysis is the key to accuracy improvement. By comparing the sensitivity indices, the structural errors that have a greater impact on machining accuracy can be extracted so that more attention can be paid to them in the process of design, manufacture, assembly, and use.

Error modeling is used to establish the relationship between the geometric error parameters and the output errors [18]. According to the different methods of mathematical description of joint motion, the error modeling methods can be divided into the matrix method [19,20], the screw method [21,22], and the vector method [23,24]. The matrix method, taking the D-H method as an example, mainly establishes a local coordinate system in each component, describes the motion relationship of adjacent components through the homogeneous transformation matrix, and constructs the relationship between each error parameter and the output error with the matrix product, differentiation, and other operations. However, the presence of multiple limbs and passive joints of parallel mechanisms makes the matrix method very difficult to model. The screw method represents the instantaneous motion of the moving and rotating joints by means of screws and then obtains the motion of the moving and rotating joints by integrating the screws. However, the complexity of the derivation, the lack of practical physical meaning of the error parameters, and the severe parameter redundancy greatly limit its use in error modeling of parallel mechanisms. The vector method is used to construct closed-loop vector equations and then differentiate each parameter and simplify them by vector operations to obtain the transfer relationship between the error parameters and the output error. The vector expression form is intuitive, and the physical meaning is clear, which is very suitable for the error modeling of parallel mechanisms with closed-loop structural features, and the closed-loop vector method is used for geometric error modeling in this paper.

Error sensitivity refers to the influence of each geometric error parameter on the output error, and sensitivity analysis can guide the extraction of the critical geometric error of the mechanism [25] and provide a basis for the optimization of mechanical performance [26,27]. The sensitivity analysis of the geometric error is mainly divided into two types: local sensitivity analysis (LSA) and global sensitivity analysis (GSA) [28]. LSA is mainly used to analyze the effect of geometric error on the output error when the tool is located at a particular pose, while GSA analyzes the average effect of error parameters on the output error in the entire workspace. Patel [29] analyzed the error sensitivity of the Hexapod parallel machine tool by approximate linear error mapping. Fan [30] analyzed the sensitivity of the 3-PRS parallel mechanism and identified its critical errors using the partial differential method based on the error transform vector. Defining the sensitivity index is the key to sensitivity analysis. Since the output error contains both position and orientation errors, multiple sensitivity indices exist when the elements in the output error vector are analyzed independently in the traditional sensitivity analysis method. And the position and orientation errors are inhomogeneous, making it difficult to compare multiple error sensitivity indices and thus difficult to identify critical geometric errors. Jiang [31] and Du [32] used the Monte Carlo algorithm to analyze the sensitivity of the 3-DOF parallel spindle head as well as the 2UPR-RPU over-constrained parallel manipulator, respectively, but they constructed two sensitivity indices for output position and orientation, respectively, which are still difficult to compare. To solve this problem, the influence of geometric error on the tool orientation error is ignored, and only the tool center point (TCP) position error is considered to construct the error sensitivity index in Refs. [33–35], which can reduce the number of sensitivity indices and facilitate the comparison. However, the tool orientation error also has a large impact on the machining accuracy, and ignoring the orientation error leads to incomplete error sensitivity analysis. Li [36] proposed a sensitivity analysis method for the position and orientation errors by introducing the average cutting length, but the method actually converts the orientation error into the positional error of a point on the tool, whereas in the actual cutting process, the actual cutting is performed not just at a point on the tool, and thus the method cannot fully reflect the impact on machining accuracy. In addition, most of the sensitivity analysis methods are based on the error values of position and orientation errors or the TCP position error modulus to model the sensitivity indices; however, when the tool is in different poses, the same position error values bring different impacts on machining accuracy, and the traditional methods cannot consider the impacts of the error values on the machining accuracy in different pose, which results in an incomplete analysis of the error sensitivity. In summary, it is necessary to propose a single sensitivity index that comprehensively considers position and orientation errors and can directly evaluate the effect of geometric errors on machining accuracy at each pose.

In this paper, a new method of sensitivity analysis is proposed. By introducing the tool radius and effective cutting length, a sensitivity index that can comprehensively consider the position and orientation errors is defined, which solves the problem that the traditional sensitivity indexes are difficult to compare and cannot fully reflect the machining accuracy.

Following this section, in Section 2, the configuration of the hybrid machine tool is introduced, and the TCP position error model is derived using the closed-loop vector method. In Section 3 the error sensitivity indices are defined. In Section 4, the error sensitivity of the 3-DOF parallel spindle head is analyzed using the proposed sensitivity indices, and the simulation results are discussed. Finally, the conclusion of this paper is given.

2. Configuration and Error Modeling

2.1. Configuration

Figure 1 shows the five-axis hybrid machine tool studied in this paper, which consists of a working platform and a 3-DOF parallel spindle head mounted on a column. A base platform and a moving platform are connected by three identical PRRU limbs to form the 3-DOF parallel spindle head, as shown in Figure 1b. Here P, R, and U denote prismatic, revolute, and universal joints, respectively. The three P joints are actuated joints. The parallel spindle head can realize Z-direction movement and rotation around the X and Y axes and cooperate with the XY platform to realize 5-DOF motion. The XY platform is essentially two perpendicular linear axes with a very simple structure, making it easy to compensate for geometric errors. The 3-DOF parallel spindle head has a complex structure with a large number of limbs and passive joints, resulting in a large manufacturing and assembly error. And the nonlinear motion characteristics of the parallel spindle head lead to difficulties in compensating for its geometric errors. Therefore, the geometric error of the 3-DOF parallel spindle head is the main source of geometric error in the five-axis hybrid machine tool. This paper mainly analyzes the error sensitivity of this 3-DOF parallel spindle head.

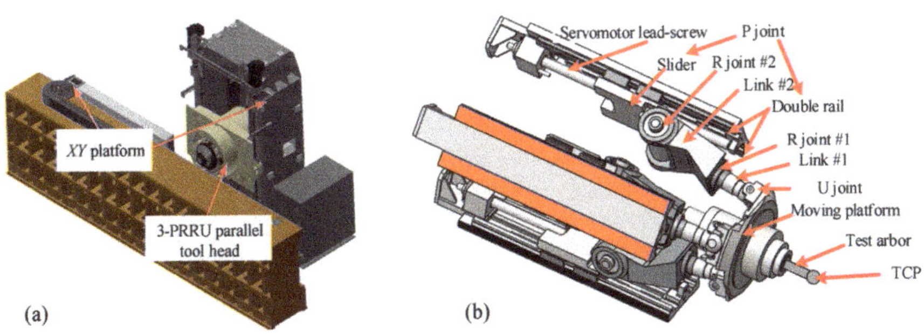

Figure 1. (**a**) Configuration of the hybrid machine tool; (**b**) structure of the 3-DOF parallel spindle head.

2.2. TCP Position Error Modeling

In kinematic studies, the 3-PRRU structure can be simplified to a 3-PRS mechanism. For each limb, its closed-loop vector equation is constructed.

$$\boldsymbol{H} + \boldsymbol{R}_{\mathrm{TT}}\boldsymbol{R}_i\boldsymbol{a}_i = \boldsymbol{R}_i\boldsymbol{b}_i + \boldsymbol{R}_i\boldsymbol{R}_{Bi}\boldsymbol{q}_i + \boldsymbol{R}_i\boldsymbol{R}_{Bi}\boldsymbol{R}_{Ci}\boldsymbol{l}_i \tag{1}$$

where $\boldsymbol{a}_i$, $\boldsymbol{b}_i$, and $\boldsymbol{l}_i$ are shown in Figure 2. $\boldsymbol{H} = [x\ y\ z]^{\mathrm{T}}$ denotes the vector $\boldsymbol{OO'}$, $\boldsymbol{q}_i$ represents the drive vector, $\boldsymbol{R}_i$ represents the rotation matrix of each limb, and $\boldsymbol{R}_{Bi}$ and $\boldsymbol{R}_{Ci}$ denote the rotation matrices of the P and R joint. $\boldsymbol{R}_{\mathrm{TT}}$ represents the orientation matrix of the moving platform, which is described using the T-T angle [37,38] to decouple the rotation around the Z' axis from the other two rotations.

$$\boldsymbol{R}_{\mathrm{TT}} = \begin{bmatrix} \cos\varphi & -\sin\varphi & 0 \\ \sin\varphi & \cos\varphi & 0 \\ 0 & 0 & 1 \end{bmatrix} \begin{bmatrix} \cos\theta & 0 & \sin\theta \\ 0 & 1 & 0 \\ -\sin\theta & 0 & \cos\theta \end{bmatrix} \begin{bmatrix} \cos\varphi & \sin\varphi & 0 \\ -\sin\varphi & \cos\varphi & 0 \\ 0 & 0 & 1 \end{bmatrix} \begin{bmatrix} \cos\psi & -\sin\psi & 0 \\ \sin\psi & \cos\psi & 0 \\ 0 & 0 & 1 \end{bmatrix} \tag{2}$$

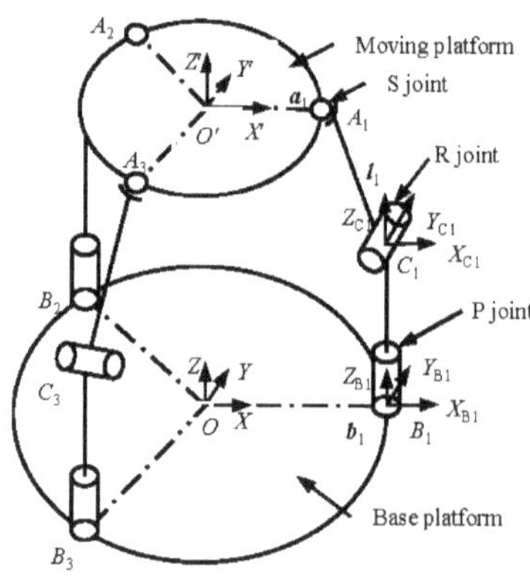

Figure 2. Equivalent mechanism diagram.

The error transfer relationship for the parallel spindle head is obtained by taking a partial derivative of Equation (1).

$$\begin{aligned} &\Delta\boldsymbol{H} + \Delta\boldsymbol{\alpha} \times (\boldsymbol{R}_{\mathrm{TT}}\boldsymbol{R}_i\boldsymbol{a}_i) + \boldsymbol{R}_{\mathrm{TT}}\boldsymbol{R}_i\Delta\boldsymbol{a}_i = \\ &\boldsymbol{R}_i\Delta\boldsymbol{b}_i + \Delta q_i\boldsymbol{R}_i\boldsymbol{R}_{\mathrm{B}i}\boldsymbol{q}_i/q_i + (\boldsymbol{R}_i\boldsymbol{R}_{\mathrm{B}i}\Delta\boldsymbol{\theta}_{Bi}) \times (\boldsymbol{R}_i\boldsymbol{R}_{\mathrm{B}i}\boldsymbol{q}_i) + \Delta l_i\boldsymbol{R}_i\boldsymbol{R}_{\mathrm{B}i}\boldsymbol{R}_{\mathrm{C}i}\boldsymbol{l}_i/l_i \\ &+(\boldsymbol{R}_i\boldsymbol{R}_{\mathrm{B}i}\Delta\boldsymbol{\theta}_{Bi}) \times (\boldsymbol{R}_i\boldsymbol{R}_{\mathrm{B}i}\boldsymbol{R}_{\mathrm{C}i}\boldsymbol{l}_i) + (\boldsymbol{R}_i\boldsymbol{R}_{\mathrm{B}i}\boldsymbol{R}_{\mathrm{C}i}\Delta\boldsymbol{\theta}_{\mathrm{C}i}) \times (\boldsymbol{R}_i\boldsymbol{R}_{\mathrm{B}i}\boldsymbol{R}_{\mathrm{C}i}\boldsymbol{l}_i) \end{aligned} \tag{3}$$

Here, Δ denotes the error vector. $\Delta\boldsymbol{a}_i$ and $\Delta\boldsymbol{b}_i$ denote the geometric error vector of $\boldsymbol{a}_i$, $\boldsymbol{b}_i$, respectively. Δq_i and Δl_i denote the input error of the P joint and the length error of the link, respectively. $\Delta\boldsymbol{\theta}_{Bi}$ and $\Delta\boldsymbol{\theta}_{Ci}$ are the orientation error vectors of orientation matrix $\boldsymbol{R}_{Bi}$ and $\boldsymbol{R}_{Ci}$, respectively. $\Delta\boldsymbol{H} = [\Delta x\ \Delta y\ \Delta z]^{\mathrm{T}}$ and $\Delta\boldsymbol{\alpha} = [\Delta\alpha_x\ \Delta\alpha_y\ \Delta\alpha_z]^{\mathrm{T}}$ denote the position and orientation error of the center of the moving platform, and a total of 42 error parameters are included in this error model, i.e., $dr = [\Delta q_i\ \Delta l_i\ \Delta a_{ix}\ \Delta a_{iy}\ \Delta a_{iz}\ \Delta b_{ix}\ \Delta b_{iy}\ \Delta b_{iz}\ \Delta\theta_{Bix}\ \Delta\theta_{Biy}\ \Delta\theta_{Biz}\ \Delta\theta_{Cix}\ \Delta\theta_{Ciy}\ \Delta\theta_{Ciz}]$ (i = 1, 2, 3).

Use $\boldsymbol{w}_i$ to denote $\boldsymbol{R}_i\boldsymbol{R}_{\mathrm{B}i}\boldsymbol{R}_{\mathrm{C}i}\boldsymbol{l}_i$ and multiply $\boldsymbol{w}_i$ on both sides of Equation (3).

$$\begin{aligned} &\boldsymbol{w}_i^{\mathrm{T}}\Delta\boldsymbol{H} + [(\boldsymbol{R}_{\mathrm{TT}}\boldsymbol{R}_i\boldsymbol{a}_i) \times \boldsymbol{w}_i]^{\mathrm{T}}\Delta\boldsymbol{\alpha} = \boldsymbol{w}_i^{\mathrm{T}}\boldsymbol{R}_i\Delta\boldsymbol{b}_i + \Delta q_i\boldsymbol{w}_i^{\mathrm{T}}\boldsymbol{R}_i\boldsymbol{R}_{\mathrm{B}i}\boldsymbol{q}_i/q_i \\ &+[(\boldsymbol{R}_i\boldsymbol{R}_{\mathrm{B}i}\boldsymbol{q}_i) \times \boldsymbol{w}_i]^{\mathrm{T}}\boldsymbol{R}_i\boldsymbol{R}_{\mathrm{B}i}\Delta\boldsymbol{\theta}_{Bi} + \Delta l_i - \boldsymbol{w}_i^{\mathrm{T}}\boldsymbol{R}_{\mathrm{TT}}\boldsymbol{R}_i\Delta\boldsymbol{a}_i \end{aligned} \tag{4}$$

Similarly, using v_i to denote $R_iR_{Bi}R_{Ci}e_2$, multiply v_i on both sides of Equation (3). Here, $e_2 = [0\ 1\ 0]^T$.

$$\begin{aligned} &v_i^T\Delta H + [(R_{TT}R_ia_i)\times v_i]^T\Delta\alpha = v_i^TR_i\Delta b_i + [(R_iR_{Bi}q_i)\times v_i]^TR_iR_{Bi}\Delta\theta_{Bi} \\ &+[w_i\times v_i]^TR_iR_{Bi}\Delta\theta_{Bi} + [w_i\times v_i]^TR_iR_{Bi}R_{Ci}\Delta\theta_{Ci} - v_i^TR_{TT}R_i\Delta a_i \end{aligned} \quad (5)$$

Combining Equations (4) and (5) gives:

$$Adp = Bdr \quad (6)$$

where $dp = [\Delta H^T\ \Delta\alpha^T]^T$ and $dr = [dr_1^T\ dr_2^T\ dr_3^T]^T$. dr_i is the geometric error vector of limb i. dr_i, A, and B can be expressed as:

$$dr_i = [\Delta q_i \quad \Delta l_i \quad \Delta a_i^T \quad \Delta b_i^T \quad \Delta\theta_{Bi}^T \quad \Delta\theta_{Ci}^T]^T$$

$$A = \begin{bmatrix} w_1^T & [(R_{TT}R_1a_1)\times w_1]^T \\ w_1^T & [(R_{TT}R_2a_2)\times w_2]^T \\ w_1^T & [(R_{TT}R_3a_3)\times w_3]^T \\ v_1^T & [(R_{TT}R_1a_1)\times v_1]^T \\ v_2^T & [(R_{TT}R_2a_2)\times v_2]^T \\ v_3^T & [(R_{TT}R_3a_3)\times v_3]^T \end{bmatrix},\quad B = \begin{bmatrix} \tau_1 & & \\ & \tau_2 & \\ & & \tau_3 \\ \rho_1 & & \\ & \rho_2 & \\ & & \rho_3 \end{bmatrix},$$

where:

$$\begin{aligned} \tau_i &= \begin{bmatrix} w_i^TR_iR_{Bi}q_i/q_i & 1 & -w_i^TR_{TT}R_i & w_i^TR_i & [(R_iR_{Bi}q_i)\times w_i]^TR_iR_{Bi} & 0_{1\times3} \end{bmatrix} \\ \rho_i &= \begin{bmatrix} 0 & 0 & -v_i^TR_{TT}R_i & v_i^TR_i & [(R_iR_{Bi}q_i)\times v_i]^TR_iR_{Bi} + [w_i\times v_i]^TR_iR_{Bi} & [w_i\times v_i]^TR_iR_{Bi}R_{Ci} \end{bmatrix} \end{aligned} \quad (7)$$

Equation (6) contains both position and orientation vectors and is not suitable for error sensitivity analysis. Since the position error of the TCP is affected by both the position and orientation errors of the moving platform, the TCP position error model is derived to avoid the non-uniformity of the position and orientation errors. When the tool length is L, the TCP position can be expressed as:

$$P = H + R_{TT}\cdot[0 \quad 0 \quad L]^T \quad (8)$$

where $P = [X\ Y\ Z]^T$. Derivation of the above equation gives:

$$dP = Cdp_{TT} \quad (9)$$

where $dp_{TT} = [\Delta x\ \Delta y\ \Delta z\ \Delta\varphi\ \Delta\theta\ \Delta\psi]^T$ represents the pose error based on the T-T angle, and we can obtain the transformation relationship between $[\Delta\varphi\ \Delta\theta\ \Delta\psi]^T$ and $[\Delta\alpha_x\ \Delta\alpha_y\ \Delta\alpha_z]^T$ as:

$$\begin{bmatrix} \Delta\alpha_x \\ \Delta\alpha_y \\ \Delta\alpha_z \end{bmatrix} = \begin{bmatrix} d\psi c\varphi s\theta - d\varphi c\varphi s\theta - d\theta s\varphi \\ d\theta c\varphi - d\varphi s\varphi s\theta + d\psi s\varphi s\theta \\ d\varphi - d\varphi c\theta + d\psi c\theta \end{bmatrix} = D\begin{bmatrix} \Delta\varphi \\ \Delta\theta \\ \Delta\psi \end{bmatrix} \quad (10)$$

Based on Equations (6), (9) and (10), the TCP position error transfer model can be expressed as:

$$dP = C\begin{bmatrix} 1 & 0 & 0 & \\ 0 & 1 & 0 & \\ 0 & 0 & 1 & \\ & & & D^{-1} \end{bmatrix}A^{-1}Bdr = \mathrm{T}dr \quad (11)$$

Observing Equation (6), the coefficients of Δb_{iz} and Δq_i, and $\Delta\theta_{Biz}$ and $\Delta\theta_{Ciz}$ are the same; they are redundant with each other and have the same effect on the output error, and the coefficient of $\Delta\theta_{Ciy}$ is 0, which does not have any effect on the terminal error. In order to simplify the analyzing process, we keep only one of them for the redundant parameters and eliminate $\Delta\theta_{Ciy}$. $\boldsymbol{T}$ can represent the transfer relationship between the geometric error parameters and the TCP position error. The use of the $\boldsymbol{T}$ for error sensitivity modeling avoids the problem of difficult comparisons caused by the non-uniformity of the output position and orientation errors.

3. Definition of Sensitivity Indices

3.1. Introduction of the Sensitivity Indices

Geometric error sensitivity indices are used to evaluate the effect of geometric error parameters on the output errors. By comparing the sensitivity indices of each error parameter, it is possible to extract the error parameters that have a high impact on the output error, i.e., the critical geometric error parameters from the error parameters. The sensitivity indices are mainly categorized into local sensitivity indices (LSI) and global sensitivity metrics (GSI). Here, the LSI is used to evaluate the effect of geometric error on the output error at a particular pose, while the GSI is used to evaluate the average effect of geometric error on the output error over the entire workspace.

In the traditional sensitivity analysis method, the definition of sensitivity indices is directly related to the output of the error model, and a total of six sensitivity indices are generally defined for three position errors and three orientation errors. And the non-uniformity of the position and orientation leads to non-uniformity between the defined sensitivity indices. The non-uniformity of the sensitivity indices as well as their excessive number lead to an inability to effectively determine the critical geometric errors that have a large impact on the output errors. For example, a certain geometric error may have a large impact on orientation error but a small impact on position error, and then it is not possible to make a comparison using six sensitivity indices.

In order to solve the problem, some scholars have simplified the sensitivity indices using only position or orientation errors [39,40], which, although convenient for comparison, cannot fully reflect the impact of error parameters on machining accuracy. Figure 3 shows the comparison of machining errors of milled slots in three different output error cases; the cutting area of the tool is represented by an envelope cylinder. Due to the presence of geometric errors, the actual tool does not overlap with the commanded tool and the non-overlapping portion leads to machining errors, resulting in overcut or undercut. The TCP position (TP) errors are the same in all three cases, and the tool orientation (TO) errors are different. The TO errors in Figure 3a,c are equal in size and opposite in direction, and the TO error in Figure 3b is zero. The TO error in Figure 3a increases the machining error and has the largest machining error, and the TO error in Figure 3c decreases the machining error and has the smallest machining error. Figure 3 illustrates that it is not reasonable to use the TP error or TO error alone to evaluate the machine error; so, to perform a comprehensive geometric error sensitivity analysis, the effects of both position error and orientation error must be considered.

In addition, most of the traditional error sensitivity analyses are based on the values of TP and TO errors (e.g., ΔX ΔY ΔZ $\Delta\varphi$ $\Delta\theta$ $\Delta\psi$) or the TP error modulus ($|\boldsymbol{dP}|$) for sensitivity indices modeling; however, when the tool is in different orientation, the same TP or TO error values bring different impacts on the machining accuracy. As shown in Figure 4, Figure 4a,b show the machine in two different poses, and the TP and TO errors are the same in both cases; however, due to the different machine poses, the impact of the same TP error on the machining is different.

Therefore, it is unscientific to use the output error values for direct sensitivity modeling under different orientations, and there is a need to propose a single sensitivity index that comprehensively considers position and orientation errors and can directly evaluate the effect of geometric errors on machining accuracy at each pose.

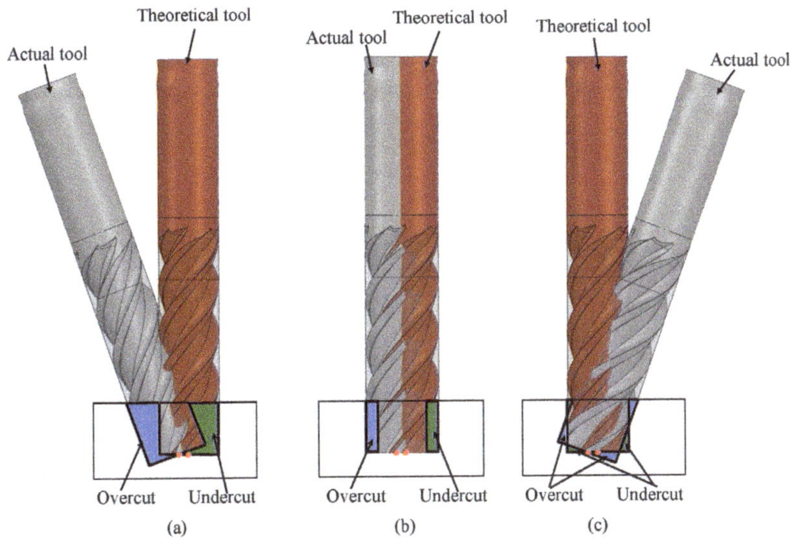

Figure 3. Comparison of machining errors of the milled slots in three different output error cases. (**a**) TO error is *do;* (**b**) TO error is 0; (**c**) TO error is $-do$.

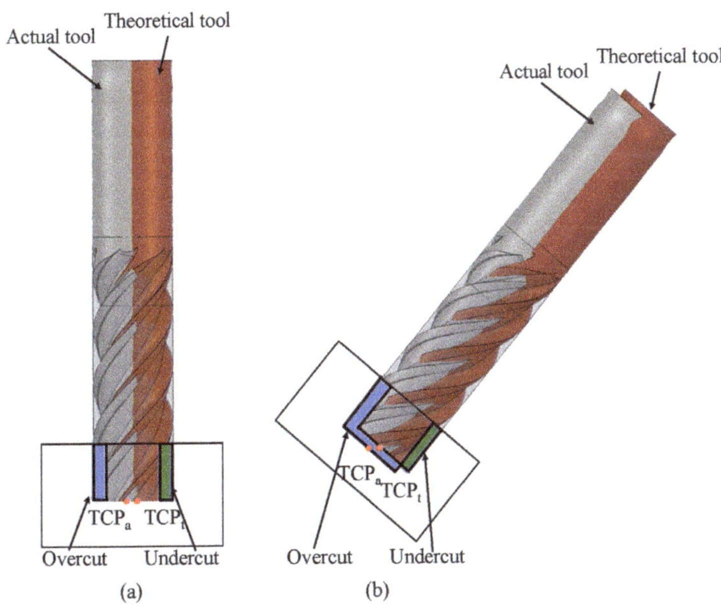

Figure 4. Comparison of the effect of the output errors on machining error at different poses. (**a**) zero orientation; (**b**) non-zero orientation.

3.2. Definition of the Sensitivity Indices

According to the analysis in Section 3.1, the consideration of output position or orientation errors cannot fully reflect the influence of geometric error on machining accuracy, and the essence of machining error is the error between the actual cutting area and the theoretical cutting area. Given the tool radius R and tool length L, define the effective cutting length for L_e. As shown in Figure 5, $\boldsymbol{A}_1\boldsymbol{B}_1$ represents the theoretical tool axis, and the actual tool axis is $\boldsymbol{A}_2\boldsymbol{B}_2$ after being affected by geometric errors. That is, the theoretical cutting area is a cylinder with $\boldsymbol{A}_1\boldsymbol{B}_1$ as the central axis and R as the radius, and the actual cutting area is a cylinder with $\boldsymbol{A}_2\boldsymbol{B}_2$ as the central axis and R as the radius after being affected by geometric errors. The portion of the theoretical and actual cutting area that fails to intersect is the machining error (the amount of overcut or undercut). The theoretical cutting area is denoted as Ω_1, and the actual cutting area is denoted as Ω_2. The volume of Ω_1 is V, and the volume of the part that fails to intersect is V_1. For each geometric error parameter, the resulting machining error V_1 can be found at each pose, and $w_i = V_1/V$ can then be used to define the LSI and GSI.

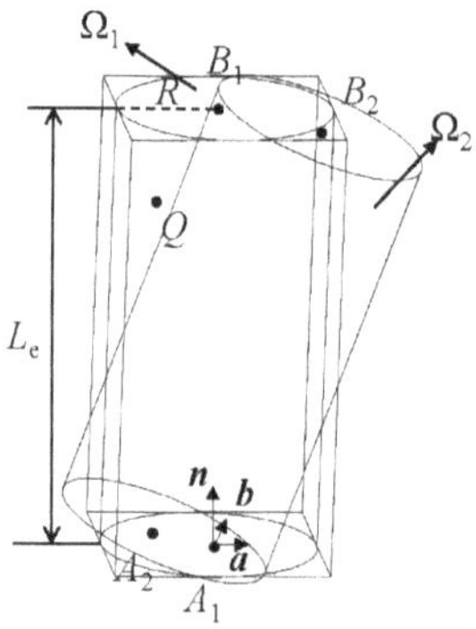

Figure 5. Definition of sensitivity indices based on the theoretical and actual cutting areas.

Thus, the sensitivity indices definition can be converted into a problem of solving the volume of the intersecting part of two spatial cylinders, which can be solved efficiently by the Monte Carlo method [41]. The Monte Carlo method is a stochastic simulation method based on probability and statistical theory methods that uses random numbers to solve computational problems. In order to randomly select points uniformly within the theoretical cutting region, a cuboid containing Ω_1 is first constructed, as shown in Figure 5. Since the cross-product result of two vectors is always perpendicular to the original two vectors as long as it is not **0**, the three normal vectors of the cuboid are solved using the characteristics of the cross-product operation.

$$\boldsymbol{n} = \frac{\boldsymbol{A}_1\boldsymbol{B}_1}{|\boldsymbol{A}_1\boldsymbol{B}_1|} \tag{12}$$

$$\boldsymbol{a} = \frac{\boldsymbol{n} \times \begin{bmatrix}1 & 0 & 0\end{bmatrix}^{\mathrm{T}}}{\left|\boldsymbol{n} \times \begin{bmatrix}1 & 0 & 0\end{bmatrix}^{\mathrm{T}}\right|} \tag{13}$$

$$if\ \boldsymbol{a} = \begin{bmatrix}0 & 0 & 0\end{bmatrix}^{\mathrm{T}}, \boldsymbol{a} = \frac{\boldsymbol{n} \times \begin{bmatrix}0 & 1 & 0\end{bmatrix}^{\mathrm{T}}}{\left|\boldsymbol{n} \times \begin{bmatrix}0 & 1 & 0\end{bmatrix}^{\mathrm{T}}\right|} \tag{14}$$

Since we need to take only any cuboid containing the theoretical cutting area, $[1\ 0\ 0]^{\mathrm{T}}$ and $[0\ 1\ 0]^{\mathrm{T}}$ are taken, and other vectors can be taken.

$$\boldsymbol{b} = \frac{\boldsymbol{n} \times \boldsymbol{a}}{|\boldsymbol{n} \times \boldsymbol{a}|} \tag{15}$$

The three unitized normal vectors of the cuboid are obtained according to Equations (12)–(15). Here, $\boldsymbol{a}$, $\boldsymbol{b}$, and $\boldsymbol{n}$ are the three unitized normal vectors of the cuboid, and random points can then be selected:

$$\boldsymbol{Q} = \boldsymbol{A}_1 + rand(-R, R) \cdot \boldsymbol{a} + rand(-R, R) \cdot \boldsymbol{b} + rand(0, L_e) \cdot \boldsymbol{n} \tag{16}$$

$\boldsymbol{Q}$ is the position vector of a randomly taken point Q within the outer cuboid of Ω_1. $rand(-R, R)$ denotes the random number selected in $[-R\ R]$. Assuming that a total of N points is randomly taken in this cuboid and assuming that there are m out of N points that fall within Ω_1 and n out of m points that fall within Ω_2, the parameter w_i can be expressed as follows according to the Monte Carlo method:

$$w_i = 1 - n/m \tag{17}$$

Specifically, the determination of whether the point Q is within the cutting area is:

$$if\ \boldsymbol{A}_1\boldsymbol{Q} \times \boldsymbol{n} < R \text{ and } 0 < \boldsymbol{A}_1\boldsymbol{Q} \cdot \boldsymbol{n} < L_e,\ Q \in \Omega_1 \tag{18}$$

$$\begin{array}{l} if\ \boldsymbol{A}_2\boldsymbol{Q} \times \boldsymbol{n}_2 < R \text{ and } 0 < \boldsymbol{A}_2\boldsymbol{Q} \cdot \boldsymbol{n}_2 < L_e,\ Q \in \Omega_2 \\ here\ \boldsymbol{n}_2 = \frac{A_2B_2}{|A_2B_2|} \end{array} \tag{19}$$

If the point Q satisfies Equation (18), then Q is located in the theoretical cutting area, $m = m + 1$. If the point Q satisfies both Equations (18) and (19), then Q is located in the intersection of the theoretical cutting area and the actual cutting area, $n = n + 1$. Thus, for a given input vector and geometric error parameter vector, the LSI solution process using the Monte Carlo method can be expressed as Algorithm 1.

Algorithm 1: LSI solving algorithm based on Monte Carlo method

INPUT: $\boldsymbol{dr}$, L, L_e, N, $\boldsymbol{q}$
Begin
Step 1. Solve the coordinates of the points A_1, B_1, A_2, and B_2 corresponding to $\boldsymbol{dr}$ and $\boldsymbol{q}$ based on the error kinematic model.
Step 2. Solve for $\boldsymbol{n}$, $\boldsymbol{a}$, and $\boldsymbol{b}$ according to Equations (12)–(15).
Step 3. Randomly generate N points inside the least outer cuboid according to Equation (16).
Step 4. Determine the number m of points within Ω_1 and the number n of points in the region where Ω_1 intersects with Ω_2 among the N points according to Equations (18) and (19).
Step 5. Then, the LSI corresponding to the error parameter $\boldsymbol{dr}$ and input vector $\boldsymbol{q}$ can be expressed as $w_i = 1 - n/m$.

The GSI can be defined as the integral of the parameter w_i over the entire workspace, indicating the average effect of each geometric error on machining accuracy for all possible poses. The GSI can be expressed as:

$$W_i = \frac{\int_W w_i dW}{\int_W dW} \tag{20}$$

where W represents the working space of the parallel spindle head, corresponding to the motion space of the three input axes. Critical geometric errors that have a significant impact on machining errors can be determined based on the LSI and GSI defined in this paper.

4. Error Sensitivity Analysis of the 3-DOF Parallel Spindle Head

In this section, the geometric error sensitivity of the 3-DOF parallel spindle head is analyzed, and the critical errors are given based on the proposed sensitivity indices. Since we are more concerned with the accuracy characteristics in the full workspace, the proposed GSI is utilized here for the GSA. The motion range of the 3-DOF parallel spindle head is $q_i \in [300\ 550]$ (i = 1, 2, 3). The tool radius is set to R = 10 mm, the tool length is set to L = 150 mm, and the effective cutting length is set to L_e = 35 mm. Since the three limbs of this

3-DOF parallel spindle head are perfectly symmetrical, the sensitivity indices of the same type of error on its different limbs are the same, e.g., GSI(Δa_{1x}) = GSI(Δa_{2x}) = GSI(Δa_{3x}). Therefore, it is sufficient to calculate the sensitivity indices corresponding to the error parameters of only one of the limbs for the actual calculation. Here for limb 1, the GSI corresponding to each geometric error parameter is calculated, and the results are shown in Figure 6.

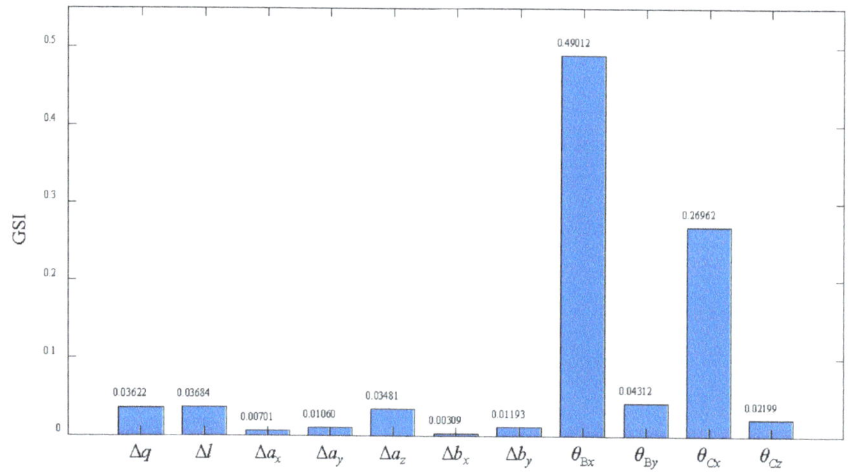

Figure 6. GSI corresponding to each geometric error parameter.

In order to show its percentage size more clearly, the GSI corresponding to each error is plotted in the form of a pie chart, as shown in Figure 7. For the accuracy design, we need to find the critical error parameters that have a large impact on the machine accuracy, and the errors with a defined percentage of more than 5% are considered the critical geometric errors, and according to the analysis results, the critical geometric errors are identified as $\Delta\theta_{Bix}$ and $\Delta\theta_{Cix}$ (i = 1, 2, 3).

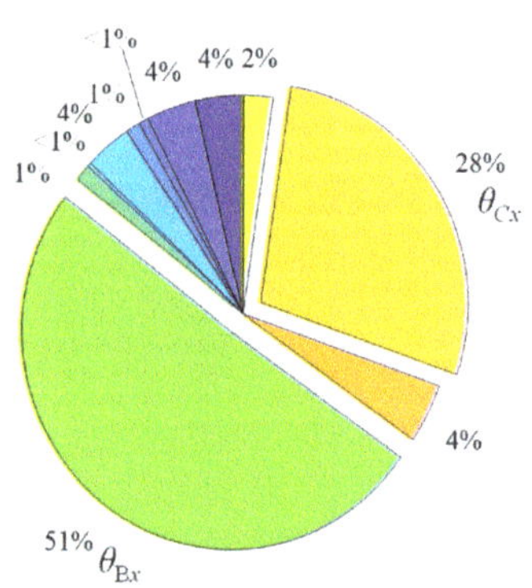

Figure 7. Pie chart of the GSI values.

In order to validate the proposed sensitivity analysis method, all the geometric position and orientation errors of the machine tool are set to 0.1 mm and 0.1°, respectively. The comparison of machine accuracy is carried out in the case of R = 10 mm and L_e = 35 mm. The maximum, mean, and root mean square (RMS) values of the actual cutting volume error across the whole workspace for the four different cases of error parameters in Table 1 are shown in Figure 8.

Table 1. The four machine accuracy design parameters used in the simulation.

Case	Critical Geometric Errors	Other Geometric Errors
1	Keep the initial value unchanged	Keep the initial value unchanged
2	Reduce by half	Keep the initial value unchanged
3	Keep the initial value unchanged	Reduce by half
4	Reduce by half	Reduce by half

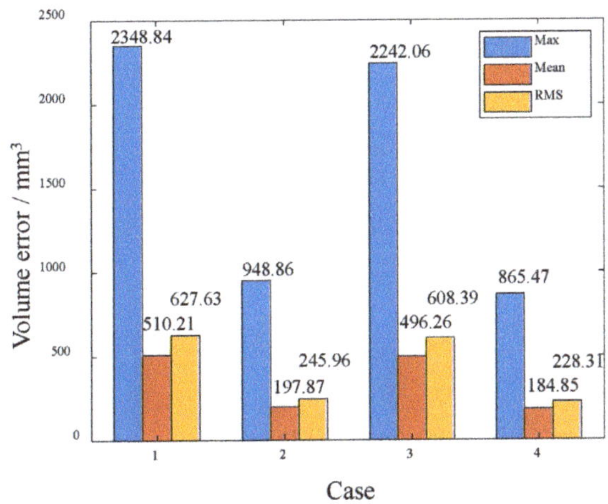

Figure 8. Maximum, mean, and root mean square (RMS) values of the actual cutting volume error across the whole workspace for the four different cases.

The relative magnitudes of the maximum, mean, and RMS values of the cutting volume error are consistent in all four cases. Taking the maximum value as an example, compared with the initial design parameters, the output accuracy is improved by 59.60% after reducing the critical geometric error, while the output accuracy is improved by only 4.55% after reducing other geometric errors. Considering the 63.15% improvement in output accuracy when reducing all geometric error parameters, a good improvement in machine accuracy can be achieved by reducing the critical geometric error. Meanwhile, since the number of critical errors (six) is much smaller than the number of remaining geometric errors, reducing the critical geometric errors achieves a better improvement of the machine accuracy, indicating that the critical geometric errors identified in this paper have a significant effect on the machine accuracy.

In addition, in order to illustrate the applicability of the proposed sensitivity analysis method more effectively, the cutting volume errors for the four different cases of error parameters in Table 1 are compared after replacing different R and L_e. Cases *a*–*f* in Table 2 correspond to the six cases of tool radius, as well as the effective cutting length, and the accuracy comparisons are shown in Figure 9. Since Figure 8 illustrates that the relative magnitudes of the maximum, mean, and RMS values are consistent across the four cases, only the mean values are compared. Simulation results show that under different machining parameters, only changing the critical geometric error can improve the machine accuracy well, which verifies the effectiveness of the sensitivity analysis method proposed in this paper.

Table 2. Six cases of the tool radius and the effective cutting length.

Case	R/mm	L_e/mm
a	6	25
b	6	45
c	8	25
d	8	45
e	12	25
f	12	45

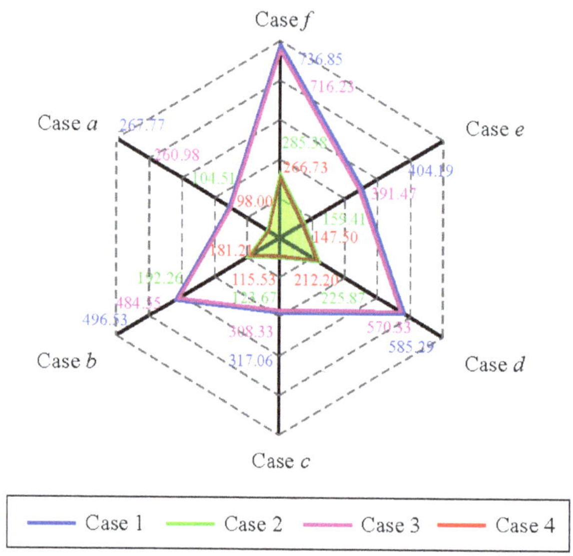

Figure 9. Accuracy comparisons of case *a*–*f* (mm^3).

5. Conclusions

A sensitivity analysis method is proposed in this paper for the geometric error of 3-DOF parallel spindle heads. By establishing the TCP position error model, the non-uniformity of the output position and orientation is avoided. By introducing the tool radius and effective cutting length, the spatial error of the machine tool is directly converted into the machining error, and a single sensitivity index that can comprehensively consider the position and orientation errors is proposed, which solves the problem of the traditional sensitivity analysis method not being able to analyze the output position and orientation error uniformly. Meanwhile, since the defined sensitivity indices do not rely on the expression of a coordinate system, they can reflect the impact of errors on machining accuracy under different poses. Based on the sensitivity analysis method proposed in this paper, quantitative information on the sensitivity of the geometric error parameters and their effect on machining errors is obtained. Six critical errors are extracted, and the validity of the extracted critical errors is verified by simulations. By improving the critical errors, the machining accuracy of the parallel spindle head can be greatly improved. The sensitivity analysis method proposed in this paper can provide important guidance for the accuracy compensation of five-axis machine tools.

In this paper, only the error sensitivity analysis method of the parallel spindle head has been investigated, but we have not carried out the accuracy improvement research on the physical prototype on the basis of error sensitivity analysis. Subsequently, optimization algorithms based on sensitivity analysis methods can be developed to improve the accuracy performance of machine tools.

Author Contributions: Conceptualization, L.W.; methodology, M.L.; software, M.L.; validation, M.L.; formal analysis, G.Y.; investigation, M.L.; resources, G.Y.; data curation, M.L.; writing—original draft preparation, M.L.; writing—review and editing, L.W.; visualization, G.Y.; supervision, G.Y.; project administration, L.W.; funding acquisition, L.W. All authors have read and agreed to the published version of the manuscript.

Funding: This research was supported by the National Natural Science Foundation of China (NSFC) under grants 52275442 and 51975319.

Data Availability Statement: All data included in this study are available upon request by contacting the corresponding author.

Conflicts of Interest: The authors declare no conflict of interest.

References

1. Slamani, M.; Mayer, R.; Balazinski, M.; Zargarbashi, S.H.; Engin, S.; Lartigue, C. Dynamic and geometric error assessment of an XYC axis subset on five-axis high-speed machine tools using programmed end point constraint measurements. *Int. J. Adv. Manuf. Technol.* **2010**, *50*, 1063–1073. [CrossRef]
2. Lee, R.S.; She, C.H. Developing a postprocessor for three types of five-axis machine tools. *Int. J. Adv. Manuf. Technol.* **1997**, *13*, 658–665. [CrossRef]
3. El-Khasawneh, B.S.; Ferreira, P.M. Computation of stiffness and stiffness bounds for parallel link manipulators. *Int. J. Mach. Tools Manuf.* **1999**, *39*, 321–342. [CrossRef]
4. Briot, S.; Bonev, I.A. Accuracy analysis of 3-DOF planar parallel robots. *Mech. Mach. Theory* **2008**, *43*, 445–458. [CrossRef]
5. Laliberté, T.; Gosselin, C.M.; Jean, M. Static balancing of 3-DOF planar parallel mechanisms. *IEEE-ASME Trans. Mechatron.* **1999**, *4*, 363–377. [CrossRef]
6. Burghardt, A.; Szybicki, D.; Kurc, K.; Muszyńska, M.; Mucha, J. Experimental study of Inconel 718 surface treatment by edge robotic deburring with force control. *Strength Mater.* **2017**, *49*, 594–604. [CrossRef]
7. Khanghah, S.P.; Boozarpoor, M.; Lotfi, M.; Teimouri, R. Optimization of micro-milling parameters regarding burr size minimization via RSM and simulated annealing algorithm. *Trans. Indian Inst. Met.* **2015**, *68*, 897–910. [CrossRef]
8. Burghardt, A.; Szybicki, D.; Kurc, K.; Muszyńska, M. Optimization of process parameters of edge robotic deburring with force control. *Int. J. Appl. Mech. Eng.* **2016**, *21*, 987–995. [CrossRef]
9. Ramesh, R.; Mannan, M.A.; Poo, A.N. Error compensation in machine tools—A review: Part I: Geometric, cutting-force induced and fixture-dependent errors. *Int. J. Mach. Tools Manuf.* **2000**, *40*, 1235–1256. [CrossRef]
10. Ramesh, R.; Mannan, M.A.; Poo, A.N. Error compensation in machine tools—A review: Part II: Thermal errors. *Int. J. Mach. Tools Manuf.* **2000**, *40*, 1257–1284. [CrossRef]
11. Khan, A.W.; Chen, W. A methodology for systematic geometric error compensation in five-axis machine tools. *Int. J. Adv. Manuf. Technol.* **2011**, *53*, 615–628. [CrossRef]
12. Chanal, H.; Duc, E.; Ray, P.; Hascoet, J.Y. A new approach for the geometrical calibration of parallel kinematics machines tools based on the machining of a dedicated part. *Int. J. Mach. Tools Manuf.* **2007**, *47*, 1151–1163. [CrossRef]
13. Majarena, A.C.; Santolaria, J.; Samper, D.; Aguilar, J.J. An overview of kinematic and calibration models using internal/external sensors or constraints to improve the behavior of spatial parallel mechanisms. *Sensors* **2010**, *10*, 10256–10297. [CrossRef]
14. Li, Q.; Wang, W.; Jiang, Y.; Li, H.; Zhang, J.; Jiang, Z. A sensitivity method to analyze the volumetric error of five-axis machine tool. *Int. J. Adv. Manuf. Technol.* **2018**, *98*, 1791–1805. [CrossRef]
15. Xu, Y.; Zhang, L.; Wang, S.; Du, H.; Chai, B.; Hu, S.J. Active precision design for complex machine tools: Methodology and case study. *Int. J. Adv. Manuf. Technol.* **2015**, *80*, 581–590. [CrossRef]
16. Li, M.; Wang, L.; Yu, G.; Li, W. A new calibration method for hybrid machine tools using virtual tool center point position constraint. *Measurement* **2021**, *181*, 109582. [CrossRef]
17. Li, M.; Wang, L.; Yu, G.; Li, W.; Kong, X. A multiple test arbors-based calibration method for a hybrid machine tool. *Robot. Comput.-Integr. Manuf.* **2023**, *80*, 102480. [CrossRef]
18. Tian, W.; Gao, W.; Zhang, D.; Huang, T. A general approach for error modeling of machine tools. *Int. J. Mach. Tools Manuf.* **2014**, *79*, 17–23. [CrossRef]
19. Lee, S.; Zeng, Q.; Ehmann, K.F. Error modeling for sensitivity analysis and calibration of the tri-pyramid parallel robot. *Int. J. Adv. Manuf. Technol.* **2017**, *93*, 1319–1332. [CrossRef]
20. Cui, H.; Zhu, Z.; Gan, Z.; Brogardh, T. Kinematic analysis and error modeling of TAU parallel robot. *Robot. Comput.-Integr. Manuf.* **2005**, *21*, 497–505. [CrossRef]
21. Tian, W.; Mou, M.; Yang, J.; Yin, F. Kinematic calibration of a 5-DOF hybrid kinematic machine tool by considering the ill-posed identification problem using regularisation method. *Robot. Comput.-Integr. Manuf.* **2019**, *60*, 49–62. [CrossRef]

22. Sun, T.; Zhai, Y.; Song, Y.; Zhang, J. Kinematic calibration of a 3-DoF rotational parallel manipulator using laser tracker. *Robot. Comput.-Integr. Manuf.* **2016**, *41*, 78–91. [CrossRef]
23. Vischer, P.; Clavel, R. Kinematic calibration of the parallel Delta robot. *Robotica* **1998**, *16*, 207–218. [CrossRef]
24. Huang, P.; Wang, J.; Wang, L.; Yao, R. Kinematical calibration of a hybrid machine tool with regularization method. *Int. J. Mach. Tools Manuf.* **2011**, *51*, 210–220. [CrossRef]
25. Niu, P.; Cheng, Q.; Liu, Z.; Chu, H. A machining accuracy improvement approach for a horizontal machining center based on analysis of geometric error characteristics. *Int. J. Adv. Manuf. Technol.* **2021**, *112*, 2873–2887. [CrossRef]
26. Sun, Y.; Lueth, T.C. Safe manipulation in robotic surgery using compliant constant-force mechanism. *IEEE Trans. Med. Robot. Bionics* **2023**, *5*, 486–495. [CrossRef]
27. Zhang, S.; He, C.; Liu, X.; Xu, J.; Sun, Y. Kinematic chain optimization design based on deformation sensitivity analysis of a five-axis machine tool. *Int. J. Precis. Eng. Manuf.* **2020**, *21*, 2375–2389. [CrossRef]
28. Guo, S.; Mei, X.; Jiang, G. Geometric accuracy enhancement of five-axis machine tool based on error analysis. *Int. J. Adv. Manuf. Technol.* **2019**, *105*, 137–153. [CrossRef]
29. Patel, A.J.; Ehmann, K.F. Volumetric error analysis of a Stewart platform-based machine tool. *CIRP Ann.* **1997**, *46*, 287–290. [CrossRef]
30. Fan, K.C.; Wang, H.; Zhao, J.W.; Chang, T.H. Sensitivity analysis of the 3-PRS parallel kinematic spindle platform of a serial-parallel machine tool. *Int. J. Mach. Tools Manuf.* **2003**, *43*, 1561–1569. [CrossRef]
31. Jiang, S.; Chi, C.; Fang, H.; Tang, T.; Zhang, J. A minimal-error-model based two-step kinematic calibration methodology for redundantly actuated parallel manipulators: An application to a 3-DOF spindle head. *Mech. Mach. Theory* **2022**, *167*, 104532. [CrossRef]
32. Du, X.; Wang, B.; Zheng, J. Geometric Error Analysis of a 2UPR-RPU Over-Constrained Parallel Manipulator. *Machines* **2022**, *10*, 990. [CrossRef]
33. Chen, G.; Liang, Y.; Sun, Y.; Chen, W.; Wang, B. Volumetric error modeling and sensitivity analysis for designing a five-axis ultra-precision machine tool. *Int. J. Adv. Manuf. Technol.* **2013**, *68*, 2525–2534. [CrossRef]
34. Cheng, Q.; Zhao, H.; Zhang, G.; Gu, P.; Cai, L. An analytical approach for crucial geometric errors identification of multi-axis machine tool based on global sensitivity analysis. *Int. J. Adv. Manuf. Technol.* **2014**, *75*, 107–121. [CrossRef]
35. Cheng, Q.; Sun, B.; Liu, Z.; Li, J.; Dong, X.; Gu, P. Key geometric error extraction of machine tool based on extended Fourier amplitude sensitivity test method. *Int. J. Adv. Manuf. Technol.* **2017**, *90*, 3369–3385. [CrossRef]
36. Li, T.; Li, F.; Jiang, Y.; Wang, H.; Zhang, J. Error modeling and sensitivity analysis of a 3-P (Pa) S parallel type spindle head with parallelogram structure. *Int. J. Adv. Robot. Syst.* **2017**, *14*, 1729881417715012. [CrossRef]
37. Bonev, I.A. *Geometric Analysis of Parallel Mechanisms*; Université Laval: Québec, QC, Canada, 2002.
38. Xie, F.; Liu, X.J.; Wang, J. A 3-DOF parallel manufacturing module and its kinematic optimization. *Robot. Comput.-Integr. Manuf.* **2012**, *28*, 334–343. [CrossRef]
39. Guo, S.; Jiang, G. Investigation of sensitivity analysis and compensation parameter optimization of geometric error for five-axis machine tool. *Int. J. Adv. Manuf. Technol.* **2017**, *93*, 3229–3243. [CrossRef]
40. Li, J.; Xie, F.; Liu, X.J. Geometric error modeling and sensitivity analysis of a five-axis machine tool. *Int. J. Adv. Manuf. Technol.* **2016**, *82*, 2037–2051. [CrossRef]
41. Liang, Y.; Chen, G.; Chen, W.; Sun, Y.; Chen, J. Analysis of volumetric error of machine tool based on Monte Carlo method. *J. Comput. Theor. Nanosci.* **2013**, *10*, 1290–1295. [CrossRef]

robotics

MDPI

Article

Instantaneous Kinematics and Free-from-Singularity Workspace of 3-XXRRU Parallel Manipulators

Henrique Simas [1], Raffaele Di Gregorio [2,*] and Roberto Simoni [3]

1 Raul Guenther Laboratory of Applied Robotics, Department of Mechanical Engineering, Federal University of Santa Catarina, Florianópolis 88040-900, SC, Brazil; henrique.simas@ufsc.br
2 Laboratory of Mechatronics and Virtual Prototyping (LaMaViP), Department of Engineering, University of Ferrara, Via Saragat, 1, 44100 Ferrara, Italy
3 Department of Mobility Engineering, Federal University of Santa Catarina, Joinville 89219-600, SC, Brazil; roberto.simoni@ufsc.br
* Correspondence: raffaele.digregorio@unife.it

Abstract: 3-XXRRU parallel manipulators (PMs) constitute a family of six-degrees-of-freedom (DOF) PMs with three limbs of type XXRRU, where R and U stand for revolute pair and universal joint, respectively, and XX indicates any actuated two-DOF mechanism that moves the axis of the first R-pair. The members of this family share the fact that they all become particular 3-RRU structures when the actuators are locked. By exploiting this feature, the present paper proposes a general approach, which holds for all the members of this family, to analyze the instantaneous kinematics, workspace, and kinetostatic performances of any 3-XXRRU PM. The results of this study include the identification of singularity conditions without reference to a specific actuation system, the proposal of two specific dimensionless performance indices ranging from 0 to 1, the determination of the optimal actuation system, and the demonstration that 3-XXRRU PMs, when appropriately sized and actuated, possess a broad singularity-free workspace that is also fully isotropic. These findings hold significance in the context of the dimensional synthesis and control of 3-XXRRU PMs. Moreover, when combined with the closed-form solutions for their positional analysis, as demonstrated in a previous publication by the same authors, 3-XXRRU PMs emerge as intriguing alternatives to other six-DOF PMs. The efficacy of the proposed approach is further illustrated through a case study.

Keywords: parallel manipulators; instantaneous kinematics; singularity analysis; workspace; motion control

Citation: Simas, H.; Di Gregorio, R.; Simoni, R. Instantaneous Kinematics and Free-from-Singularity Workspace of 3-XXRRU Parallel Manipulators. *Robotics* **2023**, *12*, 138. https://doi.org/10.3390/robotics12050138

Academic Editor: Marco Ceccarelli

Received: 30 June 2023
Revised: 27 September 2023
Accepted: 3 October 2023
Published: 5 October 2023

1. Introduction

Parallel manipulators (PMs) [1] feature an end effector (platform) connected to a frame (base) by means of a number of kinematic chains (limbs) that can be of any type (i.e., either open or closed or even hybrid chains). The multiple connections between the platform and base make PMs stiffer and more precise than their serial counterparts but, at the same time, make their kinematics more complex and their workspace smaller.

Usually, the number of limbs is equal to the degrees of freedom (DOF) of the PM, and the limbs are open chains with one actuator per limb, which is located on or near the base to reduce the mobile masses [1–4]. The higher the limb number, the smaller the workspace since the number of possible limb collisions increases. Thus, reducing the limb number without changing the DOF number and actuators' locations is of interest. Such a goal is reachable by using hybrid chains as limbs where more than one actuator per limb, located on the base, controls as many joint variables through a suitable closed-chain transmission. In the literature, a number of six-DOF PMs with only three limbs and more than one actuator per limb have been proposed (see [5–14], for instance).

In this context, the authors of the present paper together with Meneghini [15] have proposed a family of six-DOF PMs with three limbs of XXRRU type[1], where R and U stand for revolute pair and universal joint, respectively, and XX indicates any actuated two-DOF mechanism that moves the axis of the first R-pair. In addition, they have shown that both the direct and the inverse position analyses of 3-XXRRU PMs are solvable in closed form.

The presence of singularities [16–22] can further reduce PMs' useful workspace. By considering the relationship between actuated-joint rates (inputs) and platform twist (output) as an input–output relationship (hereafter named IOR) [16–20], singularities are mechanism configurations where the one-to-one correspondence, which the IOR[2] states, between actuated-joint rates and platform twist fails. With reference to the IOR, two instantaneous kinematics problems are definable: the instantaneous inverse kinematics (IIK) problem, which is the determination of one set of actuated-joint rates compatible with one assigned platform twist, and the instantaneous forward kinematics (IFK) problem, which is the determination of one platform twist compatible with one assigned set of actuated-joint rates. Accordingly, three main types of singularities are identifiable [16]: (I) serial singularities, which are mechanism configurations where the IIK is unsolvable, (II) parallel singularities, which are mechanism configurations where the IFK is unsolvable, and (III) the mechanism configurations where both the problems are unsolvable.

Serial singularities are by definition configurations from which the platform cannot exit by moving along any direction (twist), that is, there is a local reduction in the platform's DOF number. Such a condition occurs at the workspace boundaries. Differently, parallel singularities are by definition configurations where the actuated joints are not enough to fully control the platform twist, that is, somehow, there is a local increase in the platform's DOF number. This condition can occur only in PMs and, unfortunately, it usually occurs inside the workspace. The duality, the virtual work principle states between instantaneous kinematics and statics, brings one to conclude that, at a parallel singularity, a small load (even elementary) applied to the platform needs infinite generalized torques at least in one actuator to be equilibrated. Moreover, the nearer the PM configuration to a parallel singularity, the higher the internal loads in the links and the generalized torques in the actuators. In short, a PM must stay far enough from parallel singularities to avoid its breakdown. That is the reason why parallel singularities must be identified during PMs' design and must be avoided during functioning. Therefore, the existence of parallel singularities constitutes a further limitation to the useful workspace of a PM, which might be even tighter than the one due to there being a high number of limbs.

The 3-XXRRU PMs, as proposed in [15], all share a common characteristic: when the actuators are locked, they all transform into specific 3-RRU structures (refer to Figure 1). In each RRU limb of these structures, the axes of the first three R-pairs are parallel, while the axis of the fourth R-pair (i.e., the one fixed to the platform) is perpendicular to the axes of all the other R-pairs[3]. Leveraging this characteristic, this paper presents a general approach applicable to all such PMs. This approach is used to analyze the instantaneous kinematics, workspace, and kinetostatic performance of any 3-XXRRU PM.

The key findings of this study are as follows:

(i) Identification of singularity conditions without the need to reference a specific actuation system;
(ii) Introduction of two dimensionless performance indices ranging from 0 to 1;
(iii) Determination of the optimal actuation system;

[1] The sequence of capital letters indicates the types of joints or sub-mechanisms encountered by moving from the base to the platform along the limb.

[2] It is worth noting that the input–output instantaneous relationship (IOR) of manipulators is always linear and homogeneous both in the actuated-joint rates (inputs) and in the platform twist (output) [16,18] since it is the time derivative of the mechanism constraint-equation system which is holonomic and time independent for manipulators.

[3] It is worth noting that a U joint consists of two R-pairs in series with mutually orthogonal axes, which intersect one another at a point, named the center of the U joint.

(iv) Demonstration that suitably sized and actuated 3-XXRRU PMs offer a broad, singularity-free workspace that is fully isotropic.

These results hold significant relevance for the dimensional synthesis and control of 3-XXRRU PMs. Moreover, the availability of closed-form solutions for their position analysis problems positions 3-XXRRU PMs as intriguing alternatives to other six-DOF PMs. The effectiveness of the proposed approach is further illustrated through a case study.

The paper is organized as follows. Section 2 provides the necessary background materials on 3-XXRRU PMs, deduces their IOR for a general geometry, and presents their singularity analysis together with how to evaluate their kinetostatic performances. Then, Section 3 applies the found singularity conditions and performance indices to a case study, and Section 4 discusses the results. Finally, Section 5 draws the conclusions.

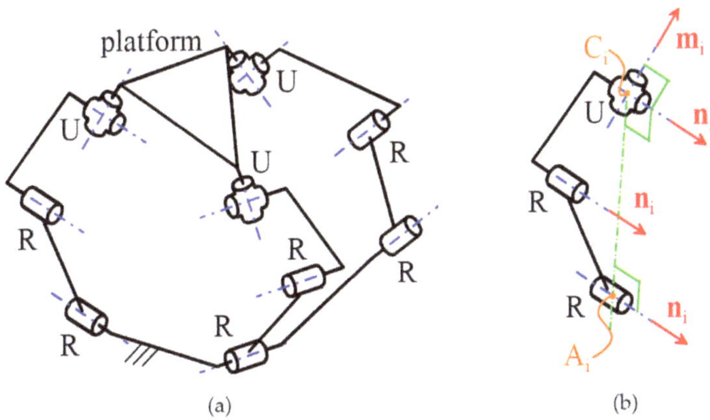

Figure 1. Notations: (**a**) a 3-RRU structure generated from a 3-XXRRU PM by locking the actuators and (**b**) the *i*-th RRU limb, for $i = 1, 2, 3$, of the 3-RRU structure.

2. Materials and Methods

By considering each limb of a PM as a unique articulated joint whose DOF number is equal to the limb connectivity[4] [21–23], the *general mobility criterion*[5] [21,22] brings one to conclude that adding any number of limbs with connectivity equal to six does not reduce the mobility of the platform with respect to the base (i.e., the resulting PM architecture always has six DOFs). Limbs of XXRRU type have connectivity equal to six when the two DOFs provided by the XX mechanism do not replicate any DOF provided by the remaining RRU kinematic chain. In particular, the first three R-pairs of the RRU chain (Figure 1b), which have parallel axes, generate a planar motion with a motion plane perpendicular to the axes of these R-pairs. Then, the fourth R-pair of the RRU chain (i.e., the one adjacent to the platform) adds a rotation around an axis that is parallel to the above-said motion plane. Therefore, in order to make the platform motion a general six-DOF motion through an XXRRU limb, the further two DOFs of the XX mechanism must provide either a general reorientation of the above-said motion plane (i.e., a general change in the direction of the axes of the first three R-pairs) or a translation perpendicular to the above-mentioned motion plane together with a particular reorientation of the same plane that involves only one DOF. If these conditions are respected in all three XXRRU limbs, the resulting 3-XXRRU PM will have six DOFs.

4 The connectivity of a limb is by definition the DOF number of the kinematic chain constituted by platform and base uniquely connected by that limb.

5 Such a criterion counts the DOF number, l, of a spatial mechanism through the formula $l = 6(m-1) - \sum_{i=1,6} (6-i)c_i$ where m is the number of rigid bodies and c_i is the number of constraints with i DOF among the m bodies.

When the actuators are locked, all of the 3-XXRRU PMs, presented in [15], become 3-RRU structures (Figure 1). In the i-th RRU limb (Figure 1b), for i = 1, 2, 3, the axis, (A_i, $\mathbf{n}_i$)[6], of the first R-pair is located through the coordinates of a point, A_i, belonging to it, and a unit vector, $\mathbf{n}_i$, parallel to it. The axis, (C_i, $\mathbf{m}_i$), of the fourth R-pair, fixed to the platform, is located through the coordinates of the center, C_i, of the U joint and a unit vector, $\mathbf{m}_i$, parallel to it. Without losing generality, point A_i is chosen lying on the plane perpendicular to $\mathbf{n}_i$ that passes through C_i.

With these notations, the actuated two-DOF mechanism, XX, of the i-th limb may change only two parameters chosen among the A_i coordinates and/or the $\mathbf{n}_i$ components by respecting the above-defined mobility conditions. Moreover, parallel singularities of 3-XXRRU PMs correspond to configurations of the 3-RRU sub-structure at which the platform can perform elementary motions.

2.1. Input–Output Instantaneous Relationship

The IOR of any 3-XXRRU PM of this family is deducible by time differentiating its constraint equation system, which is writable as follows (see Figure 1b; hereafter, a bold letter denotes a vector and a bold capital letter referable to a point denotes the position vector of that point measured in a reference system fixed to the base):

$$(\mathbf{C}_i - \mathbf{A}_i) \cdot \mathbf{n}_i = 0 \quad i = 1,2,3 \tag{1a}$$

$$\mathbf{m}_i \cdot \mathbf{n}_i = 0 \quad i = 1,2,3 \tag{1b}$$

The time derivative of system (1) is:

$$\dot{\mathbf{C}}_i \cdot \mathbf{n}_i = \dot{\mathbf{A}}_i \cdot \mathbf{n}_i + (\mathbf{A}_i - \mathbf{C}_i) \cdot \dot{\mathbf{n}}_i \quad i = 1,2,3 \tag{2a}$$

$$\dot{\mathbf{m}}_i \cdot \mathbf{n}_i = -\mathbf{m}_i \cdot \dot{\mathbf{n}}_i \quad i = 1,2,3 \tag{2b}$$

Let P and $\boldsymbol{\omega}$ be a point fixed to the platform and the angular velocity of the platform, respectively. Since points C_i and unit vectors $\mathbf{m}_i$, for i = 1, 2, 3, are fixed to the platform, the following relationships of rigid-body mechanics hold [23]:

$$\dot{\mathbf{C}}_i = \dot{\mathbf{P}} + \boldsymbol{\omega} \times (\mathbf{C}_i - \mathbf{P}) \quad i = 1,2,3 \tag{3a}$$

$$\dot{\mathbf{m}}_i = \boldsymbol{\omega} \times \mathbf{m}_i \quad i = 1,2,3 \tag{3b}$$

whose introduction into Equations (2a) and (2b), after some simple rearrangements that use the properties of the mixed product of three vectors, yields:

$$\mathbf{n}_i \cdot \dot{\mathbf{P}} + [(\mathbf{C}_i - \mathbf{P}) \times \mathbf{n}_i] \cdot \boldsymbol{\omega} = \mathbf{n}_i \cdot \dot{\mathbf{A}}_i + (\mathbf{A}_i - \mathbf{C}_i) \cdot \dot{\mathbf{n}}_i \quad i = 1,2,3 \tag{4a}$$

$$(\mathbf{n}_i \times \mathbf{m}_i) \cdot \boldsymbol{\omega} = \mathbf{m}_i \cdot \dot{\mathbf{n}}_i \quad i = 1,2,3 \tag{4b}$$

[6] Hereafter, a line will be denoted (P, **u**) where P is a point belonging to the line and u is a unit vector parallel to the line.

Equations (4a) and (4b) are writable in matrix form as follows:

$$\begin{bmatrix} \mathbf{N}^T & \mathbf{H}^T \\ 0_{3\times 3} & \mathbf{M}^T \end{bmatrix} \begin{pmatrix} \dot{\mathbf{P}} \\ \boldsymbol{\omega} \end{pmatrix} = \begin{pmatrix} \mathbf{n}_1 \cdot \dot{\mathbf{A}}_1 + (\mathbf{A}_1 - \mathbf{C}_1) \cdot \dot{\mathbf{n}}_1 \\ \mathbf{n}_2 \cdot \dot{\mathbf{A}}_2 + (\mathbf{A}_2 - \mathbf{C}_2) \cdot \dot{\mathbf{n}}_2 \\ \mathbf{n}_3 \cdot \dot{\mathbf{A}}_3 + (\mathbf{A}_3 - \mathbf{C}_3) \cdot \dot{\mathbf{n}}_3 \\ \mathbf{m}_1 \cdot \dot{\mathbf{n}}_1 \\ \mathbf{m}_2 \cdot \dot{\mathbf{n}}_2 \\ \mathbf{m}_3 \cdot \dot{\mathbf{n}}_3 \end{pmatrix} \tag{5}$$

where $\mathbf{0}_{3\times 3}$ is the 3×3 null matrix and $\hat{\$} = \begin{pmatrix} \dot{\mathbf{P}} \\ \boldsymbol{\omega} \end{pmatrix}$ is the platform twist; whereas **N**, **M**, and **H** are 3×3 matrices defined as follows:

$$\mathbf{N} = \begin{bmatrix} \mathbf{n}_1 & \mathbf{n}_2 & \mathbf{n}_3 \end{bmatrix},\ \mathbf{M} = \begin{bmatrix} (\mathbf{n}_1 \times \mathbf{m}_1) & (\mathbf{n}_2 \times \mathbf{m}_2) & (\mathbf{n}_3 \times \mathbf{m}_3) \end{bmatrix}, \tag{6a}$$

$$\mathbf{H} = \begin{bmatrix} (\mathbf{C}_1 - \mathbf{P}) \times \mathbf{n}_1 & (\mathbf{C}_2 - \mathbf{P}) \times \mathbf{n}_2 & (\mathbf{C}_3 - \mathbf{P}) \times \mathbf{n}_3 \end{bmatrix}; \tag{6b}$$

In system (5), $\dot{\mathbf{A}}_i$ and $\dot{\mathbf{n}}_i$, for i = 1, 2, 3, are linearly related to the actuated-joint rates through linear and homogeneous expressions that depend on the type of two-DOF mechanism XX, which is present in the XXRRU limb. System (5) is the IOR of a generic 3-XXRRU PM.

2.2. *Singularity Analysis*

The singularity analysis, which is the determination of the geometric/analytic conditions that identify the singular configurations (singularities), practically consists of analyzing the IOR of the PM, that is, system (5) for a 3-XXRRU PM.

Regarding the possible actuation choices for these PMs, system (5) immediately reveals that, if the XX mechanisms only move points A_i, for i = 1, 2, 3, (i.e., $\dot{\mathbf{n}}_1 = \dot{\mathbf{n}}_2 = \dot{\mathbf{n}}_3 = 0$), the last three equations, which simply become $\mathbf{M}^T\ \boldsymbol{\omega} = \mathbf{0}$, always provide $\boldsymbol{\omega} = \mathbf{0}$ out of singularities that make det(**M**) = 0. That is, with this actuation choice, the platform can only translate and the 3-XXRRU becomes a 3-DOF translational PM (TPM). The so-obtained TPM is somehow a more general geometry of the Cartesian TPM proposed by Kim and Tsai in [24,25]. Moreover, since $\dot{\mathbf{A}}_i$ only appears in the dot product $\mathbf{n}_i \cdot \dot{\mathbf{A}}_i$, the only motion direction of point A_i that causes a platform motion is the one along $\mathbf{n}_i$. In particular, since the time derivative of $\mathbf{n}_i \cdot \mathbf{n}_i = 1$ yields $\mathbf{n}_i \cdot \dot{\mathbf{n}}_i = 0$, choosing $\dot{\mathbf{A}}_i = a_i \dot{\mathbf{n}}_i$ where a_i is an arbitrary scalar constant would give $\mathbf{n}_i \cdot \dot{\mathbf{A}}_i = a_i \mathbf{n}_i \cdot \dot{\mathbf{n}}_i = 0$, that is, no effect on the platform motion. In short, the XX mechanism of the i-th XXRRU limb can devote at most one actuator to move point A_i and must move it only along $\mathbf{n}_i$, and an actuated prismatic (P) pair with sliding direction parallel to $\mathbf{n}_i$ that moves point A_i does not affect the platform translation through the term $\mathbf{n}_i \cdot \dot{\mathbf{A}}_i$ when the direction of $\mathbf{n}_i$ changes and the P pair is locked.

Differently, the same analysis, reveals that if the XX mechanisms only change the directions of unit vectors $\mathbf{n}_i$, for i = 1, 2, 3, (i.e., $\dot{\mathbf{A}}_1 = \dot{\mathbf{A}}_2 = \dot{\mathbf{A}}_3 = 0$), the platform can perform a general 6-DOF motion since $\dot{\mathbf{n}}_i$, for i = 1, 2, 3, appears in all six equations of system (5). In particular, since, in the last three equations of system (5), $\dot{\mathbf{n}}_i$ appears only in the dot product $\mathbf{m}_i \cdot \dot{\mathbf{n}}_i$, the component of $\dot{\mathbf{n}}_i$ that makes the platform orientation change is only the one along the direction of $\mathbf{m}_i$. Moreover, since, in the first three equations of system (5), $\dot{\mathbf{n}}_i$ appears only in the dot product $(\mathbf{A}_i - \mathbf{C}_i) \cdot \dot{\mathbf{n}}_i$, the component of $\dot{\mathbf{n}}_i$ that makes the platform translate is only the one along the direction of $(\mathbf{A}_i - \mathbf{C}_i)$.

The conclusion is that the most effective actuation system is a two-DOF mechanism that, firstly, controls the $\mathbf{n}_i$ direction and then, after having obtained the desired platform orientation, locks the $\mathbf{n}_i$ direction through a system of clutches and brakes, and, eventually, uses one of the two actuators previously used to change the $\mathbf{n}_i$ direction to make the platform translate toward the desired final pose by moving A_i along the $\mathbf{n}_i$ direction.

The above discussion of the possible actuation choices makes clear that when one entry of the six-tuple that appears on the right-hand side of system (5) is equal to zero, even though $\dot{\mathbf{A}}_i$ and/or $\dot{\mathbf{n}}_i$ are different from zero, some of the actuated-joint rates are indeterminate for an assigned platform twist, that is, a serial singularity occurs. Therefore, from an analytical point of view, a serial singularity occurs when, provided that $\dot{\mathbf{A}}_i$ and/or $\dot{\mathbf{n}}_i$ are different from zero, at least one of the following six conditions is satisfied:

$$\mathbf{n}_i \cdot \dot{\mathbf{A}}_i + (\mathbf{A}_i - \mathbf{C}_i) \cdot \dot{\mathbf{n}}_i = 0 \quad i = 1,2,3 \tag{7a}$$

$$\mathbf{m}_i \cdot \dot{\mathbf{n}}_i = 0 \quad i = 1,2,3 \tag{7b}$$

Differently, a parallel singularity occurs when the six-tuple on the right-hand side of system (5) is assigned, but the corresponding platform twist is not computable by solving system (5). This can happen if and only if the determinant of the 6 × 6 matrix (parallel Jacobian) that multiplies the platform twist on the left-hand side of system (5) is equal to zero. Since this parallel Jacobian is an upper triangular block matrix, the following relationship holds [26,27]:

$$\det\begin{pmatrix} \mathbf{N}^T & \mathbf{H}^T \\ 0_{3\times3} & \mathbf{M}^T \end{pmatrix} = \det(\mathbf{N})\det(\mathbf{M}) \tag{8}$$

where, by remembering definition (6a), the following geometric/analytic explicit expressions of $\det(\mathbf{N})$ and $\det(\mathbf{M})$ can be introduced:

$$\det(\mathbf{N}) = \mathbf{n}_1 \cdot (\mathbf{n}_2 \times \mathbf{n}_3), \ \det(\mathbf{M}) = (\mathbf{n}_1 \times \mathbf{m}_1) \cdot [(\mathbf{n}_2 \times \mathbf{m}_2) \times (\mathbf{n}_3 \times \mathbf{m}_3)]. \tag{9}$$

Therefore, a parallel singularity occurs if and only if one or the other of the following two conditions is satisfied:

$$\mathbf{n}_1 \cdot (\mathbf{n}_2 \times \mathbf{n}_3) = 0, \tag{10a}$$

$$(\mathbf{n}_1 \times \mathbf{m}_1) \cdot [(\mathbf{n}_2 \times \mathbf{m}_2) \times (\mathbf{n}_3 \times \mathbf{m}_3)] = 0 \tag{10b}$$

Since the left-hand sides of both Equations (10a) and (10b) are mixed products of unit vectors, Equation (10a) (Equation (10b)) is satisfied if and only if the three unit vectors $\mathbf{n}_i$, for i = 1, 2, 3, (the three unit vectors $\mathbf{n}_i \times \mathbf{m}_i$, for i = 1, 2, 3) are all parallel to a unique plane. From a kinematic point of view, the analysis of system (5) reveals that, when Equation (10a) (Equation (10b)) is satisfied, even though the actuated joints are locked, that is, even though $\dot{\mathbf{A}}_i = 0$ and $\dot{\mathbf{n}}_i = 0$ for i = 1, 2, 3, $\dot{\mathbf{P}}$ ($\boldsymbol{\omega}$) has an indeterminate component along the direction perpendicular to the unique plane the three unit vectors $\mathbf{n}_i$, for i = 1, 2, 3, (the three unit vectors $\mathbf{n}_i \times \mathbf{m}_i$, for i = 1, 2, 3) are parallel to, that is, the platform can translate along (rotate around a line parallel to) that direction.

2.3. *Evaluation of Kinetostatic Performance*

From a geometric point of view, the absolute value of the mixed product of three unit vectors, $\mathbf{u}_i$, for i = 1, 2, 3, where $\mathbf{u}_i$ can be either $\mathbf{n}_i$ or $\mathbf{n}_i \times \mathbf{m}_i$, is equal to the volume of the rhombohedron (Figure 2) whose three non-parallel edges are parallel to the three unit vectors $\mathbf{u}_i$, for i = 1, 2, 3, and have a length equal to one. Such a volume ranges from 0, when the rhombohedron is flattened, to 1, when the rhombohedron is a cube, and, in this case, can be used as an dimensionless index, say j_u ($\equiv|\mathbf{u}_1 \cdot (\mathbf{u}_2 \times \mathbf{u}_3)|$), of the distance of a non-singular configuration from a parallel singularity since the greater the volume is, the further away the configuration is from satisfying Equation (10). Such an index can be analytically expressed as a function of two angles, $\theta_{1,u}$ and $\theta_{2,u}$ (Figure 2), which range from 0° to 90°, as follows:

$$j_u \equiv |\mathbf{u}_1 \cdot (\mathbf{u}_2 \times \mathbf{u}_3)| \equiv |\lambda_1\lambda_2\lambda_3| = \cos\theta_{1,u}\sin\theta_{2,u} \quad (11)$$

where λ_i, for i = 1, 2, 3, are the three eigenvalues of matrix $\mathbf{U} = \begin{bmatrix}\mathbf{u}_1 & \mathbf{u}_2 & \mathbf{u}_3\end{bmatrix}$.

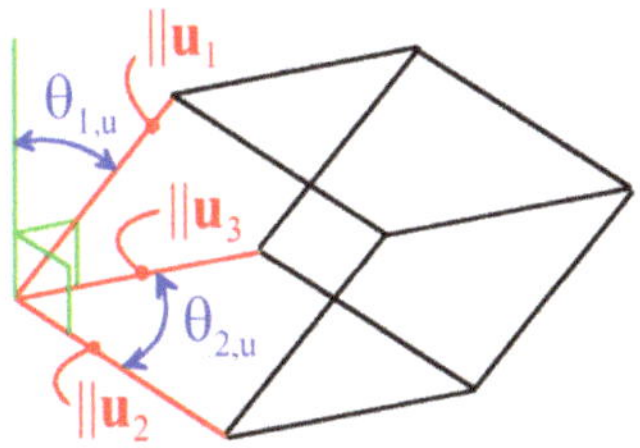

Figure 2. Rhombohedron with an edge length equal to one, which is associated with the three unit vectors $\mathbf{u}_i$, for i = 1, 2, 3.

Equations (8), (9) and (11) bring one to build the following index, J, for evaluating the kinetostatic performance of a 3-RRU configuration[7]:

$$J \equiv |\mathbf{n}_1 \cdot (\mathbf{n}_2 \times \mathbf{n}_3)||(\mathbf{n}_1 \times \mathbf{m}_1) \cdot [(\mathbf{n}_2 \times \mathbf{m}_2) \times (\mathbf{n}_3 \times \mathbf{m}_3)]| = j_n j_{n\times m} = \cos\theta_{1,n}\sin\theta_{2,n}\cos\theta_{1,n\times m}\sin\theta_{2,n\times m} \quad (12)$$

From a static point of view (Figure 3), the i-th limb applies to the platform one force, $F_i\mathbf{n}_i$, where F_i is the signed magnitude, parallel to $\mathbf{n}_i$ and with its line of action passing through C_i, and one moment, $M_i(\mathbf{n}_i \times \mathbf{m}_i)$, where M_i is the signed magnitude, parallel to $(\mathbf{n}_i \times \mathbf{m}_i)$. Thus, the equilibrium equations of the platform are ($\mathbf{F}_e$ and $\mathbf{M}_{e,P}$ are the resultant force and the resultant moment about point P of the external force system applied to the platform):

$$\sum_{i=1,3} F_i\mathbf{n}_i + \mathbf{F}_e = \mathbf{0} \quad (13a)$$

$$\sum_{i=1,3} F_i[(\mathbf{C}_i - \mathbf{P}) \times \mathbf{n}_i] + \sum_{i=1,3} M_i(\mathbf{n}_i \times \mathbf{m}_i) + \mathbf{M}_{e,P} = \mathbf{0} \quad (13b)$$

which, in matrix form, become:

$$\begin{bmatrix}\mathbf{N} & \mathbf{0}_{3\times3} \\ \mathbf{H} & \mathbf{M}\end{bmatrix}\begin{pmatrix}F_1\\F_2\\F_3\\M_1\\M_2\\M_3\end{pmatrix} = -\begin{pmatrix}\mathbf{F}_e\\ \mathbf{M}_{e,P}\end{pmatrix} \quad (14)$$

The first three equations of system (14) show that when j_n ($\equiv |\mathbf{n}_1 \cdot (\mathbf{n}_2 \times \mathbf{n}_3)|$) is equal to 1 (i.e., matrix $\mathbf{N}$ is isotropic [28]), by changing the direction of $\mathbf{F}_e$, the condition $|F_i| \leq \|\mathbf{F}_e\|$, for i = 1, 2, 3, is always satisfied (i.e., no limb must apply a force with a magnitude greater than the one of $\mathbf{F}_e$). Analogously, when a pure moment $\mathbf{M}_{e,P}$ (i.e., $\mathbf{M}_{e,P} \neq 0$ and $\mathbf{F}_e = \mathbf{0}$) is applied to the platform and $j_{n\times m}$ ($\equiv |(\mathbf{n}_1 \times \mathbf{m}_1) \cdot [(\mathbf{n}_2 \times \mathbf{m}_2) \times (\mathbf{n}_3 \times \mathbf{m}_3)]|$) is equal to 1 (i.e., matrix $\mathbf{M}$ is isotropic [28]), the last three equations of system (14) show that, by changing the direction of $\mathbf{M}_{e,P}$, the condition $|M_i| \leq \|\mathbf{M}_{e,P}\|$, for i = 1, 2, 3, is always satisfied (i.e., no limb must apply a moment with a magnitude greater than the one of $\mathbf{M}_{e,P}$). Accordingly, hereafter, a 3-RRU configuration with j_n = 1 ($j_{n\times m}$ = 1) is named

[7] With reference to Figure 1b, it is worth stressing that the positive direction, toward which the unit vectors $\mathbf{n}_i$ and $\mathbf{m}_i$, for i = 1, 2, 3, point, is arbitrarily chosen and that it does not affect the values of j_n and $j_{n\times m}$.

isotropic with respect to $\mathbf{F}_e$ ($\mathbf{M}_{e,P}$). Of course, the best redistribution of loads among the limbs occurs at configurations that are isotropic with respect to both $\mathbf{F}_e$ and $\mathbf{M}_{e,P}$, that is, when $J = j_n j_{n\times m} = 1$, hereafter named *fully isotropic*.

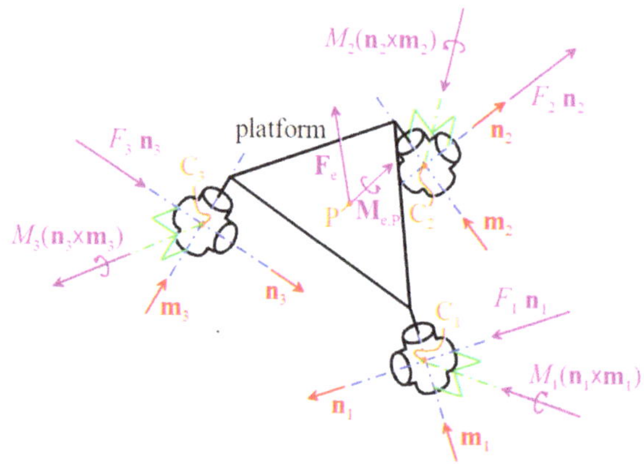

Figure 3. Free-body diagram of the platform.

3. Results

This section proves the effectiveness of the relationships deduced in Section 2 by applying them to the analysis of a 3-XXRRU PM with the geometry shown in Figure 4. In particular, the relationships deduced in Section 2 are used here for the determination of the free-from-singularity workspace of that 3-XXRRU PM and of its kinetostatic performances inside that workspace.

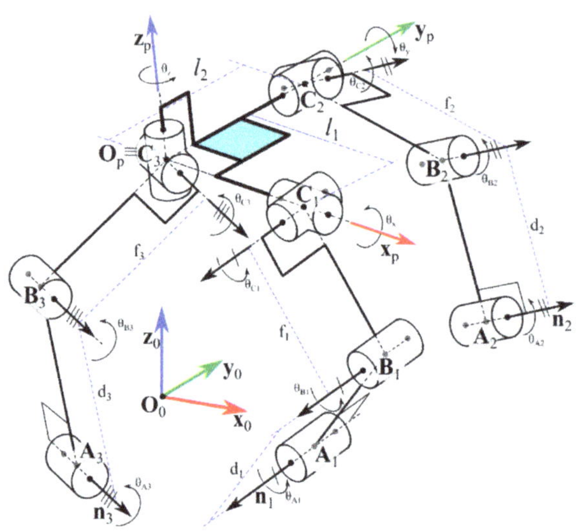

Figure 4. Geometry of the studied 3-XXRRU.

The geometry of the studied 3-XXRRU PM has the peculiarity (see Figure 4) that the axes of the three R-pairs adjacent to the platform share a common intersection, point C_3, and are mutually orthogonal. As a consequence, this particular geometry allows the introduction of a Cartesian reference system $O_p x_p y_p z_p$ (Figure 4), fixed to the platform with origin, O_p, coincident with C_3 and coordinate axes coincident with the axes of the three R-pairs adjacent to the platform.

With reference to Figure 4 and the general notations of Figure 1, the following further specific notations are introduced. $\mathbf{x}_p$, $\mathbf{y}_p$, and $\mathbf{z}_p$ are the unit vectors of the x_p, y_p, and z_p coordinate axes, respectively, of the $O_px_py_pz_p$ reference system and are chosen so that $\mathbf{x}_p \equiv \mathbf{m}_1$, $\mathbf{y}_p \equiv \mathbf{m}_2$, and $\mathbf{z}_p \equiv \mathbf{m}_3$. $O_0x_0y_0z_0$ is a Cartesian reference system fixed to the base. l_1 and l_2 are the lengths of the segments C_3C_1 and C_3C_2. In the i-th limb, for i = 1, 2, 3, point B_i is the intersection of the second R-pair's axis with the plane perpendicular to $\mathbf{n}_i$ and passing through C_i; also, d_i and f_i are the lengths of the segments A_iB_i and B_iC_i, respectively. Moreover, one assumes that the XX mechanism can, firstly, freely orientate unit vector $\mathbf{n}_i$ and then make the platform translate toward the desired final pose, which, as explained in Section 2.2, is the best actuation technique for 3-XXRRU PMs.

Since the actuation system can freely orientate unit vectors $\mathbf{n}_i$, for i = 1, 2, 3, the choice of moving them so that they are always mutually orthogonal is assumed. Such a choice makes all the reached 3-RRU configurations isotropic with respect to $\mathbf{F}_e$ (i.e., with j_n = 1). Therefore, the kinetostatic performance of each 3-RRU configuration could degrade only because of a $j_{n\times m}$ value lower than 1 and can be evaluated by using only the $j_{n\times m}$ index. Nevertheless, for this particular geometry and $\mathbf{n}_i$ motion strategy, the demonstration that follows shows that the index $j_{n\times m}$ is equal to one, too (i.e., the platform moves by keeping the 3-XXRRU configuration fully isotropic). Indeed, with reference to Figure 5, the following explicit expressions can be written.

$$\begin{cases} \mathbf{n}_1 = -\mathbf{m}_2 \sin\theta_x + \mathbf{m}_3 \cos\theta_x \\ \mathbf{n}_2 = \mathbf{m}_1 \sin\theta_y + \mathbf{m}_3 \cos\theta_y \\ \mathbf{n}_3 = -\mathbf{m}_1 \sin\theta_z + \mathbf{m}_2 \cos\theta_z \end{cases} \tag{15a}$$

$$\left\{ \begin{array}{l} \mathbf{n}_1 \times \mathbf{m}_1 = \mathbf{m}_3 \sin\theta_x + \mathbf{m}_2 \cos\theta_x \\ \mathbf{n}_2 \times \mathbf{m}_2 = \mathbf{m}_3 \sin\theta_y - \mathbf{m}_1 \cos\theta_y \\ \mathbf{n}_3 \times \mathbf{m}_3 = \mathbf{m}_2 \sin\theta_z + \mathbf{m}_1 \cos\theta_z \end{array} \right\rangle \Rightarrow j_{n\times m} = \left|\cos\theta_x \sin\theta_y \cos\theta_z - \sin\theta_x \cos\theta_y \sin\theta_z\right| \tag{15b}$$

Moreover, since unit vectors $\mathbf{n}_i$, for i = 1, 2, 3, are moved by keeping them mutually orthogonal, the following system of three equations in three unknowns must be satisfied

$$\begin{cases} \mathbf{n}_1 \cdot \mathbf{n}_2 = (-\mathbf{m}_2 \sin\theta_x + \mathbf{m}_3 \cos\theta_x) \cdot (\mathbf{m}_1 \sin\theta_y + \mathbf{m}_3 \cos\theta_y) = \cos\theta_x \cos\theta_y = 0 \\ \mathbf{n}_1 \cdot \mathbf{n}_3 = (-\mathbf{m}_2 \sin\theta_x + \mathbf{m}_3 \cos\theta_x) \cdot (-\mathbf{m}_1 \sin\theta_z + \mathbf{m}_2 \cos\theta_z) = -\sin\theta_x \cos\theta_z = 0 \\ \mathbf{n}_2 \cdot \mathbf{n}_3 = (\mathbf{m}_1 \sin\theta_y + \mathbf{m}_3 \cos\theta_y) \cdot (-\mathbf{m}_1 \sin\theta_z + \mathbf{m}_2 \cos\theta_z) = -\sin\theta_y \sin\theta_z = 0 \end{cases} \tag{16}$$

whose solutions are ($\cos\theta_x$, $\sin\theta_y$, $\cos\theta_z$) = (0, 0, 0), which implies ($\sin\theta_x$, $\cos\theta_y$, $\sin\theta_z$) = (±1, ±1, ±1) and ($\sin\theta_x$, $\cos\theta_y$, $\sin\theta_z$) = (0, 0, 0), which implies ($\cos\theta_x$, $\sin\theta_y$, $\cos\theta_z$) = (±1, ±1, ±1). Both these solutions, when introduced into Equation (15b), yield $j_{n\times m} = 1$, which demonstrates what has been declared above.

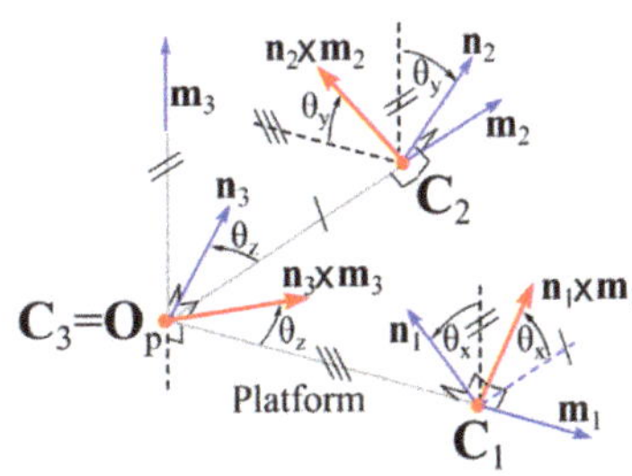

Figure 5. Geometric relationships among the unit vectors $\mathbf{n}_i$, $\mathbf{m}_i$, and $\mathbf{n}_i \times \mathbf{m}_i$, for i = 1, 2, 3, in the platform of the 3-XXRRU shown in Figure 4.

With the above-reported assumptions and choices, the free-from-singularity orientation workspace was numerically determined by using the following data (l.u. stands for arbitrary length unit; the points' coordinates are measured in $O_0x_0y_0z_0$): $A_1 = (1, 1, 0)^T$ l.u., $A_2 = (0, 3, 2)^T$ l.u., $A_3 = (0, 0, 2)^T$ l.u., $l_1 = 1.5$ l.u., $l_2 = 2$ l.u., $d_1 = d_2 = d_3 = 3$ l.u., $f_1 = f_2 = f_3 = 2.5$ l.u. Figure 6 shows the so-determined free-from-singularity orientation workspace (the parameters ϕ_1, ϕ_2, and ϕ_3 on the axes are the ZYZ Euler angles that locate the orientation of $O_px_py_pz_p$ with respect to $O_0x_0y_0z_0$). Figure 7 shows the volume inside which point C_3 moves while the directions of unit vectors $\mathbf{n}_i$, for $i = 1, 2, 3$, are modified (by keeping them mutually orthogonal) to make the platform reach the orientation shown in Figure 6 (the coordinates of point C_3 measured in $O_0x_0y_0z_0$ are reported on the axes of Figure 7).

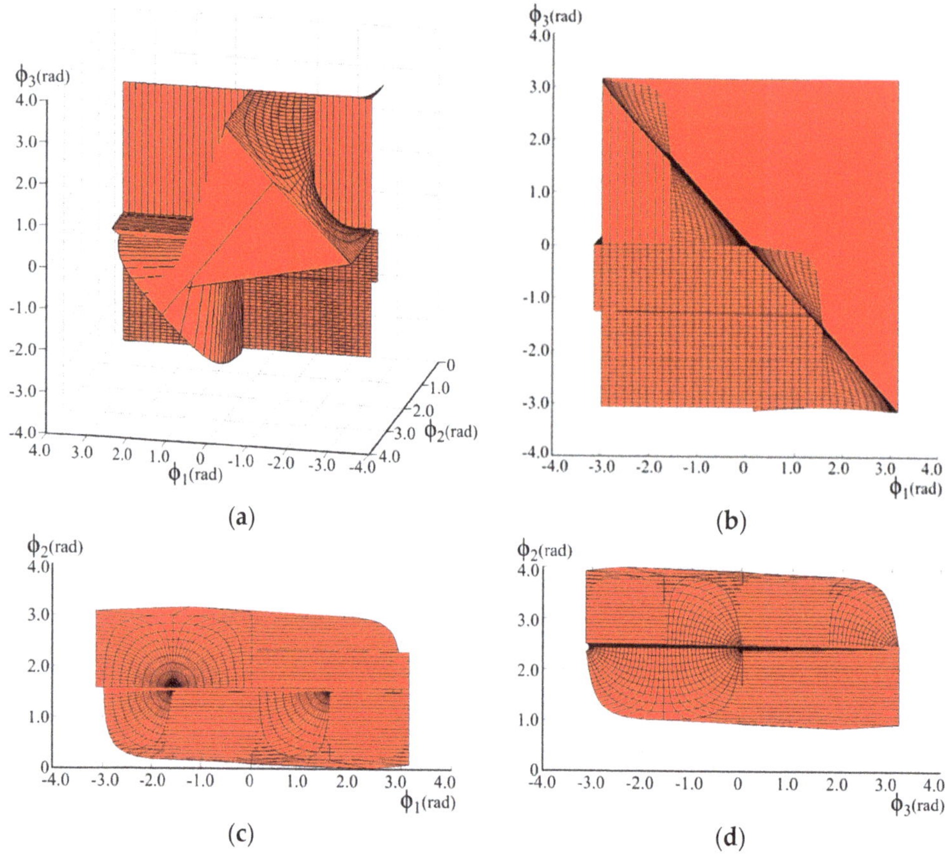

Figure 6. Computed free-from-singularity orientation workspace (the parameters ϕ_1, ϕ_2, and ϕ_3 on the axes are the ZYZ Euler angles that locate the orientation of $O_px_py_pz_p$ with respect to $O_0x_0y_0z_0$): (**a**) 3D view, (**b**) projection onto the $\phi_1\phi_3$-plane, (**c**) projection onto the $\phi_1\phi_2$-plane, and (**d**) projection onto the $\phi_2\phi_3$-plane.

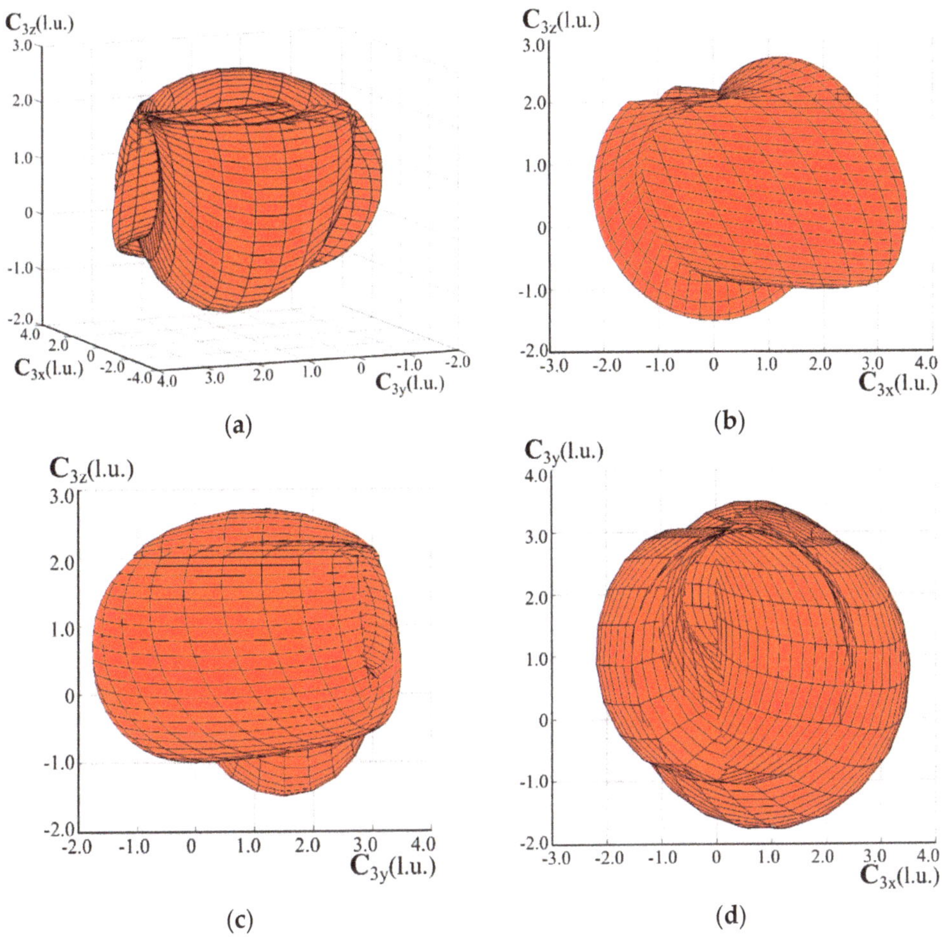

Figure 7. Volume inside which point C_3 moves while the directions of unit vectors $\mathbf{n}_i$, for $i = 1, 2, 3$, are modified (by keeping them mutually orthogonal) to make the platform reach the orientation shown in Figure 6 (the coordinates of point C_3 measured in $O_0x_0y_0z_0$ are reported on the axes): (**a**) 3D view, (**b**) projection onto the xz-plane, (**c**) projection onto the yz-plane, and (**d**) projection onto the xy-plane.

4. Discussion

The numerical procedure adopted to determine Figures 6 and 7 firstly considers matrix **N** as a rotation matrix, since unit vectors $\mathbf{n}_i$, for $i = 1, 2, 3$, must be kept mutually orthogonal, and expresses matrix **N** through the ZYZ Euler angles (ψ_1, ψ_2, ψ_3) with $\psi_1, \psi_3 \in [-\pi, \pi]$ rad and $\psi_2 \in [0, \pi]$ rad. Then, it discretizes the values of ψ_1, ψ_2, and ψ_3 by selecting a number, say k, of equally spaced values in their ranges, computes the corresponding k^3 platform poses, and keeps only those compatible with the link lengths. Eventually, the k value is increased until the shape of the determined workspace does not change; such a condition was reached for k = 40, and Figures 6 and 7 refer to k = 40. Accordingly, Figures 6 and 7 display the workspace restricted to the fully isotropic configurations, hereafter named the fully isotropic workspace.

The analysis of Figure 6 highlights that the fully isotropic orientation workspace includes an ample portion of the parallelepiped (i.e., $\phi_1, \phi_3 \in [-\pi, \pi]$ rad and $\phi_2 \in [0, \pi]$

rad) collecting all the possible orientations the platform could assume as a free rigid body. Moreover, the sizes of the excluded regions depend on the chosen link length, which brings one to conclude that, during design, the link lengths can be sized for an assigned fully isotropic orientation workspace.

Accepting non-singular configurations with $j_n \geq j_{n,min} < 1$ and $j_{n\times m} \geq j_{n\times m,min} < 1$, where $j_{n,min}$ and $j_{n\times m,min}$ are two constant values not equal to zero, is another design choice that could be adopted to obtain the desired free-from-singularity workspace. Such an approach might also be implemented by modifying the control strategy in an already manufactured machine.

Whatever the adopted design criteria happen to be, the above case study proves that, in the family of 3-XXRRU PMs, suitable combinations of geometries and control strategies can be selected which provide good kinetostatic performance in an ample free-from-singularity workspace.

The 3-RRU structure studied in Section 3 lends itself to better illustrate the parallel-singularity conditions (i.e., Equations (10a) and (10b)) through Figures 8 and 9. Figure 8 shows two singular configurations of this structure that satisfy Equation (10a). In particular, the three unit vectors $\mathbf{n}_i$, for i = 1, 2, 3, are all parallel to the x_0y_0-coordinate plane with the three R-pair axes (A_i, $\mathbf{n}_i$), for i = 1, 2, 3, that are not coplanar in Figure 8a and are coplanar in Figure 8b. The analysis of Figure 8 clearly shows that the particular disposition of the first three R-pair axes of each limb makes platform translation possible along the z_0-coordinate axis. Differently, Figure 9 shows one singular configuration of the same structure that satisfies Equation (10b). Indeed, in such a configuration, the three unit vectors $\mathbf{n}_i \times \mathbf{m}_i$, for i = 1, 2, 3, are all parallel to the x_0y_0-coordinate plane. The analysis of Figure 9 clearly shows that the particular disposition of all the R-pair axes makes the platform rotation possible around the axis (C_3, $\mathbf{z}_0$).

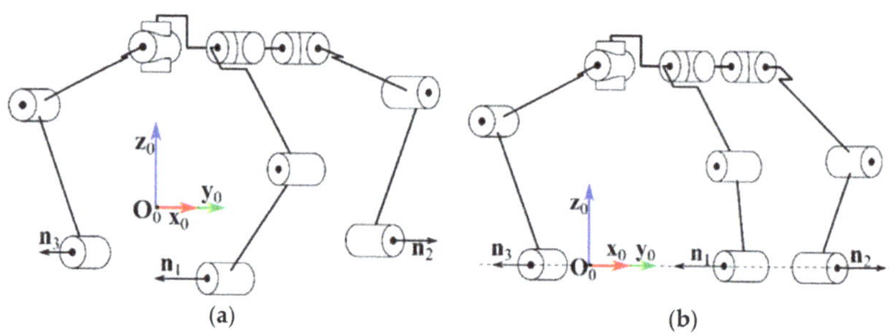

Figure 8. Two singular configurations of the 3-RRU structure of Figure 4 that satisfy Equation (10a): (**a**) the three unit vectors $\mathbf{n}_i$, for i = 1, 2, 3, are all parallel to the x_0y_0-coordinate plane with the three R-pair axes (A_i, $\mathbf{n}_i$), for i = 1, 2, 3, that are not coplanar, and (**b**) all three R-pair axes (A_i, $\mathbf{n}_i$), for i = 1, 2, 3, lie on the x_0y_0-coordinate plane indicated by the dashed line.

Eventually, the possible types of XX mechanisms that implement the optimal control strategy identified in Section 2 deserve a more detailed discussion. Firstly, the orientation of $\mathbf{n}_i$ can be obtained by using any parallel pointing system[8] (PPS) among the many proposed in the literature (see, for instance, Refs. [29,30]). Then, by placing a partially actuated cylindrical(C) pair[9] on the mobile platform of the PPS, with the translation that is alternatively actuated by one of the two actuators that moves the PPS platform and with

[8] A PPS is a 2-DOF PM that is able to freely orientate one line fixed to its platform by keeping one point of the line fixed to the base. They are employed in many applications like the motion of a telescope or of a parabolic antenna, etc.

[9] It is worth noting that a C-pair can be obtained by putting in series a prismatic (P) pair and an R-pair whose axis is parallel to the sliding direction of the P-pair and that such a PR chain is easy to actuate. In such a PR chain, the R-pair will be the first non-actuated R-pair of the remaining RRU chain of the XXRRU limb.

the C-pair axis that passes through the center of the spherical motion the PPS imposes to its platform, the desired translation of point A_i along the $\mathbf{n}_i$ direction can be added to the orientation of $\mathbf{n}_i$. The non-actuated rotation of the added C-pair plays the role of the first non-actuated R-pair of the remaining RRU chain of the XXRRU limb. In order to better illustrate this description, Figure 10 shows a possible XXRRU limb where the PPS is a spherical five-bar linkage[10]. Of course, the system of brakes and clutches that allows the actuation of the translation in the C-pair by means of one of the two actuators used to orientate the PPS platform needs an ad hoc design that depends on the chosen PPS.

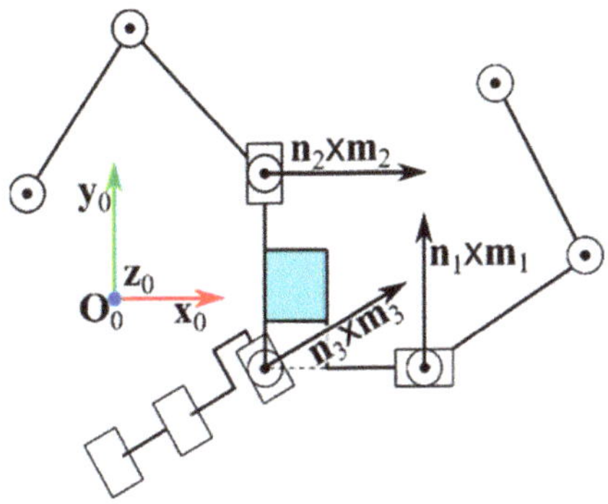

Figure 9. One singular configuration of the 3-RRU structure of Figure 4 that satisfies Equation (10b): the three unit vectors $\mathbf{n}_i \times \mathbf{m}_i$, for $i = 1, 2, 3$, are all parallel to the x_0y_0-coordinate plane.

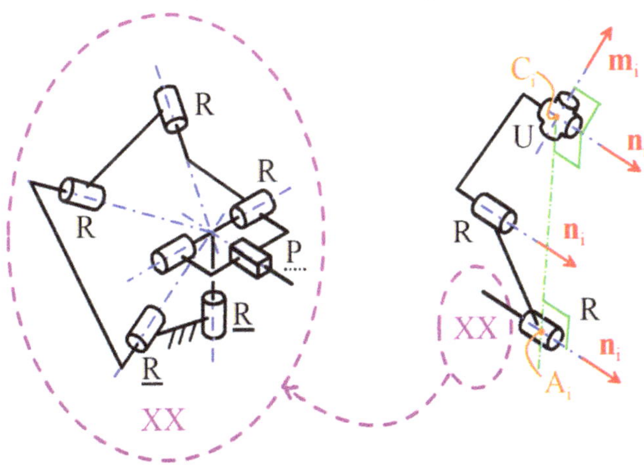

Figure 10. Example of the XXRRU limb where the PPS is a spherical five-bar linkage: the two R-pairs with a solid underscore are the two actuated pairs, adjacent to the base, that control the orientation of unit vector $\mathbf{n}_i$, whereas the P pair with a dotted underscore is the alternately actuated/locked P pair that moves point A_i along the direction of unit vector $\mathbf{n}_i$.

5. Conclusions

This study investigates the instantaneous kinematics, workspace, and kinetostatic performance of a novel family of six-DOF three-legged parallel manipulators (PMs) recently

[10] The spherical five-bar linkage is a particular PPS consisting of five binary links sequentially connected, to form a single-loop, through R-pairs whose axes share a common intersection point. This R-pair axes' arrangement guarantees that their intersection point is fixed to the frame (i.e., the links' motion is spherical) and that any line, which is fixed to a mobile link and passes through the above-mentioned intersection point, keeps that point at rest during the link motion. In Figure 10, the blue lines are the R-pair axes.

introduced by the authors in a previous publication. These PMs all share the common feature that, when their actuators are locked, they transform into 3-RRU structures and are collectively referred to as 3-XXRRU manipulators.

The examination of the instantaneous kinematics of 3-XXRRU PMs led to the derivation of a general expression for their input–output instantaneous relationship (IOR). Subsequently, the analysis of this IOR unveiled that the singularity conditions for these manipulators can be expressed in a straightforward and easily interpretable geometric manner. This insight provides valuable guidance for selecting an effective actuation system.

In combination with the static analysis of 3-XXRRU PMs, these findings led to the proposal of two dimensionless indices, ranging from 0 to 1, which measure the proximity of a non-singular configuration to the nearest parallel singularity. These indices can be employed in the dimensional synthesis of these PMs. Utilizing these indices, a suitable control strategy has been devised to maintain manipulator isotropy, particularly concerning the resultant force of external forces applied to the platform.

Lastly, a specific platform geometry has been introduced, enabling the attainment of an ample fully isotropic workspace.

6. Patents

Simas H., Simoni R., Meneghini L., Di Gregorio R.: Manipulador paralelo 3XXRRU com três pernas e seis graus de liberdade com volume de trabalho ampliado. 2023; Brazil Patent Application No. BR 10 2023 013289 8; Instituto Nacional da Propriedade Industrial -Brazil.

Author Contributions: Conceptualization, H.S., R.D.G. and R.S.; methodology, H.S., R.D.G. and R.S.; software, H.S.; validation, H.S., R.D.G. and R.S.; formal analysis, H.S., R.D.G. and R.S.; investigation, H.S., R.D.G. and R.S.; writing—original draft preparation, H.S., R.D.G. and R.S.; writing—review and editing, R.D.G.; visualization, R.S.; supervision, R.D.G.; project administration, H.S.; funding acquisition, H.S., R.D.G. and R.S. All authors have read and agreed to the published version of the manuscript.

Funding: This research was developed at the Laboratory of Mechatronics and Virtual Prototyping (LaMaViP) of Ferrara Technopole, supported by the UNIFE FIRD2023 fund, in partnership with the Laboratory of Applied Robotics of Federal University of Santa Catarina, supported by CNPq—*Conselho Nacional de Desenvolvimento Científico e Tecnológico* (National Council for Scientific and Technological Development) Project 307249/2021-2, Brazil.

Institutional Review Board Statement: Not applicable.

Data Availability Statement: This work does not use experimental data. The data necessary to replicate the computations illustrated in the paper are included in the text of the paper.

Conflicts of Interest: The authors declare no conflict of interest.

References

1. Merlet, J.-P. *Parallel Robots*; Springer: Berlin/Heidelberg, Germany, 2000; ISBN 978-1-4020-0385-1.
2. Tsai, L.-W. *Mechanism Design: Enumeration of Kinematic Structures According to Function*; CRC Press: London, UK, 2001; ISBN 978-0-8493-0901-4.
3. Tsai, L.-W. *Robot Analysis: The Mechanics of Serial and Parallel Manipulators*; John Wiley & Sons: New York, NY, USA, 1999; ISBN 978-0-4713-2593-2.
4. Meng, J.; Liu, G.; Li, Z. A geometric theory for analysis and synthesis of sub-6 DoF parallel manipulators. *IEEE Trans. Robot.* **2007**, *23*, 625–649. [CrossRef]
5. Lin, R.; Guo, W.; Gao, F. Type Synthesis of a Family of Novel Four, Five, and Six Degrees-of-Freedom Sea Lion Ball Mechanisms With Three Limbs. *ASME J. Mech. Robot.* **2016**, *8*, 021023. [CrossRef]
6. Cleary, K.; Brooks, T. Kinematic analysis of a novel 6-DOF parallel manipulator. In Proceedings of the 1993 IEEE International Conference on Robotics and Automation, Atlanta, GA, USA, 2–6 May 1993; Volume 1, pp. 708–713. [CrossRef]
7. Jin, Y.; Chen, I.-M.; Yang, G. Kinematic design of a family of 6-DOF partially decoupled parallel manipulators. *Mech. Mach. Theory* **2009**, *44*, 912–922. [CrossRef]
8. Coppola, G.; Zhang, D.; Liu, K. A 6-DOF reconfigurable hybrid parallel manipulator. *Robot. Comput. Integr. Manuf.* **2014**, *30*, 99–106. [CrossRef]

9. Monsarrat, B.; Gosselin, C.M. Workspace analysis and optimal design of a 3-leg 6-DOF parallel platform mechanism. *IEEE Trans. Robot. Autom.* **2003**, *19*, 954–966. [CrossRef]
10. Angeles, J.; Yang, G.; Chen, I.-M. Singularity analysis of three-legged, six-dof platform manipulators with URS legs. *IEEE/ASME Trans. Mechatron.* **2003**, *8*, 469–475. [CrossRef]
11. Yang, Z.; Zhang, D. Novel Design of a 3-RRUU 6-DOF Parallel Manipulator. *IOP Conf. Ser. Mater. Sci. Eng.* **2019**, *491*, 012006. [CrossRef]
12. Fu, J.; Gao, F. Optimal design of a 3-leg 6-DOF parallel manipulator for a specific workspace. *Chin. J. Mech. Eng.* **2016**, *29*, 659–668. [CrossRef]
13. Seward, N.; Bonev, I.A. A new 6-DOF parallel robot with simple kinematic model. In Proceedings of the 2014 IEEE International Conference on Robotics and Automation (ICRA), Hong Kong, China, 31 May–7 June 2014; pp. 4061–4066. [CrossRef]
14. Lu, Y.; Wang, P.; Hou, Z.; Hu, B.; Sui, C.; Han, J. Kinetostatic analysis of a novel 6-DoF 3UPS parallel manipulator with multi-fingers. *Mech. Mach. Theory* **2014**, *78*, 36–50. [CrossRef]
15. Simas, H.; Meneghini, L.; Di Gregorio, R.; Simoni, R. Position Analysis of a Novel Family of three-legged 6-DOF Parallel Manipulators of type 3-XXRRU. In *Advances in Mechanism and Machine Science: Proceedings of the 16th IFToMM World Congress 2023 [WC2023], Tokyo, Japan, 5–10 November 2023*; Okada, M., Ed.; Paper No.: 147; Springer: Cham, Switzerland, 2023.
16. Gosselin, C.M.; Angeles, J. Singularity analysis of closed-loop kinematic chains. *IEEE Trans. Robot. Automat.* **1990**, *6*, 281–290. [CrossRef]
17. Ma, O.; Angeles, J. Architecture singularities of platform manipulators. In Proceedings of the 1991 IEEE International Conference on Robotics and Automation, Sacramento, CA, USA, 9–11 April 1991; pp. 1542–1547.
18. Zlatanov, D.; Fenton, R.G.; Benhabib, B. A unifying framework for classification and interpretation of mechanism singularities. *ASME J. Mech. Des.* **1995**, *117*, 566–572. [CrossRef]
19. Zlatanov, D.; Bonev, I.A.; Gosselin, C.M. Constraint Singularities as C-Space Singularities. In Proceedings of the 8th International Symposium on Advances in Robot Kinematics (ARK 2002), Caldes de Malavella, Spain, 24–28 June 2002.
20. Zlatanov, D.; Bonev, I.A.; Gosselin, C.M. Constraint singularities of parallel mechanisms. In Proceedings of the 2002 IEEE International Conference On Robotics & Automation, Washington, DC, USA, 11–15 May 2002; pp. 496–502.
21. Hunt, K.H. *Kinematic Geometry of Mechanisms*; Oxford University Press: Oxford, UK, 1978.
22. Davidson, J.K.; Hunt, K.H. *Robots and Screw Theory: Applications of Kinematics and Statics to Robotics*; Oxford University Press: Oxford, UK, 2005.
23. Uicker, J., Jr.; Pennock, G.; Shigley, J. *Theory of Machines and Mechanisms*, 5th ed.; Oxforf University Press: Oxford, UK, 2016.
24. Kim, H.S.; Tsai, L.-W. Evaluation of a Cartesian parallel manipulator. In *Advances in Robot Kinematics*; Lenarcic, J., Thomas, F., Eds.; Kluwer Academic Publishers: London, UK, 2002; pp. 21–28.
25. Kim, H.S.; Tsai, L.-W. Design Optimization of a Cartesian Parallel Manipulator. *ASME J. Mech. Des.* **2003**, *125*, 43–51. [CrossRef]
26. Silvester, J.R. Determinants of block matrices. *Math. Gaz.* **2000**, *84*, 460–467. [CrossRef]
27. Sothanaphan, N. Determinants of block matrices with noncommuting blocks. *Linear Algebra Its Appl.* **2017**, *512*, 202–218. [CrossRef]
28. Angeles, J. *Fundamentals of Robotic Mechanical Systems*, 4th ed.; Springer: Cham, Switzerland, 2014; ISBN 978-3-319-30762-6.
29. Hervé, J.M. Uncoupled actuation of pan-tilt wrists. *IEEE Trans. Robot.* **2006**, *22*, 56–64. [CrossRef]
30. Li, Q.; Hervé, J.M.; Ye, W. *Geometric Method for Type Synthesis of Parallel Manipulators*; Springer: Singapore, 2020; pp. 223–238. ISBN 978-981-13-8754-8. [CrossRef]

robotics MDPI

Article

Applying Screw Theory to Design the Turmell-Bot: A Cable-Driven, Reconfigurable Ankle Rehabilitation Parallel Robot

Julio Vargas-Riaño [1,*], Óscar Agudelo-Varela [2,†] and Ángel Valera [1,†]

[1] Instituto Universitario de Automática e Informática Industrial (Instituto ai2), Universitat Politècnica de València, 46022 Valencia, Spain; giuprog@isa.upv.es
[2] Facultad de Ciencias Básicas e Ingeniería, Universidad de los Llanos, Villavicencio 500002, Colombia; oscar.agudelo@unillanos.edu.co
* Correspondence: julio_h_vargas_r@ieee.org
† These authors contributed equally to this work.

Abstract: The ankle is a complex joint with a high injury incidence. Rehabilitation Robotics applied to the ankle is a very active research field. We present the kinematics and statics of a cable-driven reconfigurable ankle rehabilitation robot. First, we studied how the tendons pull mid-foot bones around the talocrural and subtalar axes. We proposed a hybrid serial-parallel mechanism analogous to the ankle. Then, using screw theory, we synthesized a cable-driven robot with the human ankle in the closed-loop kinematics. We incorporated a draw-wire sensor to measure the axes' pose and compute the product of exponentials. We also reconfigured the cables to balance the tension and pressure forces using the axis projection on the base and platform planes. Furthermore, we computed the workspace to show that the reconfigurable design fits several sizes. The data used are from anthropometry and statistics. Finally, we validated the robot's statics with MuJoCo for various cable length groups corresponding to the axes' range of motion. We suggested a platform adjusting system and an alignment method. The design is lightweight, and the cable-driven robot has advantages over rigid parallel robots, such as Stewart platforms. We will use compliant actuators for enhancing human–robot interaction.

Keywords: medical and rehabilitation robotics; biomechanics; parallel manipulator; cable-driven; kinematic analysis; robot design; mechanism synthesis; compliant mechanism

Citation: Vargas-Riaño, J.; Agudelo-Varela, Ó.; Valera, Á. Applying Screw Theory to Design the Turmell-Bot: A Cable-Driven, Reconfigurable Ankle Rehabilitation Parallel Robot. *Robotics* **2023**, *12*, 154. https://doi.org/10.3390/robotics12060154

Academic Editors: Raffaele Di Gregorio and Dan Zhang

Received: 30 September 2023
Revised: 6 November 2023
Accepted: 11 November 2023
Published: 14 November 2023

1. Introduction

Over the past few years, there has been a significant expansion in the domains where robotics is employed. Beyond the conventional industrial applications such as material handling, welding, and machine tending, contemporary robotics finds its way into diverse sectors such as the construction sector [1], rescue robotics [2], agricultural [3,4] and food industry [5], waste management [6], entertainment [7], security systems [8], etc.

The medical field is another sector where there is a significant increase of the use of robots, such as robotic-assisted surgery [9], nurse care [10,11], or rehabilitation robots [12]. In the case of rehabilitation robots, the aim is to develop a system that improves the effectiveness of therapies and the facilitation of patients' daily routines and exercises. Therefore, rehabilitation robotics can be considered to a collaborative platform, uniting experts in human–robot interaction and biomedical engineering with clinicians and therapists. This synergy aims to create the essential technologies that can enhance the quality of life for patients. This work deals with the design of a ankle rehabilitation robot. Ankle injuries are prevalent in both sports and daily activities, making them among the most frequently occurring injuries. However, to aid in the recovery process, it is imperative for patients to engage in a regimen of rehabilitation exercises. These exercises are typically overseen by a therapist, aiming to strengthen the patient's ankle joint and restore its full range of motion. In an effort to ease the workload of therapists, allowing them to assist a larger number of patients in a more efficient and secure manner, different lower limb devices have been developing [13].

Designing an ankle rehabilitation device that can accommodate a large percentage of the population is challenging because its size varies depending on several factors, including sex, age, and phenotype. In addition, there is a lack of affordable motion-capture systems and databases for estimating ankle joint models, which makes it complex to develop effective rehabilitation techniques for ankle injuries.

Most ankle rehabilitation devices are for gait rehabilitation or the entire lower limb, including the hip, knee, ankle, and foot [14–18]. As a result, there is a need for more specialized devices that target the ankle joint specifically, and also a need for rehabilitation robotics that increasingly try to adapt the specific needs of patients.

This paper presents a data-driven design method that utilizes population statistics, body proportions, and anthropometry. The design is ideal for low-income countries because we can resize the 3D-printed parts by changing a few parameters. Our purpose is to use a method to easily create a resizable and reconfigurable device that can aid in the movement of the human ankle joint by restoring its range of motion.

None of the existing research focuses on developing a patient-specific ankle model for use when the patient is sitting or lying down. To address this gap-research, we propose a design method for building an affordable, reconfigurable, portable, and 3D scalable device for patients in sitting or lying positions. Instead of using a fixed device, we propose a design method for sizes depending on the target population statistics. For this reason, we created a Jupyter Notebook and put the designs on repositories for the public domain.

In this work, we have taken a human-centered approach. We propose an adjustable device for different foot sizes and be able to switch between limbs. The design is for the patients in lying down positions. Our approach is bioinspired by mimicking the tendons involved in ankle movements while avoiding any forces that could pull the foot away from the ankle. We propose reciprocal tendons that apply pressure against the plantar surface of the foot, which will induce ankle joint integration forces to generate motion. The final design is presented in this article, while the intermediate results can be found in the supplementary materials.

Instead of providing a count for the number of academic database hits, we have included the fundamental kinematics and robot design sources related to screw theory that we consulted, as well as the cited literature on ankle anatomy [19–21], where we found inspiration to create our design. We firmly believe that comprehending ankle biomechanics requires these references. Seminal works presenting the ankle model as a parallel mechanism were inspirational for our work [22]. We have developed a unique research approach that utilizes concepts such as screws, Plucker coordinates, and the product of exponentials from screw theory studied in [23–26]. We sought out works about lower-mobility parallel mechanisms [27,28]. Additionally, we used an approximated patient-specific ankle model to reconfigure the device. Our approach differs from existing patient-specific reconfigurable cable-driven ankle rehabilitation robots for laying/sitting positions. Thus, it may not be easy to compare our work with others in the field. In light of this, we searched for publications on ankle rehabilitation robots to gain a better understanding of the current state of the art. Our work is related to kinematics and robot design [29–32], and is an application for medical and service robots [33–38] . Cable-driven robotics is a trending topic research that is closely related to our work [39–45]. Screw theory has been applied to performance evaluation in cable-driven parallel robots [46]. In this work, we deal with a reconfigurable low-mobility mechanism 4P-2R. A kind of lower-mobility mechanism was also studied in [47,48]. The screw theory has also been applied to computing rotation centers in [49], and we used the screw theory approach applied to kinematics and statics. We also applied direct kinematics, but there are alternative methods [50]. Our work is related to robotics in rehabilitation. There are application for upper limbs [51–53], also using cable-driven actuators [54], and for lower and upper limbs [55]. We focus on the ankle, but other works included the complete lower limbs [56,57]. These can use exoskeletons applied in motion assistance [58], walking assistance [59], and for variable force resistance [60]. A design and performance analysis is in [61]. We found work related to our cable-driven serial chain approach analysis [62].

Related Work

We found ankle rehabilitation robots characterized by degrees of freedom and their chains. For example, a serial robot has been used for ankle rehabilitation [63]. A great number of robots use rigid kinematic chains. One of the first robots was a Stewart platform [64], which is a rigid robot with 6 DoF. Other examples are a 3-RUS/RRR [65], a redundantly actuated mechanism [66], a 3-prismatic-revolute-spherical ankle rehabilitation robot [67], a 3-PRS Parallel Robot [68], a 2-UPS/RRR ankle rehabilitation [69], a nonredundant parallel robot [70], a2-dof turntable [71], and a 2PSS platform [72]. Another approach is the use of pneumatic actuators configured as a 2-SPS Mechanism [73] and pneumatic artificial muscles [74–76]. Another alternative is a cable-driven lower-limb parallel robot [77]—optimization methods are in [78], and an innovative robot is in [79]. The most similar to our approach with four wires attached to the shank are in [80], and the CABLEankle [81]. They are designed for the plantar surface of the foot parallel to the transversal plane. As a conclusion, our design is the one designed with four cables, which are reconfigurable and ajustable and can be used in laying positions.

2. Materials and Methods

First, we studied the ankle anatomy in [19] and the International Society of Biomechanics (ISB) [20]. Following that, we used the open-source digital model (z-anatomy) [21] to identify the bones, insertions, tendons, and muscles related to the ankle movement. We introduced the dimensions and geometrical model in a Jupyter Notebook using a SageMath 10 kernel [82] in [83]. We provided the CAD model in [84] and the sources in GitHub [85]. We identified the tendons involved in ankle movement acting as an over-actuated bio-mechanism (two degrees-of-freedom and four tendon groups). Following that, we chose antagonistic tendon groups. We identified the pressure forces from the platform against the plantar surface of the foot. Next, we drew a schematic with the tension forces between the base attached to the shank and a platform attached to the foot. We adapted the robot dimensions from [86], the proportions from [87], and the statistical data from [88]. We start the design by analyzing a simple cable-driven two-axis serial chain, which helps us to understand the reciprocal products between the cable and axis screw representations. We created artificial data from the ankle model using the forward kinematics computed from the "product of exponentials" (PoE). The data serve us to validate how to get the ankle model in practice from trajectory measurements received from a modification of the Turmell-Meter system [89]. When synthesizing the robot, we enlarged the device size to prevent cable contact with the body from the platform and base. We analyzed the cable-driven antagonistic actuation in the perpendicular axes model and used it for the initial calibration. Afterwards, we propose a method for the robot reconfiguration, the workspace, and the static simulation using MuJoCo 2.3.1 [90] to validate the stability by changing the axis and tendon positions. Finally, we proposed a CAD model using SolidWorks 2022.

2.1. Robot Inspired on the Analysis of the Ankle Joint

The model has two rotational joints and four tendons involved in ankle movement. First, we analyzed the tension forces and then the compression forces. Finally, we proposed a simplified schematic for the robot.

2.1.1. Ankle-Foot Tendons

The two-axis representation and the tendon insertions in the bones at the mid-foot are in Figure 1. Figure 1a illustrates the main tendons from the lateral-anterior view of the right foot. As we show in Figure 1a, the fifth metatarsal has two insertions, the peroneus tertius, with components for dorsiflexion and eversion, and the peroneus brevis, for eversion and plantarflexion. The calcaneus has the insertion of the Achilles tendon for inversion and plantarflexion. In Figure 1b, we show the medial-bottom view of the foot, and the insertion points are between the first metatarsal and the medial cuneiform. The tibialis anterior contributes to inversion and dorsiflexion, the peroneus longus contributes to plantar flexion and eversion, and the tibialis posterior contributes to inversion and plantar flexion.

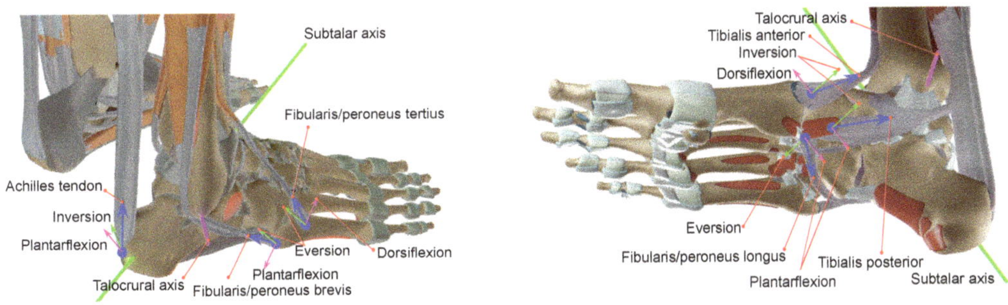

Figure 1. Lateral top and medial bottom views of the ankle-foot anatomy. (**a**) Lateral top view with components tangential to each axis. (**b**) Medial bottom view with components tangential to each axis [21].

2.1.2. Foot Compression Forces

Some requirements for ankle rehabilitation are security and comfort. The foot supports high compression forces in some regions of the plantar surface. In contrast, the dorsal skin is thin, soft, and less tolerant of compression forces. We propose a device that mainly applies pressure at the plantar surface to integrate the ankle joint and generate motion. From the bottom view of the foot, we note that the pressure points are involved in the subtalar and talocrural axes motion. Figure 2 shows the axes projection and the triangle vertices of the transversal plane. The three main points are the tubercle of the calcaneus and the first and fifth metatarsal heads. The pressure points are the projection of the triangular-shaped dome, limited by the lateral, medial-longitudinal arches, and the anterior transverse arch. These contact points transmit a compression force to the platform.

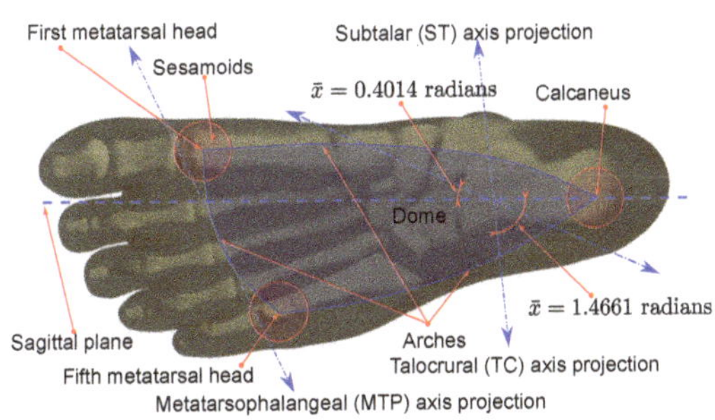

Figure 2. Right foot's bottom plantar plane projection with contact points and joint axis.

2.1.3. Robot Based on the Ankle-Foot Model

We designed a reconfigurable cable-driven robot to actuate on synergy with the ankle tendons. Cables from the base pull the platform attached to the foot. The cable anchor points have a relocation mechanism to equilibrate the rotation forces on the ankle joint. We aligned the centers to the ankle axes' intersection on the transverse plane. We show the schematic design in Figure 3. In Figure 3a, we sketch the approximated schematic and show the tendon directions regarding the subtalar (ST) and the talocrural (TC) axis. We represent the foot platform with a circle with a radius r_p, centered on PM_0, and on the same plane of the anchor points ap_1, ap_2, ap_3, and ap_4. The base is a circle with a radius r_b in the same plane as P_O, and the anchor points ab_1, ab_2, ab_3, and ab_4. We simplified the plantar surface of the foot with three contact points fs_1, fs_2, and fs_3. Finally, T_1, T_2, T_3, and T_4 are tendons. The serial kinematic chain is RR, starting at the origin P_O, followed by a rotational joint at r_1 on the TC axis, followed by the $r_1 - r_2$ link to a rotational joint at r_2 on the ST axis. A link from r_2 to PM_0 connects the platform with the ankle. The tendons from ap_1, ap_2, ap_3, and ap_4 on the platform to the corresponding ab_1, ab_2, ab_3, and ab_4 on the platform complete the parallel closed-loop structure. Each human ankle has different axis positions; thus, we propose reconfigurable cable attachments. In Figure 3b, the green arrows illustrate the concept of reconfiguration for the right foot by displacing the anchor endpoints ap_1, ap_2, ap_3, and ap_4 on the platform and the anchors ab_1, ab_2, ab_3, and ab_4 on the base. We designed a mechanism for centering pivot points P_{pb} and P_{pp}.

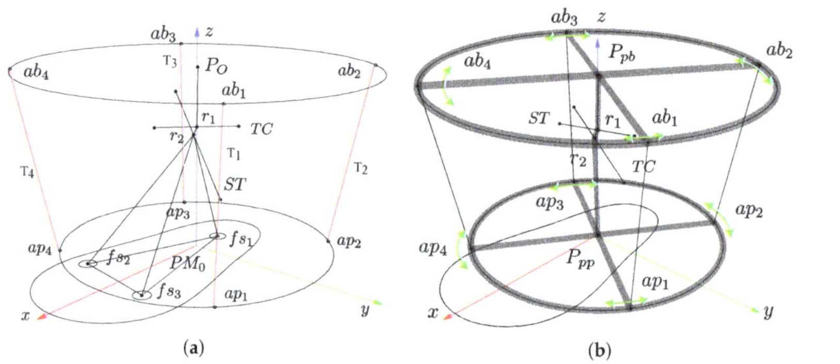

Figure 3. Ankle model and ankle rehabilitation robot sketch. (**a**) Schematic representation of the approximated tendon directions in a neutral position. (**b**) Schematic representation of reconfiguration.

We computed the center points from the intersection of the talocrural and subtalar axes projected on the transverse plane parallel to the base and the platform. Such pivot points are not the same as P_0 and PM_0, which pertain to the position sensors reference system.

2.1.4. Dimensions and Initial Configuration

In this section, we estimate the initial robot size and configuration. We used the ankle model and measurements from [86], human proportions from [87], and statistical data from [88]. We show the lower limb proportions in Figure 4a. The TC axis is dominant, and we use its statistical value for the cable-body collision analysis. We simplified the body as a sphere centered in half of the ankle's most medial point (MMP) and most lateral point (MLP); between the two malleoli. We compared this dimension with a leg model in the Figure 4b, illustrating the distance between the ankle and PM on the platform. Additionally, we show that the cables must not be in contact with the foot or the malleolus. The sphere represents the radius between the ankle and the plantar surface of the foot.

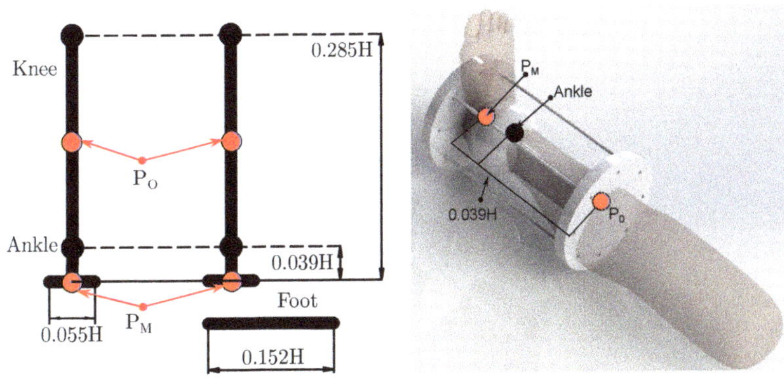

(**a**) (**b**)

Figure 4. Size design from proportions and statistics. (**a**) Proportions. (**b**) Three dimensional simplified design.

2.2. *Ankle Kinematic Model*

For the ankle kinematics representation, we used data from a two-axis model representation of the ankle joint explained in [89]. There, we described a system for platform pose capture and a method for model approximation using circle fitting, and Figure 5 shows the concept.

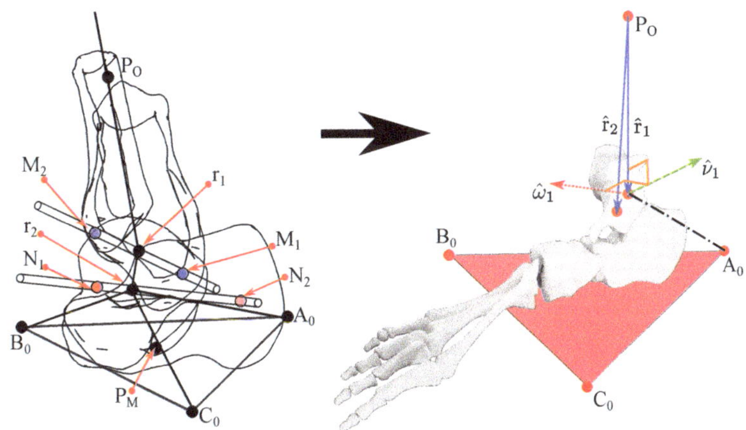

Figure 5. Product of exponentials representation of the ankle joint [89].

The points M_1, M_2, and r_1 are references on the talocrural axis. Similarly, N_1, N_2, and r_2 are in the subtalar axis. We used it to compute the following vectors and matrices:

$$\hat{r}_1 = r_1 - P_O, \qquad \hat{r}_2 = r_2 - P_O \tag{1}$$
$$\hat{\omega}_1 = \frac{M_2 - M_1}{\|M_2 - M_1\|}, \qquad \hat{\omega}_2 = \frac{N_2 - N_1}{\|N_2 - N_1\|} \tag{2}$$
$$\hat{v}_1 = -\hat{\omega}_1 \times \hat{r}_1, \qquad \hat{v}_2 = -\hat{\omega}_2 \times \hat{r}_2 \tag{3}$$
$$\hat{\omega}_1 = (\omega_{1x}, \omega_{1y}, \omega_{1z}), \quad \Omega_1 = \begin{bmatrix} 0 & -\omega_{1z} & \omega_{1y} \\ \omega_{1z} & 0 & -\omega_{1x} \\ -\omega_{1y} & \omega_{1x} & 0 \end{bmatrix} \tag{4}$$
$$\hat{\omega}_2 = (\omega_{2x}, \omega_{2y}, \omega_{2z}), \quad \Omega_2 = \begin{bmatrix} 0 & -\omega_{2z} & \omega_{2y} \\ \omega_{2z} & 0 & -\omega_{2x} \\ -\omega_{2y} & \omega_{2x} & 0 \end{bmatrix} \tag{5}$$
$$\xi_1 = \begin{pmatrix} \hat{v}_1 \\ \hat{\omega}_1 \end{pmatrix}, \qquad \xi_2 = \begin{pmatrix} \hat{v}_2 \\ \hat{\omega}_2 \end{pmatrix} \tag{6}$$

Next, we computed the Rodrigues' formulas:

$$e^{\hat{\omega}_1\theta_1} = I_{3\times3} + \Omega_1\theta_2 + \Omega_1^2(1 - \cos\theta_1) \tag{7}$$
$$e^{\hat{\omega}_2\theta_2} = I_{3\times3} + \Omega_2\theta_2 + \Omega_2^2(1 - \cos\theta_2) \tag{8}$$

representing the θ_1 and θ_2 rotations about $\hat{\omega}_1$ and $\hat{\omega}_2$, respectively. Also, we computed the matrices:

$$e^{\hat{\xi}_1\theta_1} = \begin{bmatrix} e^{\hat{\omega}_1\theta_1} & \hat{\tau}_1 \\ 0_{1\times3} & 1 \end{bmatrix}, \qquad e^{\hat{\xi}_2\theta_2} = \begin{bmatrix} e^{\hat{\omega}_2\theta_2} & \hat{\tau}_2 \\ 0_{1\times3} & 1 \end{bmatrix} \tag{9}$$

where:

$$\hat{\tau}_1 = \left(I_{3\times3} - e^{\hat{\omega}_1\theta_1}\right)\hat{\omega}_1 \times \hat{v}_1 + \hat{\omega}_1 \cdot \hat{\omega}_1^T \hat{v}_1\theta_1 \tag{10}$$
$$\hat{\tau}_2 = \left(I_{3\times3} - e^{\hat{\omega}_2\theta_2}\right)\hat{\omega}_2 \times \hat{v}_2 + \hat{\omega}_2 \cdot \hat{\omega}_2^T \hat{v}_2\theta_2 \tag{11}$$

If we define the initial pose representation as:

$$g_P(0) = \begin{bmatrix} R_0 & P_0 \\ 0_{1\times3} & 1 \end{bmatrix} \tag{12}$$

where:

$$R_0 = [\hat{s}_0 \quad \hat{n}_0 \quad \hat{a}_0] \tag{13}$$

where each column vector is:

$$\hat{s}_0 = \frac{B_0 + C_0 - 2A_0}{\|B_0 + C_0 - 2A_0\|} \tag{14}$$
$$\hat{n}_0 = \frac{(B_0 - A_0) \times (C_0 - A_0)}{\|(B_0 - A_0) \times (C_0 - A_0)\|} \tag{15}$$
$$\hat{a}_0 = \hat{s}_0 \times \hat{n}_0 \tag{16}$$

Then, we finally obtain the product of exponentials (PoE) representation for the serial chain with two hinge joints for all the P points on the platform:

$$g_P = e^{\hat{\xi}_1\theta_1} e^{\hat{\xi}_2\theta_2} g_P(0) = \begin{bmatrix} R_T & \tau_T \\ 0_{1\times3} & 1 \end{bmatrix} \tag{17}$$

where R_T is the total rotation matrix and τ_T is the total translation vector. The model formulas are in the Supplementary Materials Section S1.1.

2.3. *Synthesis of the Parallel Tendon-Driven Robot*

The cables must not be in contact with the body and must be smaller than the base and platform used by the draw-wire sensors. To simplify the collision study, we analyzed a coronal section of the foot and ankle, and the contact body is a circle containing the ankle, as illustrated in Figure 6a. We used this coronal section to approximate the platform and base sizes. In Figure 6b, two tendons drive a hinge joint, and the platform anchor points trace two concentric circular trajectories. The segments T_1 and T_2 represent two antagonistic tendons, r_s is the radius of a solid body containing the axis of rotation, and the interior circle represents a solid body. The collision contact of the cable with the body depends on the base and platform proportions and the radius r_s. By observing this, we note that if the base is larger than the platform, we can enhance the range of motion of the hinge joint. The points M_1, M_2, and r_1 are references on the talocrural axis. Similarly, N_1, N_2, and r_2 are in the subtalar axis.

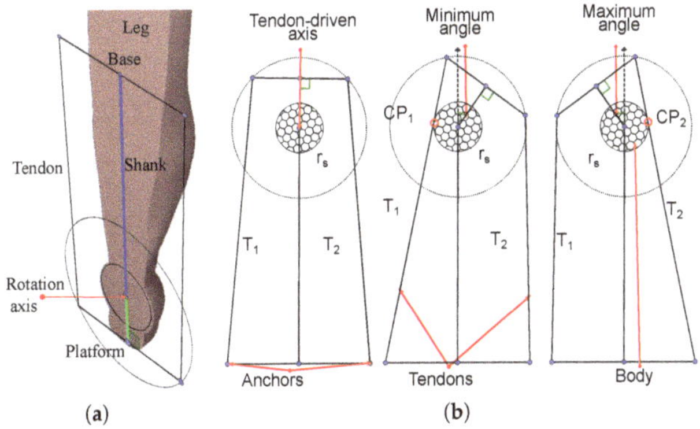

Figure 6. Base and platform sizes initial approximation. (**a**) Leg coronal section. (**b**) Collision, range of motion, and antagonistic actuation.

We used this coronal section to approximate the platform and base sizes. From the initial position, we defined a maximum and minimum reached angle. The limits are due to collision between the cables and the base body. When T_1 is extending and T_2 is contracting, Tendon T_1 touches the foot in CP_1. The minimum angle is limited for such a collision. The maximum angle occurs when T_1 is contracting and T_2 is extending. We selected a radius r_s greater than the foot width, and then we evaluated a platform radius r_p greater than the radius. By selecting a base radius r_b, we can evaluate the range of movement. The positive arc and its derivative are:

$$y = \sqrt{r_s^2 - x^2} \tag{18}$$

$$\frac{\partial y}{\partial x} = \frac{-x}{\sqrt{r_s^2 - x^2}} \tag{19}$$

The derivative is the slope of the line T_1 in the tangent point P_T:

$$\frac{-x}{\sqrt{r_s^2 - x^2}} = \frac{y - y_0}{x - x_0}, \tag{20}$$

solving for y yields:

$$y = -\frac{x^2 - x \cdot x_0 - \sqrt{r_s^2 - x^2} \cdot y_0}{\sqrt{r_s^2 - x^2}} \tag{21}$$

We found the tangential point by substituting y from Equation (18) in Equation (21), yielding the Equation (22).

$$\sqrt{r_s^2 - x_T^2} = -\frac{x_T^2 - x_T \cdot x_0 - \sqrt{r_s^2 - x_T^2} \cdot y_0}{\sqrt{r_s^2 - x_T^2}}, \tag{22}$$

and simplifying yields (23).

$$x_T \cdot x_0 - (r^2 - \sqrt{r_s^2 - x_T^2} \cdot y_0) = 0 \tag{23}$$

To solve for x_T, we used the software SageMath 10.0. Following that, we replaced the positive value in Equation (18) to find y_T. Finally, we found the intersection P_2 with the circular trajectory that has a radius r_2. To solve for x_2, we used the following equation:

$$\frac{y_T - y_0}{x_T - x_0}(x_2 - x_0) + y_0 = \sqrt{r_2^2 - x_2^2} \tag{24}$$

We found y_2 by replacing x_2 in Equation (18).

To find the arch length, we considered the initial position of anchor point P_1 on the platform. We used the following Equation:

$$d_{af} = \sqrt{r_2^2 - x_1^2} \tag{25}$$

where d_{af} is the length from the circle center to the platform central point and defines the trajectory radius regarding the positive semicircle. We solve for x_1 yields two values. By selecting the positive value, we obtain $y_1 = d_{af}$. The arc length is given by the absolute difference of the two corresponding angles. We computed the angles by using the Equation:

$$\gamma = |\arctan_2(y_2, x_2) - \arctan_2(y_1, x_1)| \tag{26}$$

In Figure 7, we illustrate the full range of motion of the platform.

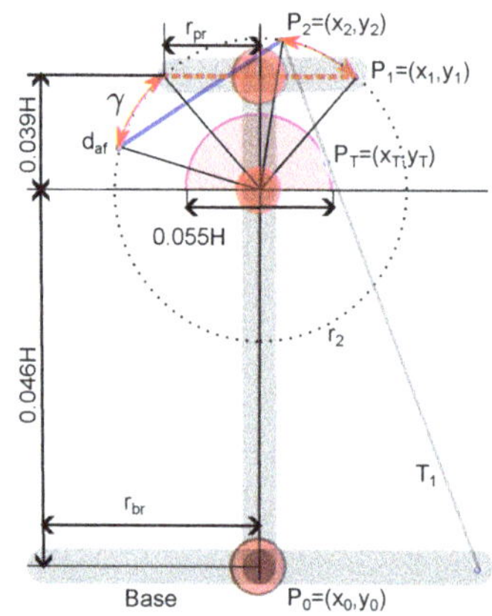

Figure 7. Maximum displacement.

Cable-Driven Two-Rotational Serial Chain

In this subsection, we studied a model in three dimensions that is analogous to the ankle joint. The model uses the screws in an antagonistic configuration to achieve tension forces at the cables. In Figure 8b, the view is normal to the proximal axis, and Figure 8a shows the normal view of the distal axis.

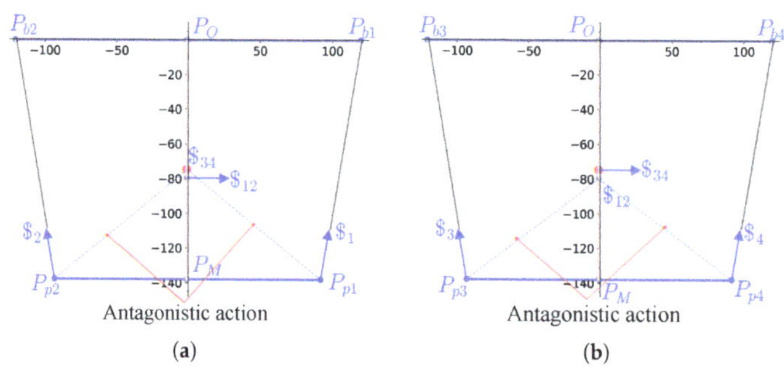

Figure 8. Normal views of the proximal and distal axis regarding the base. (**a**) Normal view of the proximal axis. (**b**) Normal view of the distal axis.

We used this simplified model to compute the cable locations. Also, we can use the model for testing and calibrating the mechanism before use in humans. With the purpose of designing a reconfigurable physical model, we represent the rotational joints as axes located coincident in opposite edges on a tetrahedral structure, as shown in Figure 9. We chose two different configurations by rotating the anchor points by 0.7854 radians (half of 1.5708 radians) from their initial position on the base and the platform. The first coincides with the rotation axis as in Figure 9a, and the second configuration is similar to that observed in the ankle, and we show this configuration in Figure 9b.

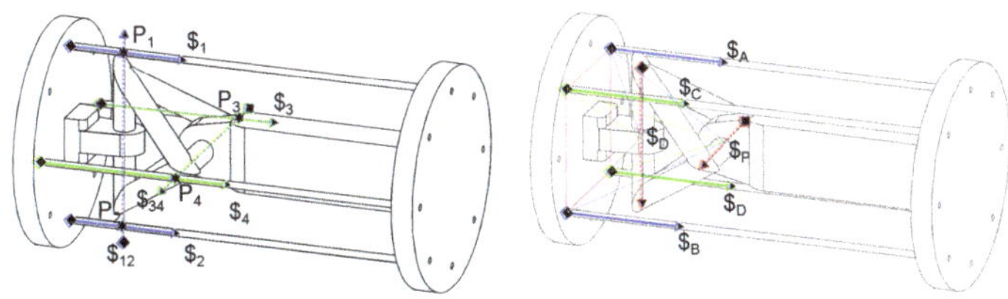

(a) (b)

Figure 9. First approximation with orthogonal axis coplanar to the tendons. (**a**) Antagonistic cables screws representation. (**b**) Screw representation for the anchor points rotated 0.7854 radians.

The first representation allows us to visually identify the intersection points between the tendons' lines of action.

$$L_{\$_1} \cap L_{\$_{12}} = P_1, \quad L_{\$_2} \cap L_{\$_{12}} = P_2 \tag{27}$$

$$L_{\$_3} \cap L_{\$_{34}} = P_3, \quad L_{\$_4} \cap L_{\$_{34}} = P_4 \tag{28}$$

Such a condition results in null reciprocal twists because they are coplanar [91]. Therefore, the reciprocal products are:

$$\$_1 \circ \$_{12} = 0, \qquad \$_2 \circ \$_{12} = 0, \tag{29}$$

$$\$_3 \circ \$_{34} = 0, \qquad \$_4 \circ \$_{34} = 0 \tag{30}$$

This configuration must be avoided because it leads to a singularity. The relations for all the tendons on the rotary joints are as follows:

$$\$_{12} \circ (\$_1 + \$_2 + \$_3 + \$_4), \quad \$_{34} \circ (\$_1 + \$_2 + \$_3 + \$_4) \tag{31}$$

In Figure 9a, we show the following:

$$\$_1 \circ \$_{34} = -\$_1 \circ \$_{34}, \qquad \$_3 \circ \$_{12} = -\$_4 \circ \$_{12} \tag{32}$$

Following that, the sum of all reciprocal products is zero, which means that the platform is in a static position. However, it is unstable because little variation in the anchor position suddenly changes the product's sign. In the second configuration, the twist pair $\$_A$ and $\$_C$ is antagonistic with regards to $\$_B$ and $\$_D$ about the proximal twist $\$_x$. Also, $\$_C$ and $\$_B$ are antagonistic with respect to $\$_A$ and $\$_D$ about the twist $\$_t$. We will use this in our design.

In summary, two conditions are important to avoid: singular configurations and cable collisions. We address the first condition by changing the angle position of the anchor points and the second condition by selecting the base and platform radiuses which are bigger than the foot standard size.

2.4. Robot Configuration

For simulation and validation, we generated trajectories and captured the axes pose using the formulas in the Supplementary Materials Sections S1.2 and S1.3. The method uses the formulas in the Supplementary Materials Section S1.4 for selecting the trajectories. The axis approximation formulas are in the Supplementary Materials Section S1.5. Finally, the range of motion and the common perpendicular line are in the Supplementary Materials Sections S1.6 and S1.7 complete the captured bi-axial model characterization. These data are important for the minimum and maximum cable lengths computation. We assume that the base plane is parallel to the transversal plane of the shank. With the reference systems aligned and P_O and PM_0 centered on the base and the platform, the shank line is perpendicular to the foot plane, and the neutral position occurs when the four tendons have similar tension. Therefore, the actuators used in antagonistic operation will have seamless forces. When the platform of the foot is parallel to the transversal plane of the shank, the talocrural and subtalar axes projected to the plane in symmetric form are:

$$\frac{x - r_{1x}}{\omega_{1x}} = \frac{y - r_{1y}}{\omega_{1y}}, \qquad \frac{x - r_{2x}}{\omega_{2x}} = \frac{y - r_{2y}}{\omega_{2y}} \tag{33}$$

Solving for y yields:

$$y = \frac{\omega_{1y}}{\omega_{1x}}x - \frac{\omega_{1y}}{\omega_{1x}}r_{1x} + \frac{r_{1y}}{\omega_{1y}} \tag{34}$$

$$y = \frac{\omega_{2y}}{\omega_{2x}}x - \frac{\omega_{2y}}{\omega_{2x}}r_{2x} + \frac{r_{2y}}{\omega_{2y}} \tag{35}$$

Subtracting (35) from (34), we have:

$$x\left(\frac{\omega_{1y}}{\omega_{1x}} - \frac{\omega_{2y}}{\omega_{2x}}\right) - \left(\frac{\omega_{1y}}{\omega_{1x}}r_{1x} - \frac{\omega_{2y}}{\omega_{2x}}r_{2x}\right) + \frac{r_{1y}}{\omega_{1y}} - \frac{r_{2y}}{\omega_{2y}} = 0 \tag{36}$$

Solving for x yields:

$$x_p = -\frac{(\omega_{2y}r_{2x} - \omega_{2x}r_{2y})\omega_{1x} + (\omega_{1x}\omega_{2x} - \omega_{1y}\omega_{2x})r_{1x}}{\omega_{1y}\omega_{2x} - \omega_{1x}\omega_{2y}} \tag{37}$$

Replacing x in (33) and solving for y yields:

$$y_p = -\frac{(\omega_{2y}r_{2x} - \omega_{2x}r_{2y})\omega_{1y} + (\omega_{1x}\omega_{2y} - \omega_{1y}\omega_{2y})r_{1x}}{\omega_{1y}\omega_{2x} - \omega_{1x}\omega_{2y}} \tag{38}$$

We created two planes parallel to the z-axis, coinciding with the intersecting point $P_{ip} = (x_p, y_p)$:

$$\Pi_1 : (P - P_{ip}) \cdot \hat{n}_{1\|z} = 0, \qquad \Pi_2 : (P - P_{ip}) \cdot \hat{n}_{2\|z} = 0 \tag{39}$$

where the normal vectors are:

$$\hat{n}_{1\|z} = \hat{\omega}_1 \times [0,0,1], \qquad \hat{n}_{2\|z} = \hat{\omega}_2 \times [0,0,1] \tag{40}$$

We computed the angle between the planes. This equation can be used to calculate the minimum and maximum angles, and the angles can be obtained as follows:

$$\gamma_{12} = \frac{1}{2}\arccos\left(\frac{\hat{n}_{1\|z} \cdot \hat{n}_{2\|z}}{|n_1||n_2|}\right), \tag{41}$$

$$\gamma_{21} = \frac{1}{2}\left[\pi - \arccos\left(\frac{\hat{n}_{1\|z} \cdot \hat{n}_{2\|z}}{|n_1||n_2|}\right)\right] \tag{42}$$

We rotated the unitary vector normal to the planes γ_{12} and γ_{21} about an axis parallel to the z-axis and passed it through P_{ip}. The resulting vectors are:

$$\omega_{12} = \mathrm{rot}(\hat{k}, -\gamma_{12}), \qquad \omega_{21} = \mathrm{rot}(\hat{k}, -\gamma_{21}) \tag{43}$$

where $\hat{k}$ is the unitary vector in the direction of the positive z-axis.

The resulting lines in symmetric form are:

$$\frac{x - x_p}{\omega_{12x}} = \frac{y - y_p}{\omega_{12y}}, \qquad \frac{x - x_p}{\omega_{21x}} = \frac{y - y_p}{\omega_{21y}} \tag{44}$$

Following that, we obtained the pivot point $P_{pp} = \left[x_{pp}, y_{pp}, z_{pp}\right]$ on the robot platform, and the pivot point $P_{pb} = \left[x_{pb}, y_{pb}, z_{pb}\right]$ on the base, corresponding to $\left[x_p, y_p\right]$ projected on the platform and the base. The platform and base attachment points are at the intersection between the circle centered on the pivot points P_{pp} and P_{pb} with the lines of Equations (44), and the circle equations are:

$$y^2 = r_p^2 - (x - x_{pp})^2 + y_{pp}, \qquad y^2 = r_b^2 - (x - x_{pb})^2 + y_{pb} \tag{45}$$

The platform radius is r_p, and the base radius is r_b. By solving for y in (44), substituting it in (45), and solving for x, we can obtain the following two values:

$$\frac{\sigma \mp \sqrt{\left(\omega_{12x}^2 + \omega_{12y}^2\right)\left(y_{pp} + r_p^2\right) - \omega_{12x}^2 y_{pp}^2}}{\omega_{12x}^2 + \omega_{12y}^2} \tag{46}$$

where:

$$\sigma = \left(\omega_{12x}^2 + \omega_{12y}^2\right)x_{pp} - \omega_{12x}\omega_{12y}y_{pp} \tag{47}$$

Substituting these values back into Equation (44) yields:

$$y_{ap_1,ap_2} = y_{pp} - \frac{\omega_{12y}}{\omega_{12x}}\left(x_{pp} - x_{ap_1,ap_2}\right) \tag{48}$$

Then, by substituting ω_{12} by ω_{21}, we can obtain the other two points. Finally, we obtained four points for the platform. Obtaining the base anchor points is similar to computing the platform anchor points, and by replacing P_{pp} by P_{bp} and r_p by r_b we computed four base corresponding points. The reconfigurable structure is easy to set up by changing the angle position given by:

$$\theta_{pi} = \arctan(y_{ap_i}, x_{ap_i}), \qquad \theta_{bi} = \arctan(y_{ab_i}, x_{ab_i}) \tag{49}$$

where θ_{pi} and θ_{bi} are the corresponding angles of each anchor point on the platform and the base related to the x direction around the intersection projected points P_{pp} and P_{bp}.

3. Workspace from the Product of Exponentials

By knowing the platform's initial configuration, we can plot the anchor point group of movements from the product of exponential matrices by applying the Equation (17) to each platform anchor point ap_i. The product of exponential matrix representing the group of movements g_{ap_i} for each anchor point ap_i on the platform is:

$$g_{ap_i} = e^{\hat{\xi}_1\theta_1}e^{\hat{\xi}_2\theta_2}g_{ap_i}(0) = \begin{bmatrix} R_{ap_i} & \tau_{ap_i} \\ 0_{1\times 3} & 1 \end{bmatrix} \tag{50}$$

Moreover, the cable lengths can be calculated as follows:

$$l_{ci} = \|\tau_{ap_i} - ab_i\| \tag{51}$$

The range of the talocrural angle is $\theta_1 \in [-0.3491, 0.3491]$ radians, and the subtalar angle range is $\theta_2 \in [-0.2618, 0.2618]$ radians. The surfaces represent the group of movements for each anchor point on the platform.

Reconfiguration and Statics Simulation

The robot configuration depends on the ankle axis location. We compared the analogous two-axis system and the ankle biaxial model on MuJoCo 2.3.1. We used this software because it is open-source and free to use. Additionally, it is used for model-based control. We provided the XML source in [85]. The simulation process is easy, and we edited the tendon lengths in the text editor and reloaded the simulation in MuJoCo 2.3.1. We stopped the simulation, we changed the hinge joint angles, and then we ran the simulation with the computed tendon lengths. The simulation stops in a static position at previously computed angles.

4. Results

4.1. Base and Platform Dimensions

First, we show the resulting base radius for mean statistic height in Figure 10 with a relation $r_b = 1.3r_p$. Every module has radius r_m.

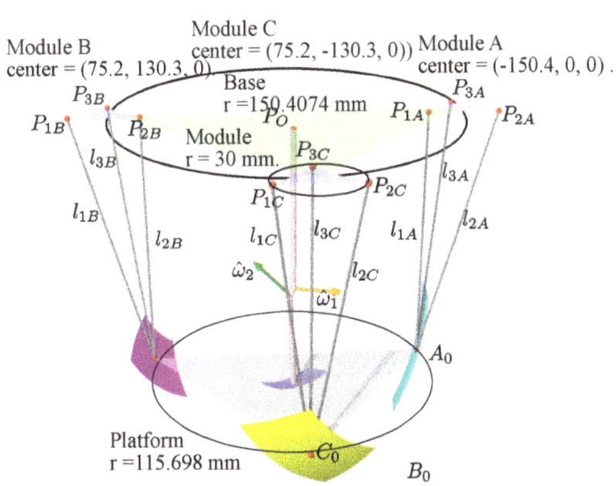

Figure 10. Resulting dimensions of the platform, base and sensor modules.

We computed the sensor lengths by subtracting points pertaining to the groups from the corresponding module vertexes. We started with the distance between the initial position from A_0 to the module P_{1A}.

$$l_{1A} = \|A_0 - P_{1A}\|, \tag{52}$$

We did the same for modules B and C, then we got the lengths from each initial point for the corresponding sensors in Table 1. The lengths graphical representation is in Figure 10.

Table 1. Distances from the initial points to the corresponding sensors.

Sensor Module	Distance to P_1	Distance to P_2	Distance to P_3
A	$l_{1A} = 176.2$	$l_{2A} = 184.9$	$l_{3A} = 184.9$
B	$l_{1B} = 176.2$	$l_{2B} = 311.1$	$l_{3B} = 269.9$
C	$l_{1C} = 293.9$	$l_{2C} = 269.9$	$l_{3C} = 311.1$

From the sensor lengths, we computed the initial position for **A**, **B**, and **C** from the Equations (S15)–(S17) on the Supplementary Materials. Following that, we compared the original with the computed sensor lengths in Table 2.

Table 2. Platform points computation.

Point	Original from the Model	Estimation from Lengths
A_0	(−115.7, 0, −176.18)	(−115.7, −0, −176.18)
B_0	(57.849, −100.2, −176.18)	(57.849, −100.2, −176.18)
C_0	(57.849, 100.2, −176.18)	(57.849, 100.2, −176.18)

The table shows that the computed positions from the lengths corresponds to the given original values.

4.2. Intermediate Results

For the sake of simplifying the results section, we put all the intermediate results on the supplementary materials. We used data generated from the ankle model forward kinematics, using the formulas in the Supplementary Materials Section S1.2. The platform position uses the formulas in the Supplementary Materials Section S1.3. The most circular trajectories selection in practice uses the formulas in the Supplementary Materials Section S1.4. The axis approximation uses the formulas in the Supplementary Materials Section S1.5. The range of motion computations uses the formulas in Supplementary Materials Section S1.6. The common perpendicular formulas are in the Supplementary Materials Section S1.7.

The trajectories generation results are in the Supplementary Materials Section S2.1. The resulting common perpendiculars computations, a unique characteristic for each ankle, are in the Supplementary Materials Section S2.2. The range of motion results are in the Supplementary Materials Section S2.3. The trajectories from the group of movements are in the Supplementary Materials Section S2.4. The subtalar axis range of motion results are in the Supplementary Materials Section S2.4. The resulting base and platform sizes are in the Supplementary Materials Section S2.6. The resulting screws initial analysis is in the Supplementary Materials Section S2.7. The horizontal planes projection of the axes intersection point are in the Supplementary Materials Section S2.8. The results of the robot configuration are in the Supplementary Materials Section S2.9. The angle computation of the anchor points is in the Supplementary Materials Section S2.10. The robot workspace is in the Supplementary Materials Section S2.11. The initial position of cable lengths is in the Supplementary Materials Section S2.12. The cables lengths at extreme poses are in the Supplementary Materials Section S2.13. The MuJoCo simulation results are in the Supplementary Materials Section S2.14.

4.3. Resulting Robot Design

In this subsection, we used the data captured from the ankle model and integrated the draw-wire sensors into the robot. The device dimensions are based on human proportions with a mean height of H = 175 cm. The objective is to adjust the measurement device and the ankle-approximated model in a configurable structure. The design is intended for the laying position. We divided the design into two main subassemblies: the platform for the foot and the base for the shank.

The resulting platform design is based on the foot anatomy by observing the Figures 1a,b and 2. The platform is adaptable to various foot sizes based on proportions [87]. The length is also adaptable, and we show the assembly in Figure 11. We added a length ruler, heel support, and three force sensing resistors (FSRs). Two FSRs are used for the forefoot, and one FSR is used for the hindfoot.

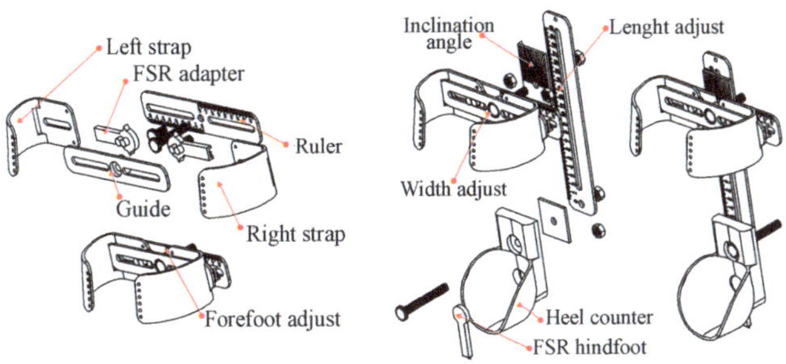

Figure 11. Foot's size adjust.

To align the center of the platform to P_{pp} at the initial position we designed perpendicular sliders, as shown in Figure 12.

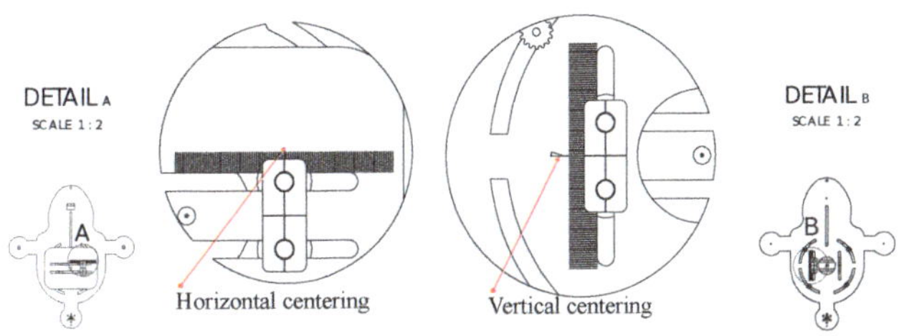

Figure 12. Centering the platform.

In Figure 13, we show the base assembly, with sensors for ankle axis estimation. We designed guides for centering the shank position P_{bp}. The anchor points are manually adjusted in different positions, depending on the ankle model.

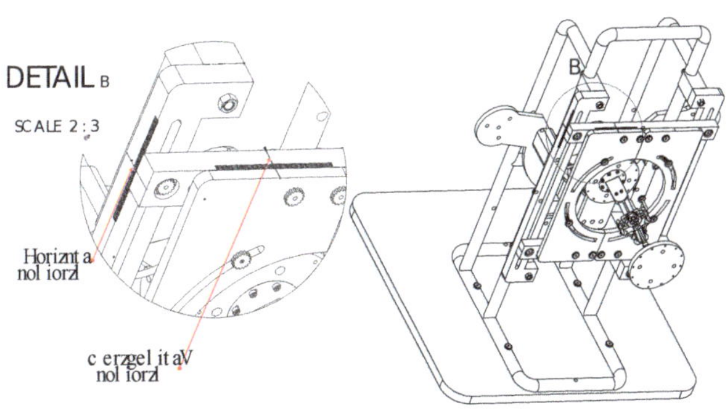

Figure 13. Centering the base.

A main tube structure is attached to a baseplate. We used spacers and 8 mm steel bolts to fasten the two plates supporting the sensors. Finally, we placed Bowden guides for the cable endpoints.

The final design includes the platform, the sensors, the base, and a possible configuration for the actuators and electronics. We also show a resulting assembly with the approximated adjustable axis mechanical model in Figure 14.

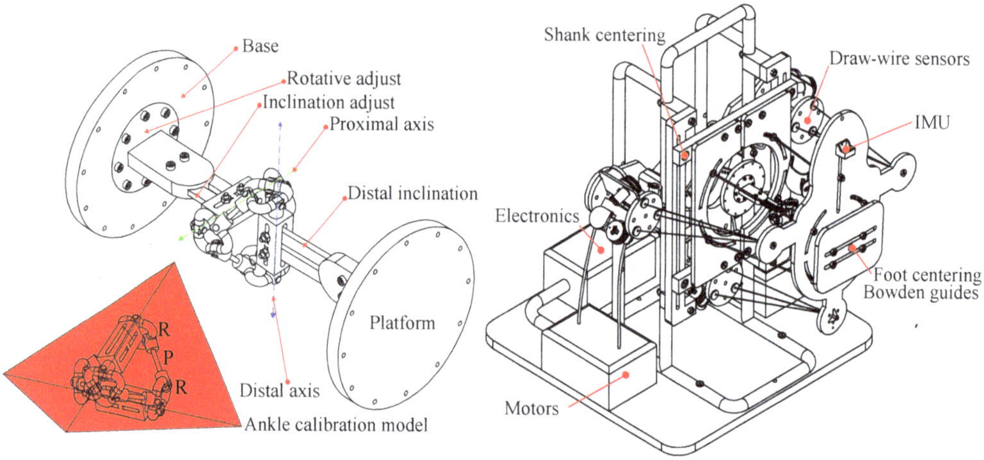

Figure 14. CAD design.

5. Discussion

In this work, we show that, by using the human-centered design of an ankle rehabilitation robot, we can apply the device to a broader group of patients.

We focus on the ankle model specific to each patient to equilibrate tension and pressure forces in an initial position. For us, ankle kinematic model identification is imperative to configure the platform and base anchor points. We show that such a design can enhance the range of motion, adapting to several sizes on both feet. Using this human-centered approach, we can limit the pressure forces acting on the plantar surface of the foot to avoid unnatural positions. We estimated the platform and base sizes. We also ensured that the cables did not touch the foot. The cable lengths can reach the ankle joint range of motion. We reconfigured the robot around the initial equilibrium position. This position is like when the human body is standing. We used projected axes on the transversal planes perpendicular to gravity. Screw theory is a powerful tool, and the results show we can effectively use it in robot geometry, kinematics, and static analysis.

The screw theory based on two axes has unique characteristics such as common perpendicular, relative angles between the axes, and relative inclination regarding the sagittal, transversal, and coronal planes. We can compare left and right foot axes for stability analysis and gait disease diagnosis.

Using MuJoCo, we simulated the statics by editing an XML file with the computed tendon lengths and anchor points, then using sensors, the ankle joint angles measures are similar to those that are initially assumed. There are various biomechanic simulators such as AnyBody, OpenSim, and Myosuite. We can translate the ankle model obtained with our method to the XML files used in these simulators.

6. Conclusions

Ankle sprains are a common injury, and there is abundant research on ankle rehabilitation robots. We found that Screw Theory is applied in many works. By observing the anatomy, we realized that the tendons associated with the ankle movement are attached to the bones at the base of the plantar dome, transmitting pressure to the foot surface in contact with the platform. We designed the Turmell-Bot to be used by patients lying in bed or sitting. The robot configuration depends on the patient-specific ankle model. The model is an approximation that we can refine by using piece-wise function approximations and machine learning. We designed a lightweight, low-cost, low-energy, portable, configurable, and comfortable device. We enhanced the device with sensors to measure the foot pressure forces involved in ankle movements. We will use compliant actuators and ratchets to hold the desired position without energy consumption. We will search for antagonistic actuation and tension control. We also plan to integrate electromyography (EMG) and functional electrical stimulation (FES) through the shank attachment to register the activation signals when the device is making rehabilitation movements.

Limitations of the Present Work and Future Development and Improvements

Before building the device, we conducted exhaustive testing on various sizes using simulations. Our proposed method for scaling the device to fit a target population involves using statistics and proportions. For example, starting with a statistical mean height, we can adjust the dimensions accordingly and use 3D printing for the necessary attachments and supports. To further refine the models, we can create datasets on the target populations and implement machine learning techniques. Our approach to mapping the neural activity related to ankle rehabilitation involves using compliant actuators and EMG sensors on the shank. By utilizing this model, we can create anatomically functional prosthetics that work effectively near the ankle. The two-axis model expressed in screw theory terms can be unique to each patient. We can use the axes' characteristics such as the attitude (relative to the sagittal, coronal, and transverse planes) and the common perpendicular segment between the two axes to study body stability and gait analysis.

Supplementary Materials: The following supporting information can be downloaded at: https://www.mdpi.com/article/10.3390/robotics12060154/s1.

Author Contributions: Conceptualization, Á.V. and J.V.-R.; methodology, J.V.-R. and Á.V.; software, J.V.-R.; validation, J.V.-R.; formal analysis, J.V.-R.; investigation, J.V.-R.; resources, Ó.A.-V. and Á.V.; data curation, J.V.-R. and Á.V.; writing—original draft preparation, J.V.-R.; writing—review and editing, J.V.-R., Ó.A.-V. and Á.V.; visualization, J.V.-R.; supervision, Á.V.; project administration, Ó.A.-V. and Á.V.; funding acquisition, Ó.A.-V. and Á.V. All authors have read and agreed to the published version of the manuscript.

Funding: This research was partially funded by Colciencias-Colfuturo PhD Scholarships Program Educational Credit Forgivable grant number 568, and by Agencia Estatal de Investigación (Ministerio de Ciencia e Innovación, Spanish Government) and Fondo Europeo de Desarrollo Regional through the Project ref. PID2021-125694OB-I00.

Institutional Review Board Statement: Not applicable.

Informed Consent Statement: Not applicable.

Data Availability Statement: Data available in https://github.com/juliohvr/Turmell-Bot (accessed on 10 November 2023). CAD in https://grabcad.com/library/turmell-bot-1 (accessed on 10 November 2023). Jupyter Notebook in https://nbviewer.org/github/juliohvr/Turmell-Bot/blob/main/TurmellBot.ipynb (accessed on 10 November 2023).

Acknowledgments: This works is supported in part by the Colombian Administrative Department of Science, Technology and Innovation, (Colciencias) under the 568 Doctorate program, the Automatics and Informatics Institute (ai2) at Universitat Politècnica de València, (UPV) and the Facultad de Ciencias Básicas e Ingeniería (FCBI) at Universidad de los Llanos (Unillanos).

Conflicts of Interest: The authors declare no conflict of interest.

Abbreviations

The following abbreviations are used in this manuscript:

ISB	International Society of Biomechanics
CAD	Computer Aided Design
PoE	Product of Exponentials
TC	Talocrural
ST	Subtalar
MTP	Metatarsophalangeal
MMP	Most Medial Point
MLP	Most Lateral Point
PM	Platform Mean
EMG	Electromyography
FES	Functional Electrostimulation

References

1. Xiao, B.; Chen, C.; Yin, X. Recent advancements of robotics in construction. *Autom. Constr.* **2022**, *144*, 104591. [CrossRef]
2. Delmerico, J.; Mintchev, S.; Giusti, A.; Gromov, B.; Melo, K.; Horvat, T.; Cadena, C.; Hutter, M.; Ijspeert, A.; Floreano, D.; et al. The current state and future outlook of rescue robotics. *J. Field Robot.* **2019**, *36*, 1171–1191. [CrossRef]
3. Lytridis, C.; Bazinas, C.; Kalathas, I.; Siavalas, G.; Tsakmakis, C.; Spirantis, T.; Badeka, E.; Pachidis, T.; Kaburlasos, V.G. Cooperative Grape Harvesting Using Heterogeneous Autonomous Robots. *Robotics* **2023**, *12*, 147. [CrossRef]
4. Tsiakas, K.; Papadimitriou, A.; Pechlivani, E.M.; Giakoumis, D.; Frangakis, N.; Gasteratos, A.; Tzovaras, D. An Autonomous Navigation Framework for Holonomic Mobile Robots in Confined Agricultural Environments. *Robotics* **2023**, *12*, 146. [CrossRef]
5. Wang, Z.; Hirai, S.; Kawamura, S. Challenges and Opportunities in Robotic Food Handling: A Review. *Front. Robot. AI* **2022**, *8*, 789107. [CrossRef]
6. Satav, A.G.; Kubade, S.; Amrutkar, C.; Arya, G.; Pawar, A. A state-of-the-art review on robotics in waste sorting: Scope and challenges. *Int. J. Interact. Des. Manuf. (IJIDeM)* **2023**, *17*, 2789–2806. [CrossRef]
7. Bogue, R. The role of robots in entertainment. *Ind. Robot* **2022**, *49*, 667–671. [CrossRef]
8. Ortega, L.D.; Loyaga, E.S.; Cruz, P.J.; Lema, H.P.; Abad, J.; Valencia, E.A. Low-Cost Computer-Vision-Based Embedded Systems for UAVs. *Robotics* **2023**, *12*, 145. [CrossRef]
9. Chopra, H.; Baig, A.A.; Cavalu, S.; Singh, I.; Emran, T.B. Robotics in surgery: Current trends. *Ann. Med. Surg.* **2022**, *81*, 104375. [CrossRef]
10. Hinrichs, P.; Seibert, K.; Gómez, P.A.; Pfingsthorn, M.; Hein, A. A Robotic System to Anchor a Patient in a Lateral Position and Reduce Nurses' Physical Strain. *Robotics* **2023**, *12*, 144. [CrossRef]
11. Soriano, G.P.; Yasuhara, Y.; Ito, H.; Matsumoto, K.; Osaka, K.; Kai, Y.; Locsin, R.; Schoenhofer, S.; Tanioka, T. Robots and Robotics in Nursing. *Healthcare* **2022**, *10*, 1571. [CrossRef]
12. Vitiello, N.; Trigili, E.; Crea, S. *Rehabilitation Robots*; Springer: London, UK, 2020; pp. 1–6. [CrossRef]
13. Gao, M.; Wang, Z.; Pang, Z.; Sun, J.; Li, J.; Li, S.; Zhang, H. Electrically Driven Lower Limb Exoskeleton Rehabilitation Robot Based on Anthropomorphic Design. *Machines* **2022**, *10*, 266. [CrossRef]
14. Warutkar, V.; Dadgal, R.; Mangulkar, U.R. Use of Robotics in Gait Rehabilitation Following Stroke: A Review. *Cureus* **2022**, *14*, e31075. [CrossRef]
15. Zhou, J.; Yang, S.; Xue, Q. Lower limb rehabilitation exoskeleton robot: A review. *Adv. Mech. Eng.* **2021**, *13*, 16878140211011862. [CrossRef]
16. Shi, D.; Zhang, W.; Zhang, W.; Ding, X. A Review on Lower Limb Rehabilitation Exoskeleton Robots. *Chin. J. Mech. Eng.* **2019**, *32*, 74. [CrossRef]
17. Mikolajczyk, T.; Ciobanu, I.; Badea, D.I.; Iliescu, A.; Pizzamiglio, S.; Schauer, T.; Seel, T.; Seiciu, P.L.; Turner, D.L.; Berteanu, M. Advanced technology for gait rehabilitation: An overview. *Adv. Mech. Eng.* **2018**, *10*, 1687814018783627. [CrossRef]
18. Callegaro, A.M.; Unluhisarcikli, O.; Pietrusinski, M.; Mavroidis, C. *Neuro-Robotics: From Brain machine Interfaces to Rehabilitation Robotics; Number 2 in Trends in Augmentation of Human Performance*; Springer: Dordrecht, The Netherlands, 2014; pp. 265–283. [CrossRef]
19. Ravella, K.C.; Ahmad, J.; Amirouche, F. *Biomechanics of the Ankle Joint*; Springer International Publishing: Cham, Switzerland, 2021; pp. 401–413. [CrossRef]
20. Wu, G.; Siegler, S.; Allard, P.; Kirtley, C.; Leardini, A.; Rosenbaum, D.; Whittle, M.; D'Lima, D.D.; Cristofolini, L.; Witte, H.; et al. ISB recommendation on definitions of joint coordinate system of various joints for the reporting of human joint motion–part I: ankle, hip, and spine. International Society of Biomechanics. *J. Biomech.* **2002**, *35*, 543–548. [CrossRef]
21. Okubo, K.; Kervyn, G.; Zielinski, M.; Vinent, L.; Bigio, A.T.; Villar., C.T. Open Source Anatomy. Available online: https://github.com/LluisV/Z-Anatomy (accessed on 13 November 2023).
22. Gregorio, R.D.; Parenti-Castelli, V.; O Connor, J.J.; Leardini, A. Mathematical models of passive motion at the human ankle joint by equivalent spatial parallel mechanisms. *Med. Biol. Eng. Comput.* **2007**, *45*, 305–313. [CrossRef]

23. Lynch, K.M.; Park, F.C. (Eds.) *Modern Robotics: Mechanics, Planning, and Control*; Cambridge University Press: Cambridge, MA, USA, 2019.
24. Gallardo-Alvarado, J. *Kinematic Analysis of Parallel Manipulators by Algebraic Screw Theory*; Springer International Publishing: Cham, Switzerland, 2016. [CrossRef]
25. Zhao, J.; Feng, Z.; Chu, F.; Ma, N. *Advanced Theory of Constraint and Motion Analysis for Robot Mechanisms*; Academic Press: Oxford, UK, 2014. [CrossRef]
26. Ball, R.S. *A Treatise on the Theory of Screws*; Cambridge University Press: Cambridge, UK, 1998.
27. Liao, Z.; Yao, L.; Lu, Z.; Zhang, J. Screw theory based mathematical modeling and kinematic analysis of a novel ankle rehabilitation robot with a constrained 3-PSP mechanism topology. *Int. J. Intell. Robot. Appl.* **2018**, *2*, 351–360. [CrossRef]
28. Di, R. Parallel Manipulators with Lower Mobility. In *Industrial Robotics: Theory, Modelling and Control*; Pro Literatur Verlag: Mammendorf, Germany; ARS: Linz, Austria, 2006. [CrossRef]
29. Gregorio, R.D. (Ed.) *Kinematics and Robot Design I, KaRD2018*; MDPI: Basel, Switzerland, 2021. [CrossRef]
30. Gregorio, R.D. (Ed.) *Kinematics and Robot Design IV, KaRD2021*; MDPI: Basel, Switzerland, 2022. [CrossRef]
31. Gregorio, R.D. *Kinematics and Robot Design II (KaRD2019) and III (KaRD2020)*; MDPI AG: Basel, Switzerland, 2022.
32. Gregorio, R.D. (Ed.) *Kinematics and Robot Design V, KaRD2022*; MDPI: Basel, Switzerland, 2023. [CrossRef]
33. Tarnita, D.; Dumitru, N.; Pisla, D.; Carbone, G.; Geonea, I. (Eds.) *New Trends in Medical and Service Robotics: MESROB 2023*; Mechanisms and Machine Science; Springer Nature: Cham, Switzerland, 2023; Volume 133. [CrossRef]
34. Rauter, G.; Carbone, G.; Cattin, P.C.; Zam, A.; Pisla, D.; Riener, R. (Eds.) *New Trends in Medical and Service Robotics: MESROB 2021*; Mechanisms and Machine Science; Springer International Publishing: Cham, Switzerland, 2021; Volume 106. [CrossRef]
35. Rauter, G.; Cattin, P.C.; Zam, A.; Riener, R.; Carbone, G.; Pisla, D. (Eds.) *New Trends in Medical and Service Robotics: MESROB 2020*; Mechanisms and Machine Science; Springer International Publishing: Cham, Switzerland, 2020; Volume 93. [CrossRef]
36. Carbone, G.; Ceccarelli, M.; Pisla, D. (Eds.) *New Trends in Medical and Service Robotics: Advances in Theory and Practice*; Mechanisms and Machine Science; Springer International Publishing: Cham, Switzerland, 2019; Volume 65. [CrossRef]
37. Bleuler, H.; Bouri, M.; Mondada, F.; Pisla, D.; Rodic, A.; Helmer, P. (Eds.) *New Trends in Medical and Service Robots: Assistive, Surgical and Educational Robotics*; Mechanisms and Machine Science; Springer International Publishing: Cham, Switzerland, 2016; Volume 38. [CrossRef]
38. Rodić, A.; Pisla, D.; Bleuler, H. (Eds.) *New Trends in Medical and Service Robots: Challenges and Solutions*; Mechanisms and Machine Science; Springer International Publishing: Cham, Switzerland, 2006; Volume 20. [CrossRef]
39. Caro, S.; Pott, A.; Bruckmann, T. (Eds.) *Cable-Driven Parallel Robots: Proceedings of the 6th International Conference on Cable-Driven Parallel Robots*; Mechanisms and Machine Science; Springer Nature Switzerland: Cham, Switzerland, 2023; Volume 132. [CrossRef]
40. Gouttefarde, M.; Bruckmann, T.; Pott, A. (Eds.) *Cable-Driven Parallel Robots: Proceedings of the 5th International Conference on Cable-Driven Parallel Robots*; Mechanisms and Machine Science; Springer International Publishing: Cham, Switzerland, 2021; Volume 104. [CrossRef]
41. Pott, A.; Bruckmann, T. (Eds.) *Cable-Driven Parallel Robots: Proceedings of the 4th International Conference on Cable-Driven Parallel Robots*; Mechanisms and Machine Science; Springer International Publishing: Cham, Switzerland, 2019; Volume 74. [CrossRef]
42. Pott, A. *Cable-Driven Parallel Robots: Theory and Application*; Springer Tracts in Advanced Robotics; Springer International Publishing: Cham, Switzerland, 2019; Volume 120. [CrossRef]
43. Gosselin, C.; Cardou, P.; Bruckmann, T.; Pott, A. (Eds.) *Cable-Driven Parallel Robots*; Mechanisms and Machine Science; Springer International Publishing: Cham, Switzerland, 2018; Volume 53. [CrossRef]
44. Pott, A.; Bruckmann, T. (Eds.) *Cable-Driven Parallel Robots: Proceedings of the Second International Conference on Cable-Driven Parallel Robots*; Mechanisms and Machine Science; Springer International Publishing: Cham, Switzerland, 2017; Volume 32. [CrossRef]
45. Bruckmann, T.; Pott, A. (Eds.) *Cable-Driven Parallel Robots*; Mechanisms and Machine Science; Springer: Cham, Switzerland, 2014; Volume 12. [CrossRef]
46. Boschetti, G.; Trevisani, A. Cable Robot Performance Evaluation by Wrench Exertion Capability. *Robotics* **2018**, *7*, 15. [CrossRef]
47. Di Gregorio, R. A Review of the Literature on the Lower-Mobility Parallel Manipulators of 3-UPU or 3-URU Type. *Robotics* **2020**, *9*, 5. [CrossRef]
48. Di Gregorio, R. A Novel 3-URU Architecture with Actuators on the Base: Kinematics and Singularity Analysis. *Robotics* **2020**, *9*, 60. [CrossRef]
49. Valderrama-Rodríguez, J.I.; Rico, J.M.; Cervantes-Sánchez, J.J.; García-García, R. A Screw Theory Approach to Computing the Instantaneous Rotation Centers of Indeterminate Planar Linkages. *Robotics* **2022**, *11*, 6. [CrossRef]
50. Flores-Salazar, E.D.; Arias-Montiel, M.; Lugo-González, E.; Gallardo-Alvarado, J.; Tapia-Herrera, R. Alternative Methods for Direct Kinematic Analysis of a Parallel Robot for Ankle Rehabilitation. In *New Trends in Medical and Service Robotics*; Rauter, G., Cattin, P.C., Zam, A., Riener, R., Carbone, G., Pisla, D., Eds.; Mechanisms and Machine Science; Springer International Publishing: Cham, Switzerland, 2021; pp. 53–61. [CrossRef]
51. Metcalf, A.G.; Gallagher, J.F.; Jackson, A.E.; Levesley, M.C. Multi-Domain Dynamic Modelling of a Low-Cost Upper Limb Rehabilitation Robot. *Robotics* **2021**, *10*, 134. [CrossRef]
52. Zakaryan, N.; Harutyunyan, M.; Sargsyan, Y. Bio-Inspired Conceptual Mechanical Design and Control of a New Human Upper Limb Exoskeleton. *Robotics* **2021**, *10*, 123. [CrossRef]

53. Khan, M.M.R.; Swapnil, A.A.Z.; Ahmed, T.; Rahman, M.M.; Islam, M.R.; Brahmi, B.; Fareh, R.; Rahman, M.H. Development of an End-Effector Type Therapeutic Robot with Sliding Mode Control for Upper-Limb Rehabilitation. *Robotics* **2022**, *11*, 98. [CrossRef]
54. Zhou, Y.; Zhang, B.; Shang, W.; Cong, S. Configuration Optimization of an Auto-reconfigurable Cable-Driven Upper-Limb Rehabilitation Robot. In *Cable-Driven Parallel Robots*; Gouttefarde, M., Bruckmann, T., Pott, A., Eds.; Mechanisms and Machine Science; Springer International Publishing: Cham, Switzerland, 2021; pp. 145–157. [CrossRef]
55. Ceresoli, F.; Aggogeri, F.; Amici, C.; Borboni, A.; Faglia, R.; Pellegrini, N.; Tiboni, M.; Antonini, M.; Fausti, D.; Mor, M.; et al. Differential System for Limb Rehabilitation. In *New Trends in Medical and Service Robotics*; Carbone, G., Ceccarelli, M., Pisla, D., Eds.; Mechanisms and Machine Science; Springer International Publishing: Cham, Switzerland, 2018; pp. 3–10. [CrossRef]
56. Bouri, M.; Abdi, E.; Bleuler, H.; Reynard, F.; Deriaz, O. *Lower Limbs Robotic Rehabilitation Case Study with Clinical Trials*; Mechanisms and Machine Science; Springer International Publishing: Cham, Switzerland, 2014; pp. 31–44. [CrossRef]
57. Badi, A.; Saad, M.; Gauthier, G.; Archambault, P. Inverse Kinematics for a Novel Rehabilitation Robot for Lower Limbs. In *Cable-Driven Parallel Robots*; Gosselin, C., Cardou, P., Bruckmann, T., Pott, A., Eds.; Mechanisms and Machine Science; Springer International Publishing: Cham, Switzerland, 2018; pp. 376–389. [CrossRef]
58. Geonea, I.; Tarnita, D.; Carbone, G.; Ceccarelli, M. Design and Simulation of a Leg Exoskeleton Linkage for Human Motion Assistance. In *New Trends in Medical and Service Robotics*; Carbone, G., Ceccarelli, M., Pisla, D., Eds.; Mechanisms and Machine Science; Springer International Publishing: Cham, Switzerland, 2019; pp. 93–100. [CrossRef]
59. Copilusi, C.; Ceccarelli, M.; Dumitru, S.; Margine, A.; Geonea, I. A Leg Exoskeleton Mechanism for Human Walking Assistance. In *New Trends in Medical and Service Robotics*; Tarnita, D., Dumitru, N., Pisla, D., Carbone, G., Geonea, I., Eds.; Mechanisms and Machine Science; Springer Nature: Cham, Switzerland, 2023; pp. 160–167. [CrossRef]
60. Pramod, A.S.; Palani, P.; Mohan, S.; Thondiyath, A. Development of a Passive Ankle-Foot Exoskeleton for Variable Force Resistance Training. In *New Trends in Medical and Service Robotics*; Tarnita, D., Dumitru, N., Pisla, D., Carbone, G., Geonea, I., Eds.; Mechanisms and Machine Science; Springer Nature: Cham, Switzerland, 2023; pp. 144–151. [CrossRef]
61. Nursultan, Z.; Ceccarelli, M.; Balbayev, G. Design and Performance Analysis of Ankle Joint Exoskeleton. In *New Trends in Medical and Service Robotics*; Tarnita, D., Dumitru, N., Pisla, D., Carbone, G., Geonea, I., Eds.; Mechanisms and Machine Science; Springer Nature: Cham, Switzerland, 2023; pp. 152–159. [CrossRef]
62. Nakka, S.; Vashista, V. Manipulability Analysis of Cable-Driven Serial Chain Manipulators. In *Cable-Driven Parallel Robots*; Caro, S., Pott, A., Bruckmann, T., Eds.; Mechanisms and Machine Science; Springer Nature: Cham, Switzerland, 2023; pp. 16–29. [CrossRef]
63. Tucan, P.; Ulinici, I.; Pop, N.; Puskas, F.; Carbone, G.; Gherman, B.; Luchian, I.; Pisla, D. Ankle Rehabilitation of Stroke Survivors Using Kuka LBR Iiwa. In *New Trends in Medical and Service Robotics*; Rauter, G., Cattin, P.C., Zam, A., Riener, R., Carbone, G., Pisla, D., Eds.; Mechanisms and Machine Science; Springer International Publishing: Cham, Switzerland, 2020; pp. 29–36. [CrossRef]
64. Girone, M.J.; Burdea, G.C.; Bouzit, M. The "Rutgers Ankle" Orthopedic Rehabilitation Interface. In Proceedings of the ASME 1999 International Mechanical Engineering Congress and Exposition, Dynamic Systems and Control, Nashville, TN, USA, 14–19 November 1999; Dynamic Systems and Control; American Society of Mechanical Engineers Digital Collection: New York, NY, USA, 2021; pp. 305–312. [CrossRef]
65. Wang, C.; Fang, Y.; Guo, S.; Chen, Y. Design and Kinematical Performance Analysis of a 3-RUS/RRR Redundantly Actuated Parallel Mechanism for Ankle Rehabilitation. *J. Mech. Robot.* **2013**, *5*, 041003. [CrossRef]
66. Saglia, J.A.; Dai, J.S. Geometry and Kinematic Analysis of a Redundantly Actuated Parallel Mechanism for Rehabilitation. In Proceedings of the ASME 2007 International Design Engineering Technical Conferences and Computers and Information in Engineering Conference, Volume 8: 31st Mechanisms and Robotics Conference, Parts A and B, Las Vegas, NV, USA, 4–7 September 2007; American Society of Mechanical Engineers Digital Collection: New York, NY, USA, 2009; pp. 1081–1090. [CrossRef]
67. Chen, G.; Mao, Z.; Zhou, H.; Yang, P. Design and control strategy of 3-prismatic-revolute-spherical ankle rehabilitation robot. *Aust. J. Mech. Eng.* **2023**, *21*, 1079–1092. [CrossRef]
68. Escarabajal, R.J.; Abu-Dakka, F.J.; Pulloquinga, J.L.; Mata, V.; Vallés, M.; Valera, Á. Development of lower-limb rehabilitation exercises using 3-PRS Parallel Robot and Dynamic Movement Primitives. *Multidiscip. J. Educ. Soc. Technol. Sci.* **2020**, *7*, 30–44. [CrossRef]
69. Dong, M.; Kong, Y.; Li, J.; Fan, W. Kinematic Calibration of a Parallel 2-UPS/RRR Ankle Rehabilitation Robot. *J. Healthc. Eng.* **2020**, *2020*, 3053629. [CrossRef]
70. Li, J.; Zuo, S.; Zhang, L.; Dong, M.; Zhang, Z.; Tao, C.; Ji, R. Mechanical Design and Performance Analysis of a Novel Parallel Robot for Ankle Rehabilitation. *J. Mech. Robot.* **2020**, *12*, 051007. [CrossRef]
71. Wang, L.; Chang, Y.; Zhu, H. Internal Model Control and Experimental Study of Ankle Rehabilitation Robot. *Robotica* **2020**, *38*, 940–956. [CrossRef]
72. Doroftei, I.; Cazacu, C.M. Developments in the Design of an Ankle Rehabilitation Platform. In *New Trends in Medical and Service Robotics*; Tarnita, D., Dumitru, N., Pisla, D., Carbone, G., Geonea, I., Eds.; Mechanisms and Machine Science; Springer Nature: Cham, Switzerland, 2023; pp. 179–187. [CrossRef]
73. Wang, C.; Wang, L.; Wang, T.; Li, H.; Du, W.; Meng, F.; Zhang, W. Research on an Ankle Joint Auxiliary Rehabilitation Robot with a Rigid-Flexible Hybrid Drive Based on a 2-SPS Mechanism. *Appl. Bionics Biomech.* **2019**, *2019*, 7071064. [CrossRef] [PubMed]
74. Zhong, J.; He, D.; Zhao, C.; Zhu, Y.; Zhang, Q. An rehabilitation robot driven by pneumatic artificial muscles. *J. Mech. Med. Biol.* **2020**, *20*, 2040008. [CrossRef]

75. Zhang, M.; Cao, J.; Xie, S.Q.; Zhu, G.; Zeng, X.; Huang, X.; Xu, Q. A Preliminary Study on Robot-Assisted Ankle Rehabilitation for the Treatment of Drop Foot. *J. Intell. Robot. Syst.* **2018**, *91*, 207–215. [CrossRef]
76. Zhang, M.; Xie, S.Q.; Li, X.; Zhu, G.; Meng, W.; Huang, X.; Veale, A.J. Adaptive Patient-Cooperative Control of a Compliant Ankle Rehabilitation Robot (CARR) with Enhanced Training Safety. *IEEE Trans. Ind. Electron.* **2018**, *65*, 1398–1407. [CrossRef]
77. Sales Gonçalves, R.; Carvalho, J.; Rodrigues, L.A.; Marques Barbosa, A. Cable-Driven Parallel Manipulator for Lower Limb Rehabilitation. *Appl. Mech. Mater.* **2013**, *459*, 535–542. [CrossRef]
78. Wang, Y.L.; Wang, K.Y.; Chai, Y.J.; Mo, Z.J.; Wang, K.C. Research on mechanical optimization methods of cable-driven lower limb rehabilitation robot. *Robotica* **2022**, *40*, 154–169. [CrossRef]
79. Vaida, C.; Birlescu, I.; Pisla, A.; Carbone, G.; Plitea, N.; Ulinici, I.; Gherman, B.; Puskas, F.; Tucan, P.; Pisla, D. RAISE—An Innovative Parallel Robotic System for Lower Limb Rehabilitation. In *New Trends in Medical and Service Robotics*; Carbone, G., Ceccarelli, M., Pisla, D., Eds.; Mechanisms and Machine Science; Springer International Publishing: Cham, Switzerland, 2019; pp. 293–302. [CrossRef]
80. Yu, R.; Fang, Y.; Guo, S. Design and Kinematic Analysis of a Novel Cable-Driven Parallel Robot for Ankle Rehabilitation. In *Advances on Theory and Practice of Robots and Manipulators*; Ceccarelli, M., Glazunov, V.A., Eds.; Mechanisms and Machine Science; Springer International Publishing: Cham, Switzerland, 2014; pp. 293–301. [CrossRef]
81. Venkata Sai Prathyush, I.; Ceccarelli, M.; Russo, M. Control Design for CABLEankle, a Cable Driven Manipulator for Ankle Motion Assistance. *Actuators* **2022**, *11*, 63. [CrossRef]
82. The Sage Developers. *SageMath, the Sage Mathematics Software System (Version 10.0)*; 2023. Available online: https://www.sagemath.org (accessed on 20 July 2023).
83. Vargas-Riaño, J.H. Turmell-Bot-Nbviewer. Available online: https://nbviewer.org/github/juliohvr/Turmell-Bot/blob/main/TurmellBot.ipynb (accessed on 10 November 2023).
84. Vargas-Riaño, J.H. Turmell-Bot | 3D CAD Model Library | GrabCAD. Available online: https://grabcad.com/library/turmell-bot-1 (accessed on 10 November 2023).
85. Vargas-Riaño, J.H. Juliohvr/Turmell-Bot: Turmell-Bot. Available online: https://github.com/juliohvr/Turmell-Bot (accessed on 10 November 2023). [CrossRef]
86. Isman, R.E.; Inman, V.T. *Anthropometric Studies of the Human Foot and Ankle*; Biomechanics Laboratory, University of California: Riverside, CA, USA, 1969.
87. Drillis, R.; Contini, R. *Body Segment Parameters*; New York University, School of Engineering and Science: New York, NY, USA, 1966.
88. Fryar, C.D.; Carroll, M.D.; Gu, Q.; Afful, J.; Ogden, C.L. Anthropometric Reference Data for Children and Adults: United States, 2015–2018. 2021; pp. 1–44. Available online: https://stacks.cdc.gov/view/cdc/100478 (accessed on 10 November 2023).
89. Agudelo-Varela, Ó.; Vargas-Riaño, J.; Valera, Á. Turmell-Meter: A Device for Estimating the Subtalar and Talocrural Axes of the Human Ankle Joint by Applying the Product of Exponentials Formula. *Bioengineering* **2022**, *9*, 199. [CrossRef]
90. Todorov, E.; Erez, T.; Tassa, Y. MuJoCo: A physics engine for model-based control. In Proceedings of the 2012 IEEE/RSJ International Conference on Intelligent Robots and Systems, Vilamoura-Algarve, Portugal, 7–12 October 2012; pp. 5026–5033. [CrossRef]
91. Zhao, J.; Li, B.; Yang, X.; Yu, H. Geometrical method to determine the reciprocal screws and applications to parallel manipulators. *Robotica* **2009**, *27*, 929–940. [CrossRef]

robotics

Article

Posture Optimization of the TIAGo Highly-Redundant Robot for Grasping Operation

Albin Bajrami †,‡, Matteo-Claudio Palpacelli *,†, Luca Carbonari and Daniele Costa

Department of Industrial Engineering and Mathematical Sciences, Polytechnic University of Marche, 60131 Ancona, Italy; a.bajrami@pm.univpm.it (A.B.); l.carbonari@univpm.it (L.C.); d.costa@univpm.it (D.C.)
* Correspondence: m.palpacelli@univpm.it
† These authors contributed equally to this work.
‡ The Italian Doctorate in Robotics and Intelligent Machines.

Abstract: This study explores the optimization of the TIAGo robot's configuration for grasping operation, with a focus on the context of aging. In fact, featuring a mobile base and a robotic arm, the TIAGo robot can conveniently aid individuals with disabilities, including those with motor and cognitive impairments in both domestic and clinical settings. Its capabilities include recognizing visual targets such as faces or gestures using stereo cameras, as well as interpreting vocal commands through acoustic sensors to execute tasks. For example, the robot can grasp and lift objects such as a glass of water and navigate autonomously in order to fulfill a request. The paper presents the position and differential kinematics that form the basis for using the robot in numerous application contexts. In the present case, they are used to evaluate the kinematic performance of the robot relative to an assigned pose in the search for the optimal configuration with respect to the higher-order infinite possible configurations. Ultimately, the article provides insight into how to effectively use the robot in gripping operations, as well as presenting kinematic models of the TIAGo robot.

Keywords: posture optimization; kinematic manipulability; kinematic redundancy; robotic grasping; aging; elderly support

Citation: Bajrami, A.; Palpacelli, M.-C.; Carbonari, L.; Costa, D. Posture Optimization of the TIAGo Highly-Redundant Robot for Grasping Operation. *Robotics* **2024**, *13*, 56. https://doi.org/10.3390/robotics13040056

Academic Editor: Raffaele Di Gregorio

Received: 24 February 2024
Revised: 15 March 2024
Accepted: 17 March 2024
Published: 23 March 2024

1. Introduction

The aging population presents significant challenges in eldercare. As the world ages, there is a growing need for technology to support healthy aging, as Tian and Li [1] point out. Technologies like personal mobility aids, vehicle modifications, healthcare access technologies, electronic assistive devices, and smart home solutions are becoming vital for helping the elderly maintain their independence and quality of life [2]. Peek et al. [3] highlight that technology can help older adults stay independent and active, stressing the importance of considering psychological and environmental factors. In the field of robotics for eldercare, several studies have been performed. Krishnan and Pugazhenthi [4] reviewed various mobility and self-transfer systems as assistive devices for the elderly, focusing on improving their quality of life and independence through intelligent robotic systems. Broekens et al. [5] provided a comprehensive review of assistive social robots, which are designed for social interaction and can play a significant role in eldercare, especially in providing companionship and functioning as an interface with digital technology. Ienca et al. [6] discussed the integration of robotics in formal and informal dementia care, outlining the ethical, legal, and social implications that should be considered early in the development of assistive and social robots for dementia care. Coşar et al. [7] introduced the ENRICHME (Enabling Robot and assisted living environment for Independent Care and Health Monitoring of the Elderly) project's robotic platform, which aims to enrich the day-to-day experience of elderly people with mild cognitive impairments by offering health monitoring, complementary care, and social support. Similar objectives are sought in the FOCAAL project, an acronym for FOg Computing in Ambient Assisted Living, funded by the Italian MIMIT, Ministry of Business and Made in Italy. The project is aimed at integrating artificial intelligence, assistive robotics, home automation, and sensors through the fog computing network architecture, which allows data to be processed locally, in the context of the patient's life, and to implement personalized actions for different contexts and assistance scenarios, such as home care, rehabilitation clinics, and geriatric medical clinics. Although robots have the potential to assist in eldercare, their integration into this field is not without challenges. According to Remmers and Fischer's research [8], care workers often perceive robots as mere tools, which can impede their widespread use in caregiving. Therefore, there is a challenge in convincing people to accept and appreciate the benefits of robots in this field. In order for robots to function effectively in eldercare, they must be designed to move and interact safely and efficiently in these settings. This requires a thorough examination of robotic movement, or kinematics, to ensure that robots can meet the diverse needs of elderly individuals without causing disruptions. Overcoming these challenges necessitates a collaborative effort from the fields of robotics, healthcare, and social science to ensure that robots can provide effective support for the elderly.

1.1. The Problem of an Aging Population in Europe

The aging of the population is due to two factors: increasing life expectancy and falling birth rates. On the one hand, life expectancy in the EU (European Union) fell slightly in 2021 due to the pandemic, but has generally increased, leading to an aging population. Huang et al. [9] analyzed the impact of COVID-19 on life expectancy and found a temporary decline in 2020 with a partial recovery in some countries by 2021. This aging trend is clear when comparing the population structure of Europe in 2007 and 2022, which shows a shift towards an older population [10]. On the other hand, the decline in birth rates lead to demographic aging. Figure 1b elucidates this trend by contrasting the population composition of Europe in 2007 with that of 2022, illustrating a demographic shift towards an older population. The hollow rectangles refer to data from 2007 and the filled rectangles to 2022.

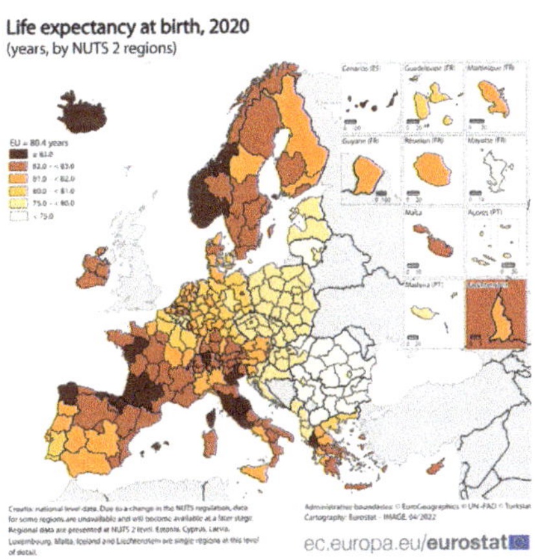

(**a**)

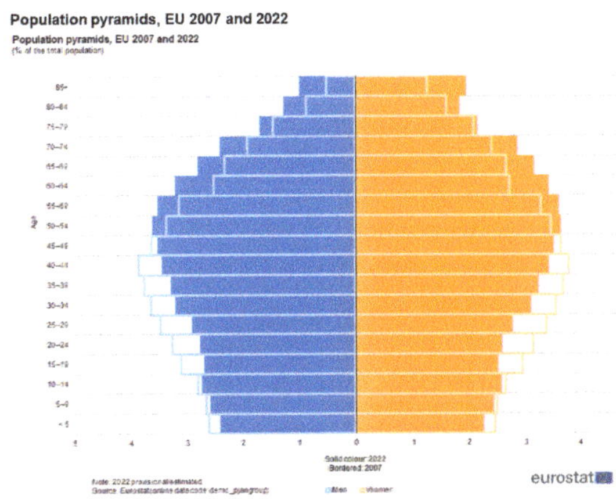

(**b**)

Figure 1. (**a**) Depicts a graphical representation of average life expectancy across Europe. (**b**) Provides a comparative view of Europe's population structure in 2007 and 2022. It shows the shifting age pyramid, with a bulge in the older age brackets, signifying the demographic aging process.

1.2. TIAGo in the Literature

TIAGo, a versatile assistive robot by PAL Robotics, comes in various versions (see Figure 2) and is ideal for research and applications in healthcare, assistance, and light industry. It features advanced perception, autonomous navigation, object manipulation, and human–robot interaction (HRI), with a modular design for customization. Telembici et al. in [11] demonstrate how TIAGo can be integrated into everyday life based on its audio capabilities, achieving a high classification rate of audio events. Grama and Rusu enhanced TIAGo's audio signal task handling and interaction, introducing three modules for better home environment interaction [12,13]. Lach et al. [14] present tactile-equipped end-effectors for TIAGo aimed at improving the reliability and success of mobile manipulation, contributing to its physical interaction capabilities. Muscar et al. [15] show how TIAGo can issue real-time warnings using its audio skills, identifying dangerous and helpful sounds and issuing alerts via email or SMS depending on the sound category identified. Miguel et al. [16] propose a fault-tolerant approach for service robots that uses a Robust Unknown-Input Observer (RUIO) and state feedback control strategy to compensate for fault effects and handle non-linearities.

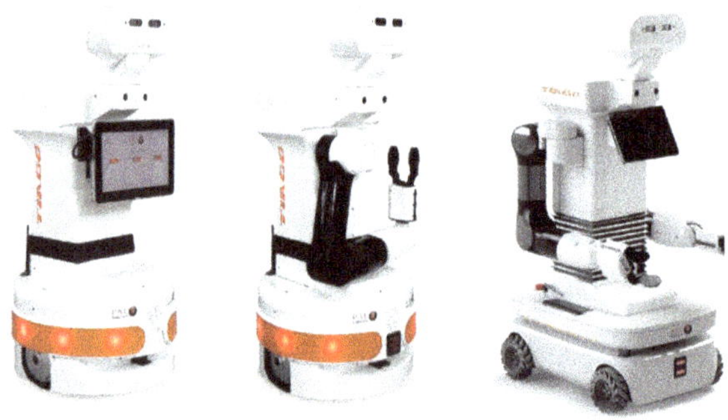

Figure 2. Different versions of the TIAGo robot marketed by Pal Robotics.

Using VOSviewer 1.6.19 [17], a bibliometric tool, we analyzed TIAGo robot research through a keyword search of *tiago* and *robot**, identifying 71 articles. These were visualized in three main clusters in VOSviewer, indicating research themes (see Figure 3):

- Red: Social Robotics and Human Interaction
- Blue: Mobile Robotics and Programming
- Green: Machine Learning and Robotic Services

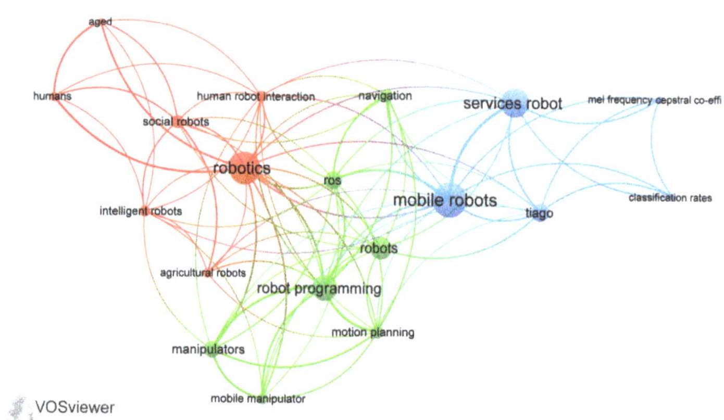

Figure 3. This figure, created with VOSviewer, depicts the bibliometric relationships in robotics research, particularly around the TIAGo robot. The terms, color-coded into three clusters (Red: Social Robotics and Human Interaction, Blue: Mobile Robotics and Programming, Green: Machine Learning and Robotic Services), represent different research themes. The size of each term indicates its prominence in the field. The visualization highlights the current research landscape and potential areas for future exploration.

The links between the various words demonstrate the co-occurrence between the keywords. The most-used words are robotics, robot programming, mobile robots, and services robot. There is no reference about kinematics. This visualization not only outlines the research landscape of TIAGo but also emphasizes the importance of each theme, with the size of the item representing its prominence in the field. Continuing, when the term *kinema** was added to refine the search, we obtained only seven results on Scopus. Among these results, we found an absence of articles that explore the kinematics of TIAGo. Some works, such as that of Sulaiman et al. [18], where they integrate different robotic arms like the UR3 into the mobile platform, or the Emiko Franka Panda robotic manipulator by Baumgartner et al. [19], show a trend to study applications through performance analysis.

1.3. Summary

In the course of this literature review and subsequent analysis, the authors of this paper have found a scarcity of articles in the literature on the kinematics of the TIAGo robotic arm. The formal definition of its kinematics is considered useful to make the most of the features offered by such a robot, providing developers working on the ROS (robot operating system) platform and TIAGo robot drivers with detailed equations to improve its motion planning logic, for example in the case of obstacle avoidance problems [20], while also providing useful information for determining optimal postures.

The concept of pose optimization is extensively discussed in scientific circles. For instance, Cheah and Caverly's research [21] focuses on the pose regulation of a six-degree-of-freedom cable-driven parallel robot (CDPR), analyzing it from a passivity perspective. Similarly, Gao et al.'s work [22] delves into dynamic Jacobian identification for robots operating in unstructured environments. Their studies illustrate how solving Jacobian identification as state estimation problems allows for the optimal estimation of the desired Jacobian, ensuring precise robotic alignment with the intended pose. Jens et al. [23] propose a kinematic redundancy in order to improve the achievable pose accuracy of a parallel robot's traveling platform. Guojian et al. [24] propose a tightly coupled laser-inertial pose estimation and map building method that uses B-spline curves to represent a continuous-time trajectory and achieve high robustness of the registration steps. These are only some scientific papers that address this topic in different ways.

In this paper, the pose optimization is more related to investigating the manipulability ellipsoids in understanding a robot's performance capabilities. This is pivotal not only for the TIAGo arm's efficiency but also for enhancing its functionality across various tasks requiring precision and finesse. The examination of how the Jacobian matrix responds during grasping tasks is essential for determining the most effective methods for robots to interact with objects. This is particularly helpful in healthcare and assistance settings, where high dexterity and agility are demanded.

This article is structured as follows: Section 2 sets the context for this work, outlining the robot's deployment in complex tasks, such as navigating through environments with obstacles and retrieving objects for assistance purposes. It illustrates the robot's capability to interpret commands and plan its motion, highlighting the practical applications of autonomous robots in supporting daily activities. Section 3 focuses on the position kinematics of the TIAGo robot, exploring its optimal configurations for interacting with objects. Section 4 investigates the differential kinematics, detailing the relationship between the robot's joint velocities and its end-effectors' velocities. In Section 5, the routines used to optimize the robot posture are presented. Some results of their application are gathered in Section 6, where the optimal configuration that maximizes a kinematic index is discussed. This paper aims to contribute to the robotics field by providing the kinematics of a highly redundant mobile robot that shows a great potential in enhancing the quality of life for people with mobility challenges.

2. Advanced Tasks in an Unstructured Environment

Featuring a mobile base, namely, an Autonomous Mobile Robot (AMR), and a 7 dof robotic arm, the TIAGo robot has a redundant number of degrees of freedom that can be exploited in complex applications. Advanced tasks in supporting elderly people or people with motor impairments require movement of objects and/or equipment within a large environment, typically characterized by fixed but also mobile obstacles, namely, those whose position is not necessarily known a priori. A complex task taken as reference in this study can be outlined as follows:

1. The person in need of support asks the robot, either vocally or with a hand gesture, to go and retrieve an object, suppose a glass of water, located on a support surface that cannot be reached by the person;
2. The robot either directly interprets the voice command or points its stereoscopic camera system towards the visual target, triggered by the voice command, interpreting its relative meaning;
3. The robot scans the surrounding environment through its sensors, namely, sonars and laser range-finders, to map the space in which it operates, necessary for planning the motion of the mobile base. Furthermore, through the vision system positioned on its head, it searches for the visual target to reach and take. The target search itself is a complex task that may require moving the robot randomly if there is visual coverage of the target to be reached;
4. A trajectory is automatically planned for the robot to follow to avoid colliding with fixed obstacles; along the path, the robot continues to acquire information from its surroundings through its sensors by stopping or changing its trajectory based on obstacle avoidance algorithms [20];
5. Once the proximity of the target object is reached, the optimal configuration of the robot with which to grasp the target is evaluated, taking into account all the available degrees of freedom as a whole;
6. Finally, the robot's motion is planned to bring the object to and interact with the person who requested it.

The present work aims to answer step 5 of the above outline, focusing on the practical task of grasping a glass from a table while enhancing the TIAGo robot dexterity and precision. In order to identify the optimal configuration of the robot on the basis of an appropriate performance index, it is necessary to investigate its position kinematics in both the direct and inverse problems, as well as determine and analyze its Jacobian matrix.

3. Position Kinematics of the TIAGo Robot

The TIAGo robot is available in different versions, as already mentioned in the introductory section. The setup taken as reference in this work is the one shown in the middle of Figure 2, where a mobile base at the bottom is driven by two differential wheels, with four caster wheels used to ensure stable support on the ground. The robot trunk can be moved vertically through the presence of an actuated prismatic pair, which allows the robot to operate at different heights. The head is connected to the upper part of the robot, where a stereoscopic vision system resides, which is equipped with two revolute joints that allow the vision system to be pointed towards the visual target. Under the head, the stereo microphone system is rigidly mounted to the top part of the trunk. The robotic arm equipped with seven revolute joints is connected to the front part of the robot trunk, under the microphones. The redundancy of the arm compared to a 6 dof Cartesian task makes it suitable for highly dexterous operations, being able to have an additional degree of freedom, without taking into account the other degrees of freedom made available by the mobile base and the trunk itself. A grasping tool is then installed at the end of the arm. In the present study, a collaborative two-finger electric gripper is considered.

In order to study the TIAGo position kinematics, the positioning of several reference frames is required, attached rigidly to the movable rigid bodies of the robot. The main reference systems are shown in Figure 4, whereas the others related to the robotic arm and the TIAGo head are shown in Figures 5 and 6, respectively.

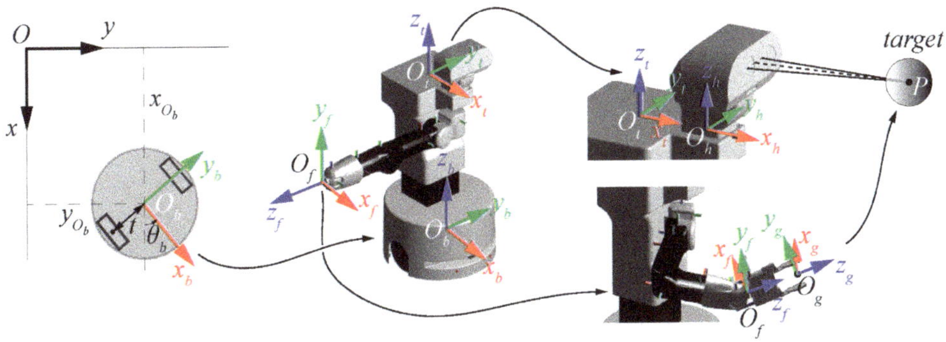

Figure 4. Main reference systems for the TIAGo robot.

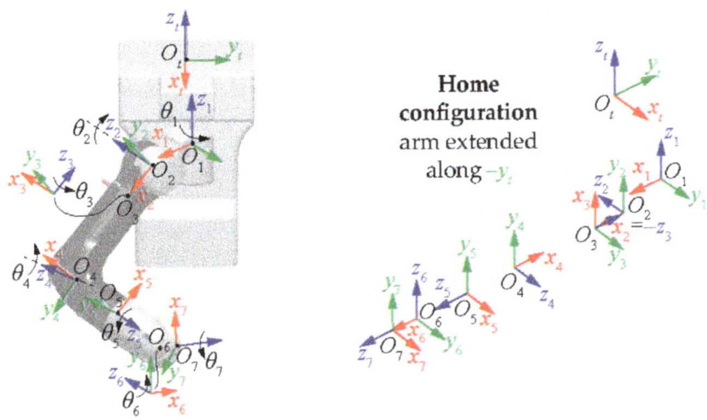

Figure 5. Reference frames of the TIAGo robotic arm.

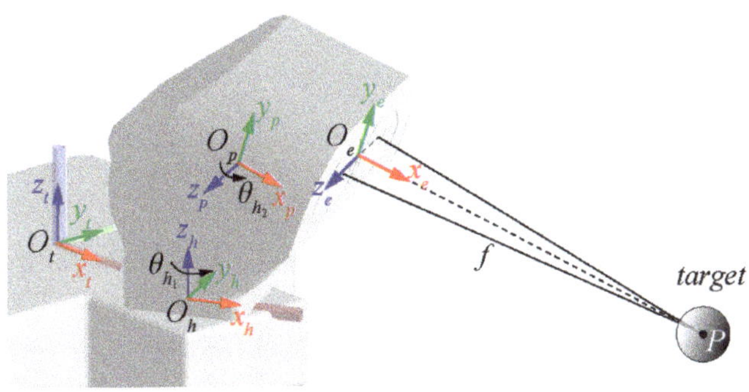

Figure 6. Reference frames of the TIAGo head.

In more detail, $O_b - x_b y_b z_b$ is the reference system placed at the center of the mobile platform, between the centers of the two actuated wheels, $O_t - x_t y_t z_t$ refers to the trunk and is placed on its top surface, $O_h - x_h y_h z_h$ is the frame at the root of the head, where the latter is physically connected, and finally, $O_f - x_f y_f z_f$ is located at the end of the robotic arm, on the flange where several tools can be assembled according to the user's needs.

In Figure 4, on the bottom right, a simple electric gripper is considered, with a further reference frame $O_g - x_g y_g z_g$ in the middle of the two actuated fingers.

The mentioned frames are placed in appropriate positions along the mechanical structure of the robot, respecting the conventions adopted in the Robotics Systems Toolbox 4.0 made available by Mathworks Matlab. Referring to the robotic arm, the frames are placed so that the joint rotations occur around the local unit vector z. The Denavit–Hartenberg (DH) convention is not convenient for studying the kinematics of this robot because of the numerous offsets between the robotic arm joints, resulting in an unintelligible home configuration when the arm joint axes have zero rotation. It is clear from Figure 5 on the right that with the chosen convention all the frames assume a simple relative orientation in the home configuration. The simulation environment offered by Matlab is used, on the one hand, as a verification tool and, on the other hand, so that the kinematic equations provided in this study are compatible with the toolbox in view of further developments on the robot.

Focusing on the head of the robot, two further reference systems allow movement from the base of the head toward the stereo vision system: $O_p - x_p y_p z_p$ is the intermediate system responsible of the head pointing direction, whereas $O_e - x_e y_e z_e$ is placed conveniently between the sensors of the two RGB-D cameras. Each camera has a wide field of view, 58° horizontally, 45° vertically, and 70° diagonally, with a depth range from 0.4 m to 8 m. Actually, the vision system can be handled by means of the Python scripts and drivers that Pal Robotics makes available in the programming environment, allowing for the evaluation of distance f and other physical distances required to place $O_e - x_e y_e z_e$ accurately.

3.1. Direct Position Kinematics

The *direct position kinematics* of the TIAGo robot is handled with conventional homogeneous transformations, according to the following modular operation:

$$ {}^{A}_{B}\mathbf{T} = \begin{bmatrix} {}^{A}_{B}\mathbf{R}\mathbf{R}_z(\vartheta) & {}^{A}\mathbf{p}_B \\ \mathbf{0}_{1\times 3} & 1 \end{bmatrix}, \tag{1} $$

where ${}^{A}\mathbf{p}_B = \left[{}^{A}x_B, {}^{A}y_B, {}^{A}z_B\right]^T$ represents the position vector between the two origins A and B of two generic reference systems and ${}^{A}_{B}\mathbf{R}$ is the rotation matrix between the two frames, always given by the same sequence of elementary rotations around local x, y, and z axes, namely,

$$ {}^{A}_{B}\mathbf{R} = \mathbf{R}_x(\alpha)\mathbf{R}_y(\beta)\mathbf{R}_z(\gamma). \tag{2} $$

In (2) the conventional elementary rotations have the form given by

$$ \mathbf{R}_x = \begin{bmatrix} 1 & 0 & 0 \\ 0 & \cos\alpha & -\sin\alpha \\ 0 & \sin\alpha & \cos\alpha \end{bmatrix} \quad \mathbf{R}_y = \begin{bmatrix} \cos\beta & 0 & \sin\beta \\ 0 & 1 & 0 \\ -\sin\beta & 0 & \cos\beta \end{bmatrix} \quad \mathbf{R}_z = \begin{bmatrix} \cos\gamma & -\sin\gamma & 0 \\ \sin\gamma & \cos\gamma & 0 \\ 0 & 0 & 1 \end{bmatrix}. \tag{3} $$

The last row in (1) consists of three zeros and one one, according to the convention for homogeneous transformations. Moreover, it can be observed that the rotation ϑ associated with the degree of freedom provided locally by a revolute joint, when focusing on the robotic arm, is explicitly involved in a rotation matrix around a z-axis. In fact, matrix ${}^{A}_{B}\mathbf{R}$ is used first to orient the local z-axis along the revolute joint, then an elementary $\mathbf{R}_z(\vartheta)$ rotation can be driven by the actuated joint. This choice has allowed for a more compact notation.

All the geometric data of the TIAGo robot required in the computation of the Homogeneous transformations in (1) can be retrieved from Table 1 by selecting the required reference frames. Each line in the table is related to a specific couple of frames, with origins labeled A and B. In Table 1, the distances ${}^{A}x_B$ and ${}^{A}y_B$ in the last row are hypothesized by comparison with other known dimensions.

Table 1. Kinematic and geometric parameters of the TIAGo robot.

A	B	${}^{A}x_B[mm]$	${}^{A}y_B[mm]$	${}^{A}z_B[mm]$	$\alpha[rad]$	$\beta[rad]$	$\gamma[rad]$	$\vartheta[rad]$
O	O_b	x_{O_b}	y_{O_b}	98.5	0	0	0	ϑ_b
O_b	O_t	0	−62.0	d	0	0	0	0
O_t	O_1	155.1	14.0	−151.0	0	0	$-\pi/2$	ϑ_1
O_1	O_2	125.0	16.5	−31.0	$\pi/2$	0	0	ϑ_2
O_2	O_3	89.5	0	1.5	$-\pi/2$	$-\pi/2$	0	ϑ_3
O_3	O_4	−20.0	−27.0	−222.0	$-\pi/2$	0	$-\pi/2$	ϑ_4
O_4	O_5	−162.0	20.0	27.0	0	$-\pi/2$	0	ϑ_5
O_5	O_6	0	0	150.0	$-\pi/2$	0	$-\pi/2$	ϑ_6
O_6	$O_7 = O_f$	66.0	0	0	$\pi/2$	$\pi/2$	0	ϑ_7
O_f	O_g	0	0	0.2	0	0	0	0
O_t	O_h	182.0	0	0	0	0	0	ϑ_{h_1}
O_h	O_p	5.0	0	98.0	$\pi/2$	0	0	ϑ_{h_2}
O_p	O_e	120	106	0	0	0	0	0

The joints of the robot, representing its degrees of freedom, can change in the ranges collected in Table 2 due to mechanical limitations. Given the configuration vector of the robot in terms of the coordinates $\mathbf{q} = \left[\mathbf{q}_b^T, d, \mathbf{q}_r^T\right]^T$ with $\mathbf{q}_b = \left[x_{O_b}, y_{O_b}, \vartheta_b\right]^T$ related to the AMR and $\mathbf{q}_r = [\vartheta_1, \vartheta_2, \vartheta_3, \vartheta_4, \vartheta_5, \vartheta_6, \vartheta_7]^T$ to the robotic arm, in order to obtain the pose ${}^{O}_{O_g}\mathbf{T}$ of the end-effector of the robotic arm with respect to the absolute reference frame $O - xyz$ on the left upper corner of Figure 4, a matrix post-multiplication can be carried out:

$$ {}^{O}_{O_g}\mathbf{T} = \begin{bmatrix} {}^{O}_{O_g}\mathbf{R} & {}^{O}\mathbf{p}_g \\ \mathbf{0}_{1\times 3} & 1 \end{bmatrix} = {}^{O}_{O_b}\mathbf{T}\,{}^{O_b}_{O_t}\mathbf{T}\,{}^{O_t}_{O_7}\mathbf{T}\,{}^{O_f}_{O_g}\mathbf{T}, \tag{4} $$

where ${}^{O}\mathbf{p}_g$ and ${}^{O}_{O_g}\mathbf{R}$ represent, respectively, the absolute position of point O_g at the end-effector and the rotation matrix that provides its orientation with respect to the absolute reference system in O. The last row in (4) consists of three zeros and one one, as usual. Moreover, looking at the member on the right of (4), the two reference frames in O_7 and O_f coincide and matrix $\mathbf{T}^{O}_{O_b}$ refers to the planar motion of the TIAGo mobile base, whose degrees of freedom can be assumed to vary in the ranges $x_{O_b} \in]-\infty, \infty[$, $y_{O_b} \in]-\infty, \infty[$ and $\vartheta_b \in [-180°, 180°]$. Obviously, the extremes with infinite value mean that the translation of the base is free within the available space, namely, the limits of the room in which the robot moves for indoor applications. If an orientation in terms of a minimum set of angles was required instead of via the rotation matrix ${}^{O}_{O_g}\mathbf{R}$ in (4), one of the most common choices in robotics applications would be the ZYZ convention of Euler angles. In this case, three angles (ϕ, θ, ψ), conveniently gathered in a vector called $\mathbf{e}$, can be used to underscore the three degrees of freedom related to the spatial rotation of the robot end-effector following the sequence ZYZ, namely, ${}^{O}_{O_g}\mathbf{R} = \mathbf{R}_z(\phi)\mathbf{R}_y(\theta)\mathbf{R}_z(\psi)$, by using the elementary rotations in (3). The conventional inverse solution to make them explicit can be used:

$$ \mathbf{e} = \begin{bmatrix} \phi \\ \theta \\ \psi \end{bmatrix} = \begin{bmatrix} \arctan 2(\mathbf{R}_{2,3}, \mathbf{R}_{1,3}) \\ \arctan 2\left(\sqrt{\mathbf{R}_{3,1}^2 + \mathbf{R}_{3,2}^2}, \mathbf{R}_{3,3}\right) \\ \arctan 2(\mathbf{R}_{3,2}, -\mathbf{R}_{3,1}) \end{bmatrix}, \tag{5} $$

where the subscripts (i, j) in $\mathbf{R}_{i,j}$ point at the ith row and j^{th} column, respectively, of matrix ${}^{O}_{O_g}\mathbf{R}$. Particular attention must be paid to $\theta = \pm\pi/2$, namely, when singularities of representation occur.

Table 2. Motion range of the actuated kinematic joints.

Joint	Type	Lower Limit	Upper Limit
d	P	0 mm	350 mm
θ_1	R	0°	157.5°
θ_2	R	−90°	62.5°
θ_3	R	−202.5°	90°
θ_4	R	−22.5°	135°
θ_5	R	−120°	120°
θ_6	R	−90°	90°
θ_7	R	−120°	120°
θ_{h_1}	R	−75°	75°
θ_{h_2}	R	−60°	45°

The planar motion of the TIAGo base, namely, a two-wheeled nonholonomic AMR, is regulated by the following well-known differential problem:

$$\dot{\mathbf{q}}_b = \begin{bmatrix} \dot{x}_{O_b} \\ \dot{y}_{O_b} \\ \dot{\theta}_b \end{bmatrix} = \begin{bmatrix} \frac{r}{2}\cos\theta_b & \frac{r}{2}\cos\theta_b \\ \frac{r}{2}\sin\theta_b & \frac{r}{2}\sin\theta_b \\ -\frac{r}{2t} & \frac{r}{2t} \end{bmatrix} \begin{bmatrix} \omega_L \\ \omega_R \end{bmatrix}, \tag{6}$$

where $r = 98.5$ mm, which is also the coordinate z of the local frame in O_b with respect to the absolute fixed frame $O - xyz$, $t = 202.2$ mm is the distance of the wheel center from the origin O_b shown in Figure 4, and ω_L and ω_R are, respectively, the left and right angular velocities of the AMR actuated wheels. Therefore, the absolute position coordinates of the TIAGo base result from a time integration of (6).

3.2. *Differential Kinematics*

The relationship between the velocity vector $\dot{\mathbf{q}} = \left[\dot{\mathbf{q}}_b^T, \dot{d}, \dot{\mathbf{q}}_r^T\right]^T$ related to the actuated joints of the robot, involving directly the planar velocities of the moving base $\dot{\mathbf{q}}_b^T$ in place of the wheel angular velocities, and the velocity vector of the robot end-effector $\dot{\mathbf{p}} = \left[{}^{O}\dot{\mathbf{p}}_g^T, {}^{O}\boldsymbol{\omega}_g^T\right]^T = \left[{}^{O}\dot{x}_g, {}^{O}\dot{y}_g, {}^{O}\dot{z}_g, {}^{O}\omega_{g,x}, {}^{O}\omega_{g,y}, {}^{O}\omega_{g,z}\right]^T$ can be obtained by parting the linear and angular velocities according to the following procedure:

- A time derivative of vector ${}^{O}\mathbf{p}_g$ gives directly the expression of the first three rows of the Jacobian matrix. Isolating the coordinates gathered in vector $\dot{\mathbf{q}}$, a 3×11 Jacobian matrix related to the linear velocities is obtained:

$${}^{O}\dot{\mathbf{p}}_g = \mathbf{J}_l\dot{\mathbf{q}} \tag{7}$$

- The angular velocity ${}^{O}\boldsymbol{\omega}_g$ results from the vector sum of all the angular velocities gained along the kinematic structure of the robot, by driving the planar rotation of the mobile base and actuating the robot joints. $\hat{\mathbf{z}}_i$ is the unit vector along the local i^{th} z-axis around which rotations occur; thus, it follows that

$${}^{O}\boldsymbol{\omega}_g = \dot{\vartheta}_b^{O}\hat{\mathbf{z}}_O + \dot{\vartheta}_1{}^{O}\hat{\mathbf{z}}_1 + \dot{\vartheta}_2^{O}\hat{\mathbf{z}}_2 + \dot{\vartheta}_3^{O}\hat{\mathbf{z}}_3 + \dot{\vartheta}_4{}^{O}\hat{\mathbf{z}}_4 + \dot{\vartheta}_5^{O}\hat{\mathbf{z}}_5 + \dot{\vartheta}_6^{O}\hat{\mathbf{z}}_6 + \dot{\vartheta}_7^{O}\hat{\mathbf{z}}_7 \tag{8}$$

which, in compact form, provides the expression of a 3×11 Jacobian matrix, this time related to rotations:

$${}^{O}\boldsymbol{\omega}_g = [\mathbf{0}_{3\times1}, \mathbf{0}_{3\times1}, \hat{\mathbf{z}}_O, \mathbf{0}_{3\times1}, \hat{\mathbf{z}}_1, \hat{\mathbf{z}}_2, \hat{\mathbf{z}}_3, \hat{\mathbf{z}}_4, \hat{\mathbf{z}}_5, \hat{\mathbf{z}}_6, \hat{\mathbf{z}}_7]\dot{\mathbf{q}} = \mathbf{J}_a\dot{\mathbf{q}} \tag{9}$$

From Equations (7) and (9), the complete 6×11 Jacobian $\mathbf{J}_G$ that allows one to solve the *direct velocity kinematics* of the TIAGo robot can be found:

$$\dot{\mathbf{p}} = \begin{bmatrix} {}^{O}\dot{\mathbf{p}}_g \\ {}^{O}\boldsymbol{\omega}_g \end{bmatrix} = \mathbf{J}_G\dot{\mathbf{q}} = \begin{bmatrix} \mathbf{J}_l \\ \mathbf{J}_a \end{bmatrix}\dot{\mathbf{q}} \tag{10}$$

Matrix $\mathbf{J}_G$ is the geometric Jacobian, but sometimes the analytical version is required. $\mathbf{J}_A$ is the analytical Jacobian, which can be derived by knowing the relationship between the angular velocity vector ${}^{O}\boldsymbol{\omega}_g$ and the time derivative of the Euler angles in (5):

$${}^{O}\boldsymbol{\omega}_g = \boldsymbol{\Theta}\dot{\mathbf{e}} = \begin{bmatrix} 0 & -\sin\phi & \cos\phi\sin\theta \\ 0 & \cos\phi & \sin\phi\sin\theta \\ 1 & 0 & \cos\theta \end{bmatrix} \begin{bmatrix} \dot{\phi} \\ \dot{\theta} \\ \dot{\psi} \end{bmatrix} \tag{11}$$

Since $\mathbf{p}_e = \left[{}^{O}\mathbf{p}_g^T, \mathbf{e}^T\right]^T$ and $\dot{\mathbf{p}}_e$ are, respectively, the pose and the velocity vector of the robot's end-effector as a function of Euler angles, it follows that

$$\dot{\mathbf{p}}_e = \begin{bmatrix} {}^{O}\dot{\mathbf{p}}_g \\ \dot{\mathbf{e}} \end{bmatrix} = \mathbf{J}_A\dot{\mathbf{q}} = \begin{bmatrix} \mathbf{J}_l \\ \boldsymbol{\Theta}^{-1}\mathbf{J}_a \end{bmatrix}\dot{\mathbf{q}} \tag{12}$$

Both $\mathbf{J}_G$ and $\mathbf{J}_A$ are rectangular matrices and share the same size; therefore, the *inverse velocity kinematics* presents the complexity of matrix inversion, typical of redundant manipulators. A well-known solution is to solve a constrained optimal problem by using the pseudoinverse $\mathbf{J}^{\dagger}$ of the Jacobian matrix, also called the Moore–Penrose inverse of the Jacobian matrix, which gives the best possible solution in terms of least squares [25]. Matrix $\mathbf{J}^{\dagger}$ is defined as $\mathbf{J}^T\left(\mathbf{J}\mathbf{J}^T\right)^{-1} = \mathbf{V}\boldsymbol{\Sigma}^{\dagger}\mathbf{U}^T$, also taking advantage of singular value decomposition, namely, $\mathbf{J} = \mathbf{U}\boldsymbol{\Sigma}\mathbf{V}^T$. It is recalled from linear algebra that both $\mathbf{U}$ and $\mathbf{V}$ are orthogonal matrixes, that $\boldsymbol{\Sigma}$ gathers diagonally the singular values of matrix $\mathbf{J}$, and, finally, that the pseudoinverse $\boldsymbol{\Sigma}^{\dagger}$ of $\boldsymbol{\Sigma}$ requires the substitution of the non-zero diagonal terms with their reciprocal, as well as a transposition of the matrix obtained. A damped version of the pseudoinverse $\mathbf{J}^{\dagger}$ can be formulated in order to avoid singular configurations, namely, $\mathbf{J}^{\dagger} = \mathbf{J}^T\left(\mathbf{J}\mathbf{J}^T + k^2\mathbf{I}_6\right)^{-1}$ with $\mathbf{I}_6$, the 6×6 identity matrix, and k a damping coefficient, whose value can be chosen according to optimization techniques [26]. This choice is important in the context of motion planning, proposing a control-managed variable damping coefficient that operates in the proximity of singular configurations and reverts to zero outside these conditions, avoiding a loss of accuracy in trajectory tracking [27]. Furthermore, a weighting matrix $\mathbf{W}$ can also be used in order to favor some motors with respect to others [28], for instance, the motion of the moving base with respect to the motion of the robotic arm, according to the expression $\mathbf{J}^{\dagger} = \mathbf{W}^{-1}\mathbf{J}^T\left(\mathbf{J}\mathbf{W}^{-1}\mathbf{J}^T + k^2\mathbf{I}_6\right)^{-1}$.

3.3. Inverse Position Kinematics

The *inverse position kinematics* of the TIAGo robot, meaning the search for a configuration $\mathbf{q}$ corresponding to an assigned Cartesian pose of the robot end-effector, can be solved only numerically by choosing an optimization strategy, as already pointed out in the previous section for velocities, since TIAGo is a highly redundant robot. Therefore, with reference to the expressions of $\mathbf{J}^{\dagger}$ presented above, the iterative Gauss–Newton algorithm is proposed to solve the inverse kinematics problem:

$$\mathbf{q}_{k+1} = \mathbf{q}_k + \mathbf{J}^{\dagger}\left(\mathbf{p}_{e,f} - \mathbf{p}_k\right) \tag{13}$$

where $\mathbf{p}_{e,f} = \left[{}^{O}x_g, {}^{O}y_g, {}^{O}z_g, \phi, \theta, \psi\right]^T$ is the assigned pose of the TIAGo end-effector, $\mathbf{q}_k$ and $\mathbf{p}_k$ are, respectively, the robot configuration vector with all eleven variables and the corresponding end-effector pose vector related to the kth iteration, and $\mathbf{J}^{\dagger}$ is the pseudoinverse already mentioned. It is shown in the following section that the algorithm in (13) is employed with some modifications to find the optimal posture of the robot for a given Cartesian pose, making use of the analytical Jacobian matrix $\mathbf{J}_A$ presented in (12) obtained through the use of Euler angles. In order to solve Equation (13), a first known configuration of the robot joint variables and their corresponding Cartesian pose must be determined, possibly not too far from the configuration searched by means of the inverse kinematics algorithm. Some attempts are sometimes required to find a good starting configuration to allow the algorithm to converge quickly to a solution, which must also meet the allowable ranges shown in Table 2. The singular configurations possibly encountered by the iterative algorithm in (13) is easily overcome by verifying the rank of $\mathbf{J}_A$ for each step, updating the pseudoinverse $\mathbf{J}^{\dagger}$ with its damped version to avoid numerical indeterminacy.

4. Posture Optimization of the Grasping Task

In this section, a study is proposed to identify the optimal robot configuration for the grasping task from the infinite possibilities theoretically available (actually ∞^5, due to the excess of degrees of freedom). As already mentioned, the grasping operation can be carried out by adjusting all the independent variables for both the mobile base and the robotic arm to enhance the robot's performance in accomplishing the task. Since the robot configuration is managed with eleven variables and the task is defined by assigning a Cartesian pose to the end-effector corresponding to six degrees of freedom, there is a surplus of five degrees of freedom that can be freely selected.

In the field of robotics, an index commonly used in the search for optimal kinematic configurations of a robot is the condition number κ of its Jacobian matrix [29,30]. The condition number offers insights into the sensitivity of the robot's motion to variations in control inputs, with lower values indicating more uniform control sensitivity across different directions. The 2-norm condition number for inversion has been chosen as the reference for the kinematic optimization proposed in this work, using the singular value decomposition already mentioned to find the ratio of the largest to smallest singular values. This method highlights configurations where the robot demonstrates similar performance in every direction, either in terms of generalized velocity or force, depending on the specific task requirements. Actually, an inversion is conveniently applied to the condition number to obtain a finite index. In this way, the highest possible theoretical value is one, corresponding to the isotropic configuration, where the robot gains analogous performance in every direction. Ultimately, in the most comprehensive case, the index expression is

$$I = \frac{1}{\kappa(\mathbf{J}\mathbf{J}^T)} = \frac{\lambda_{\min}(\mathbf{J}\mathbf{J}^T)}{\lambda_{\max}(\mathbf{J}\mathbf{J}^T)} \tag{14}$$

where $\mathbf{J}$ represents the geometric Jacobian $\mathbf{J}_G$ in (10), $\lambda_{\min}$ and $\lambda_{\max}$ are the minimum and maximum eigenvalues, respectively, of matrix $\mathbf{J}\mathbf{J}^T$, and κ is the condition number of $\mathbf{J}\mathbf{J}^T$. To specifically focus on the linear velocities of the TIAGo end-effector, expression (14) can be updated by replacing $\mathbf{J}$ with $\mathbf{J}_l$, which is a 3×11 Jacobian matrix. Similarly, for angular velocities, the 3×11 Jacobian matrix $\mathbf{J}_a$ should be considered. If required, more complex indexes that allow us to overcome the non-homogeneous feature of the full Jacobian matrix can be taken into account [31].

The degrees of freedom that have been spanned in their possible range of motion are the three planar ones of the mobile base, gathered in vector $\mathbf{q}_b$, the vertical displacement d of the robot trunk, and finally, the spin ϑ_7 around the last revolute joint of the robotic arm, related to the axial rotation of the gripper. The choice of considering these five variables as redundant is justified by the fact that the moving base and the vertical prismatic axis are the most influential in achieving a generic Cartesian configuration for the robot end-effector, the former being able to move freely in the horizontal plane unless encountering obstacles, and the latter being able to adjust the robot's height relative to the gripping target. The rotation around the robot's last joint was chosen with a pitching rotation of the grasped object in mind, allowing for a rotation of an object, such as a glass or bottle, under the assumption of spilling its contents. Other choices can be made, for instance, the rotation around the third axis of the robotic arm, considered typically the redundant axis of a 7 dof robot. These five specific degrees of freedom were selected for exploration to systematically evaluate how variations in each could impact the robot's performance in the grasping task. In particular, their ranges vary as shown below:

$$\begin{aligned} -200\,\text{mm} &\le x_b \le 200\,\text{mm} \\ -200\,\text{mm} &\le y_b \le 200\,\text{mm} \\ -\frac{\pi}{4} &\le \vartheta_b \le \frac{\pi}{4} \\ 0\,\text{mm} &\le d \le 350\,\text{mm} \\ -\frac{2\pi}{3} &\le \vartheta_7 \le \frac{2\pi}{3} \end{aligned} \quad . \tag{15}$$

By assigning values to each of the variables listed above, a finite number of solutions for the inverse kinematics problem of the remaining 6 dof robotic arm can be determined, excluding singular configurations. Therefore, given a reference initial posture $\mathbf{q}_{r6,0} = [\vartheta_1, \vartheta_2, \vartheta_3, \vartheta_4, \vartheta_5, \vartheta_6]^T$ related to a pose $\mathbf{p}_e$ not too far from the required pose $\mathbf{p}_{e,f}$ for grasping the target object, the numerical algorithm presented in (13) efficiently identifies the robot configuration that matches the assigned final pose $\mathbf{p}_{e,f}$. The algorithm consists of two routines: the first one revolves the inverse kinematics for an assigned value of the five variables in $(x_b, y_b, \vartheta_b, d, \vartheta_7)$; the second one ranges all possible values for the variables in $(x_b, y_b, \vartheta_b, d, \vartheta_7)$, providing information about the index I in (14).

4.1. Routine 1—Inverse Kinematics Routine

1. A Cartesian pose $\mathbf{p}_{e,f}$ is assigned, namely, the gripper is located on the target in the gripping configuration. In this case, as shown in Figure 7, the pose is

$$\mathbf{p}_{e,f} = \begin{bmatrix} 0 & 0 & 1 & 0.9\,\text{m} \\ 0 & 1 & 0 & 0\,\text{m} \\ -1 & 0 & 0 & 0.7\,\text{m} \\ 0 & 0 & 0 & 1 \end{bmatrix}; \tag{16}$$

2. A vector $\mathbf{q}_{ind} = [x_b, y_b, \vartheta_b, d, \vartheta_7]^T$ is chosen so that its elements fall into the intervals mentioned above. For instance, $\mathbf{q}_{ind} = [0,0,0,0,\pi/2]^T$;
3. A reference initial posture $\mathbf{q}_{r_6,0}$ is defined. Without loss of generality, the following non-singular configuration has been chosen for the robotic arm joints, also verified visually through the Matlab Robotics Systems Toolbox, as shown in Figure 8:

$$\mathbf{q}_{r_6,0} = [20°, 10°, -70°, 70°, 70°, 30°]^T; \tag{17}$$

4. Equation (4), together with the relations in (5), is used to find the initial pose $\mathbf{p}_{e,0}$ associated with the overall vector $\mathbf{q}$ that gathers $\mathbf{q}_{ind}$ and $\mathbf{q}_{r_6,0}$;
5. The 6 × 6 portion $\mathbf{J}_{A,r}$ of the analytical Jacobian matrix in (12) related only to the rotations in $\mathbf{q}_{r_6,0}$, more specifically, taking from $\mathbf{J}_A$ the third column and the columns from the fifth to the tenth, is evaluated for the $\mathbf{q}$ of step 4;
6. The algorithm in (13) is updated with the 6 × 6 Jacobian of the previous step to solve the inverse kinematics problem:

$$\mathbf{q}_{r_6,k+1} = \mathbf{q}_{r_6,k} + \mathbf{J}_{A,r}^{\dagger}\left(\mathbf{p}_{e,f} - \mathbf{p}_{e,k}\right), \tag{18}$$

where the integer k starts from 0 and gradually increases in the iterative process toward convergence, achieved only when a tolerance for both position and orientation is observed: $\|\mathbf{p}_{g,f} - \mathbf{p}_{g,k}\| \leq \epsilon_l$ and $\|\mathbf{e}_f - \mathbf{e}_k\| \leq \epsilon_a$, with sufficiently small values for the tolerances ϵ_l and ϵ_a;

7. The new position $\mathbf{p}_{e,k+1}$ is computed by means of the direct kinematics in (4) applied to $\mathbf{q}_{k+1}$, namely, the vector that gathers $\mathbf{q}_{r_6,k+1}$ and $\mathbf{q}_{ind}$;
8. The procedure continues iteratively until convergence, when reached. An example is presented in Figure 7, where the shown $\mathbf{q}_{r_6}$ results from a given $\mathbf{q}_{ind}$. On the contrary, the choice for $\mathbf{q}_{ind}$ is discarded from the routine and another vector is evaluated, starting again from step 2. The routine outputs the value of $\mathbf{q}_{r_6,k}$ that verifies the tolerances of step 6.

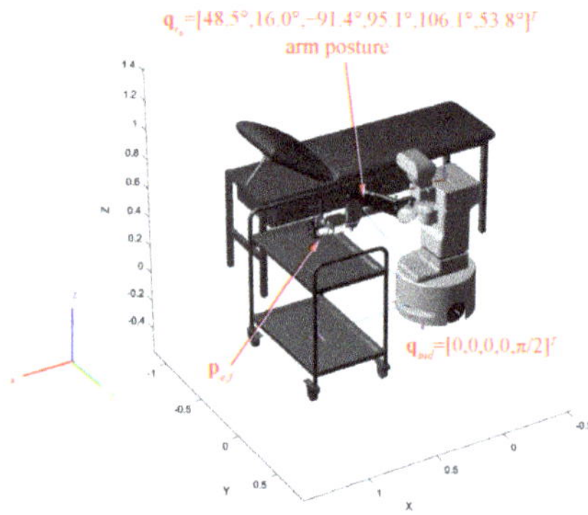

Figure 7. TIAGo robot while grasping a glass of water from a table.

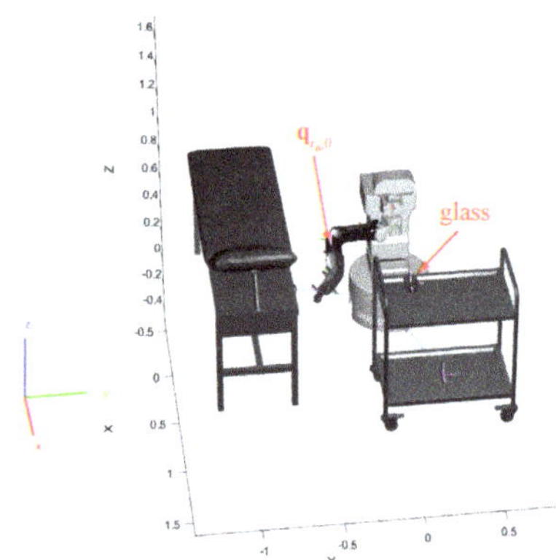

Figure 8. Simulation environment in Matlab where a reference initial posture is defined for the robotic arm.

4.2. Routine 2—Posture Optimization

1. Five nested loops sweep the values of the variables in $\mathbf{q}_{ind}$;
2. For each $\mathbf{q}_{ind}$ determined in the previous step, Routine 1 is executed in order to find the final $\mathbf{q}_{r_6,k}$ associated with $\mathbf{q}_{ind}$;
3. The eigenvalues and eigenvectors of $\mathbf{J}_l\mathbf{J}_l{}^T$ are determined, recording this information so as to know the associated manipulability ellipsoid. It is known that the eigenvectors represent the principal axes of the ellipsoid and the eigenvalues the respective dimensions;
4. The index I in (14) is evaluated and recorded for that particular posture of the robot;

5. The highest value obtained for I allows us to intercept the best posture in terms of velocity manipulability, to which the optimal ellipsoid corresponds;
6. Similarly, for rotations, steps 3 to 5 can be repeated using matrix $\mathbf{J}_a$ instead of matrix $\mathbf{J}_l$.

5. Results

The routines proposed in the previous section have been conveniently implemented in Matlab to find the best posture of the TIAGo robot, according to index I values, when the Cartesian pose in (16) is assigned. On the left of Figure 9, the entire set of ellipsoids is shown, numbering 961 from the scanning of vector $\mathbf{q}_{ind}$. It can be noted that most of the ellipsoids must have a low index I because of their flattened shape, with semi-axes in the yz-plane much larger than that along the x-axis. On the right of Figure 9, the best ellipsoid in terms of linear velocities of the robot end-effector corresponding to the highest value of $I = 0.41$ is shown, meaning that all the Cartesian directions have a similar linear velocity performance. The volume of such an ellipsoid is clearly smaller than many others that are visible in the left image, but there was a desire to favor a closer proximity to the isotropic condition over having a strongly anisotropic behavior. When an anisotropic behavior is accepted, a different posture of the robot can be chosen in order to have higher velocities in some directions.

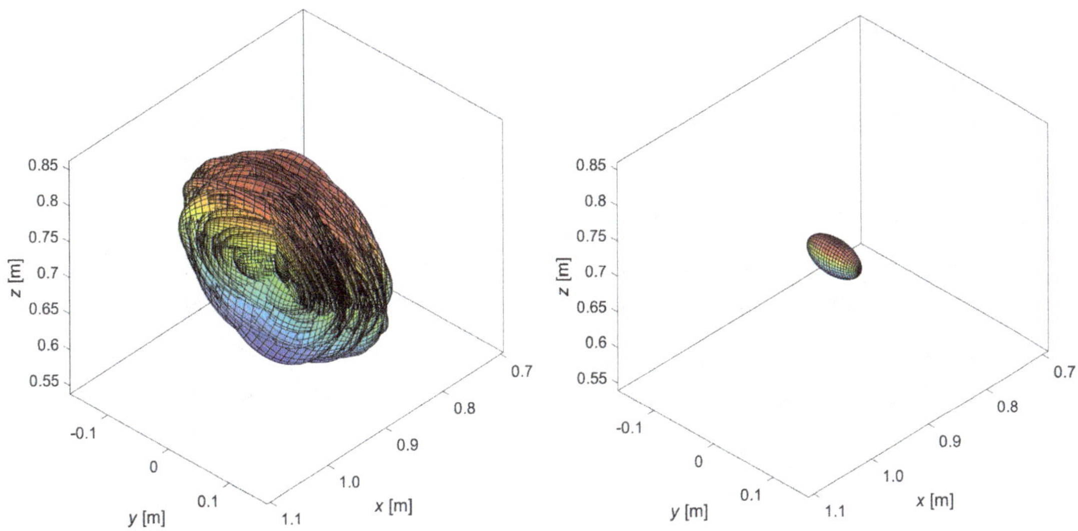

Figure 9. Entire set of linear velocity ellipsoids on the left and optimum velocity ellipsoid with maximum index I on the right.

Figure 10 shows the optimized posture that maximized index I. In this configuration, the mobile base of the robot is closer to the table and the arm is less extended, as well as positioned to lie on a horizontal plane xy. Such a condition is allowed by an extension of the vertical prismatic joint of the trunk. A comparison with the generic configuration already presented in Figure 7 points out these differences in posture, as Figure 11 clearly demonstrates.

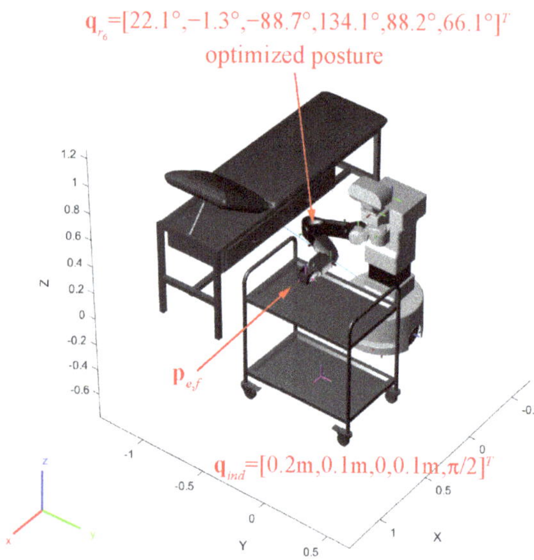

Figure 10. Optimized posture that maximizes index I.

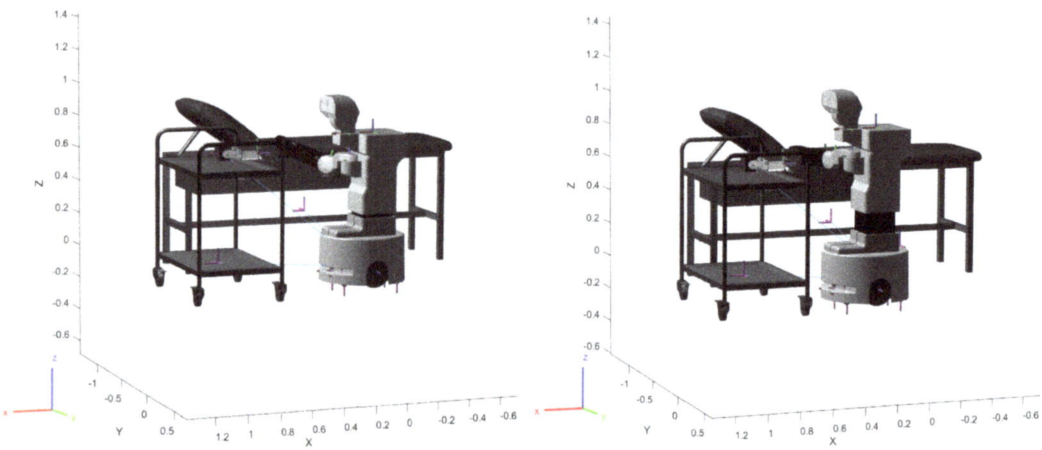

Figure 11. Generic posture on the left and optimized posture on the right for a given Cartesian pose where the TIAGo robot performs a grasping operation.

This study has been also extended to angular velocities by using matrix $\mathbf{J}_a$ instead of $\mathbf{J}_l$. It results in a set of angular velocity ellipsoids similar to the one presented in Figure 9, but more extended along the x-axis, almost representing a sphere. In this case, the best ellipsoid is related to a higher value of the index I defined in (14), namely, $I = 0.58$. It means that the robot end-effector can rotate with a similar angular velocity in every direction. Even in this case, the best ellipsoid has a shape similar to the one shown in Figure 9 on the right, but with a higher volume.

As is well known, ellipsoids in force have orthogonal behavior compared to those in velocity presented in this paper. These become useful for evaluating the robot's performance in terms of static lifting capacity or thrust/resistance.

6. Conclusions

This study has systematically explored the impact of varying degrees of freedom on the performance of the TIAGo robot's end-effector for grasping tasks, with a particular focus on the manipulability of linear and angular velocities. By employing a comprehensive methodology that includes the calculation of condition numbers through the geometric Jacobian and the execution of two distinct routines (inverse kinematics routine and posture optimization), optimal postures that maximize the robot's efficiency in simulated grasping tasks have been identified.

Optimal Postures Enhance Performance: Adjusting specific degrees of freedom can significantly improve the robot's efficiency in grasping tasks, underscoring the importance of precise control in robotic design.

Implications and Future Directions: The presented methodology provides a foundation for improving motion planning and control of the TIAGo robot, with potential applications across various platforms and tasks characterized by redundant degrees of freedom. Future research should explore these techniques in different contexts and integrate advanced predictive algorithms to further optimize performance.

Limitations: While insightful, the findings proposed in this paper are based on simulations with the TIAGo robot and may require validation in real-world scenarios.

In summary, this research highlights the critical role of posture optimization in robotic grasping tasks, offering a pathway toward more responsive and capable robotic systems.

Author Contributions: Conceptualization, M.-C.P. and A.B.; methodology, M.-C.P.; software, A.B.; validation, A.B. and D.C.; formal analysis, M.-C.P.; investigation, A.B.; resources, L.C.; data curation, D.C.; writing—original draft preparation, M.-C.P. and A.B.; writing—review and editing, M.-C.P. and A.B.; visualization, L.C. and A.B.; supervision, L.C.; project administration, M.-C.P. All authors have read and agreed to the published version of the manuscript.

Funding: This research received external funding from the project FOCAAL, FOg Computing in Ambient Assisted Living, funded by MIMIT, the Ministry of Enterprise and Made in Italy, under the Life Sciences Innovation Agreement DM 05/03/2018.

Data Availability Statement: All data are contained within this article.

Conflicts of Interest: The authors declare no conflicts of interest.

Abbreviations

The following abbreviations are used in this manuscript:

FOCAAL	FOg Computing in Ambient Assisted Living
MIMIT	Ministry of Enterprises and Made in Italy
EU	European Union
COVID	COronaVIrus Disease
HRI	Human–Robot Interaction
SMS	Short Message Service
RUIO	Robust Unknown-Input Observer
ROS	Robot Operating System
CDPR	Cable-Driven Parallel Robot
AMR	Autonomous Mobile Robot
dof	Degrees of Freedom
DH	Denavit–Hartenberg
RGB-D	Red–Green–Blue and Depth

References

1. Tian, Y.; Li, D. The Impact of Aging on Economic Growth Based on Health Investment and Technological Innovation: An Empirical Study from China. In Proceedings of the 2020 Management Science Informatization and Economic Innovation Development Conference (MSIEID), Guangzhou, China, 18–20 December 2020. [CrossRef]
2. Science, N.; Council, T. *Emerging Technologies to Support an Aging Population*; Technical Report NSTC-2019-1; Executive Office of the President of the United States: Washington, DC, USA, 2019.

3. Peek, S.T.; Luijkx, K.G.; Rijnaard, M.D.; Nieboer, M.E.; van der Voort, C.S.; Aarts, S.; van Hoof, J.; Vrijhoef, H.J.; Wouters, E.J. Older Adults' Reasons for Using Technology while Aging in Place. *Gerontology* **2015**, *62*, 226–237. [CrossRef] [PubMed]
4. Krishnan, R.H.; Pugazhenthi, S. Mobility assistive devices and self-transfer robotic systems for elderly, a review. *Intell. Serv. Robot.* **2013**, *7*, 37–49. [CrossRef]
5. Bemelmans, R.; Gelderblom, G.J.; Jonker, P.; de Witte, L. Socially Assistive Robots in Elderly Care: A Systematic Review into Effects and Effectiveness. *J. Am. Med. Dir. Assoc.* **2012**, *13*, 114–120.e1. [CrossRef]
6. Ienca, M.; Jotterand, F.; Vică, C.; Elger, B. Social and Assistive Robotics in Dementia Care: Ethical Recommendations for Research and Practice. *Int. J. Soc. Robot.* **2016**, *8*, 565–573. [CrossRef]
7. Coşar, S.; Fernandez-Carmona, M.; Agrigoroaie, R.; Pages, J.; Ferland, F.; Zhao, F.; Yue, S.; Bellotto, N.; Tapus, A. ENRICHME: Perception and Interaction of an Assistive Robot for the Elderly at Home. *Int. J. Soc. Robot.* **2020**, *12*, 779–805.[CrossRef]
8. Remmers, P.; Fischer, N. Social Robots in Care Facilities: Reflections on Current Research and the Potential of Ethical Visions. In *Frontiers in Artificial Intelligence and Applications*; IOS Press: Amsterdam, The Netherlands, 2023; ISBN 9781643683751, ISSN 1879-8314. [CrossRef]
9. Huang, G.; Guo, F.; Liu, L.; Taksa, L.; Cheng, Z.; Tani, M.; Zimmermann, K.F.; Franklin, M.; Silva, S.S.M. Changing impact of COVID-19 on life expectancy 2019–2023 and its decomposition: Findings from 27 countries. *SSM Popul. Health* **2024**, *25*, 101568. [CrossRef] [PubMed]
10. Eurostat. Life Expectancy at Birth Down to 80.1 Years in 2021. 2023. Available online: https://ec.europa.eu/eurostat/web/products-eurostat-news/w/DDN-20230316-1#:~:text=In%202021%2C%20the%20life%20expectancy,of%20the%20COVID%2D19%20pandemic (accessed on 22 February 2024).
11. Telembici, T.; Grama, L.; Rusu, C. Integrating Service Robots into Everyday Life Based on Audio Capabilities. In Proceedings of the 2020 International Symposium on Electronics and Telecommunications (ISETC), Timisoara, Romania, 5–6 November 2020. [CrossRef]
12. Grama, L.; Rusu, C. Adding audio capabilities to TIAGo service robot. In Proceedings of the 2018 International Symposium on Electronics and Telecommunications (ISETC), Timisoara, Romania, 8–9 November 2018. [CrossRef].
13. Grama, L.; Rusu, C. Extending Assisted Audio Capabilities of TIAGo Service Robot. In Proceedings of the 2019 International Conference on Speech Technology and Human-Computer Dialogue (SpeD), Timisoara, Romania, 10–12 October 2019; pp. 1–8. [CrossRef]
14. Lach, L.; Haschke, R.; Ferro, F.; Pagès, J. Leveraging Touch Sensors to Improve Mobile Manipulation. *arXiv* **2020**, arXiv:2010.10810.
15. Muscar, L.; Grama, L.; Rusu, C. A Real-Time Warning Based on TIAGo's Audio Capabilities. In Proceedings of the 2022 International Symposium on Electronics and Telecommunications (ISETC), Timisoara, Romania, 10–11 November 2022; pp. 1–4. [CrossRef]
16. Miguel, A.; Puig, V.; Alenyà, G. Fault-tolerant Control of a Service Robot using a LPV Robust Unknown Input Observer. In Proceedings of the 2019 4th Conference on Control and Fault Tolerant Systems (SysTol), Casablanca, Morocco, 18–20 September 2019; pp. 207–212. [CrossRef]
17. van Eck, N.J.; Waltman, L. Software survey: VOSviewer, a computer program for bibliometric mapping. *Scientometrics* **2009**, *84*, 523–538. [CrossRef] [PubMed]
18. Sulaiman, S.; H, A.; Apurin, A.; Martinez-Garcia, E.; Magid, E. Trajectory planning and simulation of a custom mobile manipulator for door opening tasks using ROS. In Proceedings of the Advances in Robotics—6th International Conference of The Robotics Society, Ropar, India, 5–8 July 2023. [CrossRef]
19. Baumgartner, J.; Petrič, T.; Klančar, G. Potential Field Control of a Redundant Nonholonomic Mobile Manipulator with Corridor-Constrained Base Motion. *Machines* **2023**, *11*, 293. [CrossRef]
20. Neri, F.; Forlini, M.; Scoccia, C.; Palmieri, G.; Callegari, M. Experimental Evaluation of Collision Avoidance Techniques for Collaborative Robots. *Appl. Sci.* **2023**, *13*, 2944. [CrossRef]
21. Cheah, S.K.; Caverly, R.J. Passivity-Based Pose Regulation and Jacobian-Based Force Distribution of a Cable-Driven Parallel Robot. In Proceedings of the 2021 American Control Conference (ACC), New Orleans, LA, USA, 25–28 May 2021. [CrossRef].
22. Zhong, X.G.; Peng, X.F.; Zhong, X.Y.; Lin, L.X. Dynamic Jacobian Identification Based on State-Space for Robot Manipulation. *Appl. Mech. Mater.* **2013**, *475–476*, 675–679. [CrossRef]
23. Kotlarski, J.; Abdellatif, H.; Heimann, B. Improving the pose accuracy of a planar 3RRR parallel manipulator using kinematic redundancy and optimized switching patterns. In Proceedings of the 2008 IEEE International Conference on Robotics and Automation, Pasadena, CA, USA, 19–23 May 2008. [CrossRef]
24. He, G.; Liu, Y.; Li, C. Tightly coupled laser-inertial pose estimation and map building based on B-spline curves. *Meas. Sci. Technol.* **2023**, *34*, 125130. [CrossRef]
25. Campbell, S.L.; Meyer, C.D. *Generalized Inverses of Linear Transformations*; Dover Publications: New York, NY, USA, 1991.
26. Deo, A.S.; Walker, I.D. Overview of damped least-squares methods for inverse kinematics of robot manipulators. *J. Intell. Robot. Syst.* **1995**, *14*, 43–68. [CrossRef]
27. Buss, S.R.; Kim, J.S. Selectively Damped Least Squares for Inverse Kinematics. *J. Graph. Tools* **2005**, *10*, 37–49. [CrossRef]
28. Tringali, A.; Cocuzza, S. Globally Optimal Inverse Kinematics Method for a Redundant Robot Manipulator with Linear and Nonlinear Constraints. *Robotics* **2020**, *9*, 61. [CrossRef]

29. Angeles, J.; López-Cajún, C.S. Kinematic Isotropy and the Conditioning Index of Serial Robotic Manipulators. *Int. J. Robot. Res.* **1992**, *11*, 560–571. [CrossRef]
30. Merlet, J.P. Jacobian, Manipulability, Condition Number, and Accuracy of Parallel Robots. *J. Mech. Des.* **2005**, *128*, 199–206. [CrossRef]
31. Mansouri, I.; Ouali, M. The power manipulability—A new homogeneous performance index of robot manipulators. *Robot. Comput.-Integr. Manuf.* **2011**, *27*, 434–449. [CrossRef]

robotics

Article

Driving Strategies for Omnidirectional Mobile Robots with Offset Differential Wheels

Joan Badia Torres [1,†], Alba Perez Gracia [1,*,†] and Carles Domenech-Mestres [2,†]

[1] Department of Mechanical Engineering, Polytechnic University of Catalonia (UPC), 08034 Barcelona, Spain; joan.badia.torres@upc.edu
[2] Centre de Disseny d'Equips Industrials (CDEI), Polytechnic University of Catalonia (UPC), 08034 Barcelona, Spain; carles.domenech@upc.edu
* Correspondence: alba.perez.gracia@upc.edu
† These authors contributed equally to this work.

Abstract: In this work, we present an analysis of, as well as driving strategies and design considerations for, a type of omnidirectional mobile robot: the offset-differential robot. This system presents omnidirectionality while using any type of standard wheel, allowing for applications in uneven and rough terrains, as well as cluttered environments. The known fact that these robots, as well as simple differential robots, have an unstable driving zone, has mostly been dealt with by designing driving strategies in the stable zone of internal dynamics. However, driving in the unstable zone may be advantageous when dealing with rough and uneven terrains. This work is based on the full kinematic and dynamic analysis of a robot, including its passive elements, to explain the unexpected behaviors that appear during its motion due to instability. Precise torque calculations taking into account the configuration of the passive elements were performed for better torque control, and design recommendations are included. The stable and unstable behaviors were characterized, and driving strategies were described in order to achieve the desired performance regarding precise positioning and speed. The model and driving strategies were validated through simulations and experimental testing. This work lays the foundation for the design of better control strategies for offset-differential robots.

Keywords: OMR; omnidirectional mobile robots

Citation: Badia Torres, J.; Perez Gracia, A.; Domenech-Mestres, C. Driving Strategies for Omnidirectional Mobile Robots with Offset Differential Wheels. *Robotics* **2024**, *13*, 19. https://doi.org/10.3390/robotics13010019

Academic Editor: Raffaele Di Gregorio

Received: 15 December 2023
Revised: 15 January 2024
Accepted: 16 January 2024
Published: 18 January 2024

1. Introduction

Omnidirectional mobile robots are a desirable option in several robotics applications, especially when maneuvering in tight spaces or when extra dexterity is required [1,2].

Many of the current omnidirectional mobile robots are based on Swedish or omni wheels; see the recent reviews of [3,4]. These wheels, however, are not as robust as standard wheels in their application on rough terrains and outdoor applications; see, for instance [5], in which Swedish wheels were modified to overcome some of their limitations. Active caster wheels are an attractive option because of their holonomy; the design of an actuated caster wheel was described in [6], with a mobile platform using two independent steerable caster wheels. In a similar development, ref. [7] designed a dual, steerable caster wheel with a differential gear mechanism. Other solutions have accomplished omnidirectionality by adding redundant degrees of freedom; see, for instance [8]. This, in turn, increases the cost and complexity of path planning and control.

One solution for omnidirectionality while using standard wheels is offset-differential robots. This type of robot has been called *OmniKitty* by [9] and *Otbot* by [10,11]. Our implementation is the *MOBY* platform, which was developed and patented at the Polytechnic University of Catalonia (UPC) [12] and is currently being used for logistics and other industrial applications. This kind of omnidirectional robot uses two standard wheels offset from the center and a third rotational degree of freedom about a vertical, centered axis. For practical considerations, the robot usually has a third omnidirectional point of support, usually a passive caster wheel, which is often overlooked in analyses.

Because of its application in real tasks, MOBY presents a unique opportunity to study its behavior. The kinematics and dynamics of this robot were studied in [9,11]. The stability of this robot has many points in common with that of the differential robots for which the centers of the robot, or the point to track, is not located at the midpoint of the wheels, for instance, when a *look-ahead* control is used. This was studied by Yun and Yamamoto in [13], in which Lyapunov analysis was undertaken in order to identify stable and unstable configurations for a look-ahead control strategy. Pulling the robot, which the authors called *driving backwards*, became an unstable configuration.

Because of the widespread use of differential robots, the topic of their instability has drawn certain attention over the years. Shojaei [14] proposed a tracking controller with an adaptive feedback linearization control based on the same look-ahead strategy, which they extended to an inverse dynamics control with an adaptive PID in [15]. One of the few recent references on the topic of unstable regions was [16], which showed that, for a differential robot, driving with the center of mass behind the wheels leads to better directional maneuverability. In general, even works that have identified the stable and unstable driving regions have focused on creating control strategies for driving in stable regions; see [17] for a review of stability issues in navigation for wheeled robots.

As mobile robots extend their field of application to uneven and complex terrains and cluttered environments (such as agricultural fields [18] or rescue missions), new designs (such as omnidirectional robots), and new driving strategies (such as driving backwards) will become more relevant. This research aims to help identify the characteristics of the unstable driving zone and to define some of the driving methods that take into account the internal dynamics of the system, regardless of the control strategy.

In this work, we expanded the kinematic and dynamic analysis presented in [11] by including the effect of the auxiliary caster wheel, and we applied it to the MOBY design to study the characteristics of the kinematic and dynamic behavior of this robot, which appears when navigating in real terrains and which is not intuitive. Unlike most of the publications for these robots, which have aimed to characterize and control only the stable internal -pushing- dynamics identified in [13], we optimized the behavior of this type of robot by also developing pull-driving strategies and quantifying their differences. We performed both simulations and real experiments, and we compared the theoretical results to the experimental results in order to propose techniques and design solutions for smooth and precise navigation.

2. Description of the Robot

The MOBY robot has three actuated degrees of freedom with the aim of obtaining an omnidirectional mobile robot. The general structure of the robot is shown in the sketch in Figure 1. Unlike the differential wheeled robot, in which the two actuated wheels can spin independently in both directions and the difference between the speeds of the wheels steers the machine, MOBY has an extra actuated rotation about an axis that is perpendicular to the plane of motion and offset from the wheels. It is, basically, an offset differential robot plus a vertical rotation about a centered axis.

For the differential robot motion, the instant center of rotation is always found on the line of the axes of the wheels, as shown in Figure 2. The third vertical axis of MOBY allows for the orientation of this line arbitrarily while keeping an independent orientation on the body of the robot. A caster wheel is also mounted to the base of the robot for stability.

This simple kinematic structure was implemented in the MOBY robot that can be seen in Figure 3. It is important to notice that it is very simple to modify the type of wheels and the type of platform to adapt this robot to indoor or outdoor applications, as shown in the figure.

The robot's movement is controlled using platform velocity commands (three degrees of freedom in the plane) that are translated into velocity commands for each axis of the robot. In a code written in the C language, the commands for each axis are calculated based on the movement requirements. These data are sent to the robot every 4 ms using the UDP protocol. The robot is equipped with a PLC (programmable logic controller) that translates the velocity values for each axis into commands for the drivers of each motor to execute the requested movement. To follow a specific trajectory, the required velocity is calculated for each moment, and by integrating it, the position and orientation of the robot are determined.

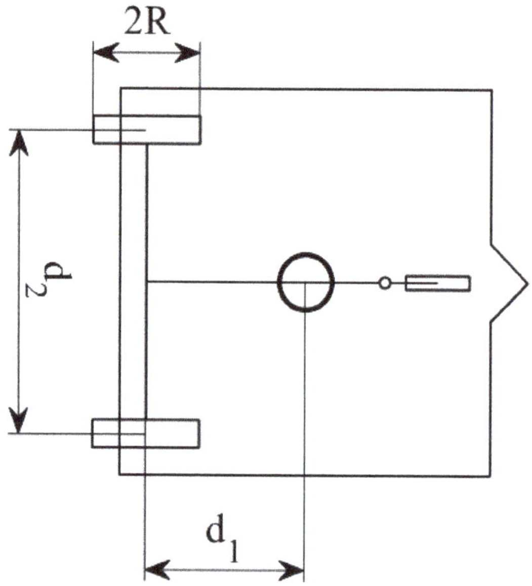

Figure 1. The structure of MOBY robot. The actuated wheels have radius R and the distance between the wheels is d_2. The vertical axis of rotation is located at a distance d_1 from the axis of the wheels.

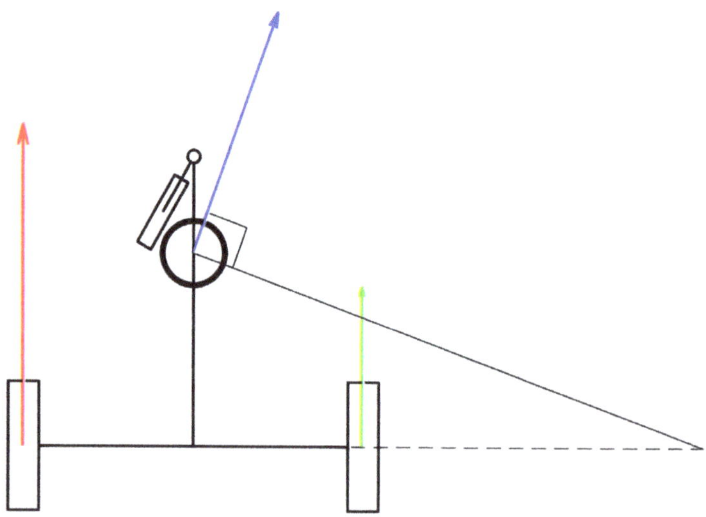

Figure 2. The instant velocity of the center of the robot (in blue) is a function of the velocities of the wheels (in red and green).

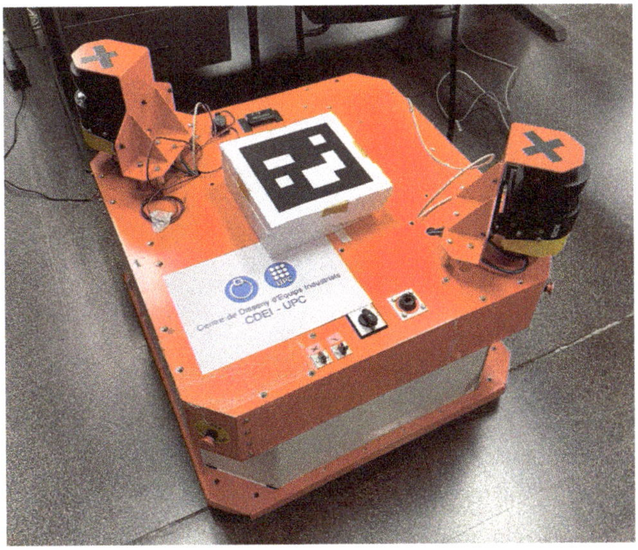

Figure 3. The real robot used for this study's experiments, MOBY.

3. Kinematic Analysis

A kinematic analysis of this type of robot can also be found in [9,11]. In the current work, we took a very similar approach, adapting it to the particular geometry of the MOBY platform and adding the caster wheel and fork.

Let us consider the active wheels of the robot as subjected to the no-slip condition to simplify the initial analysis. We defined the chassis as the differential wheels with their common axis and the structure that connects it to the platform.

The following generalized velocities describe the controlled velocities of MOBY:

- The angular velocity of the right wheel, $\dot{\alpha}_R$, denoting the angular position of the right wheel as α_R.
- The angular velocity of the left wheel, $\dot{\alpha}_L$, denoting α_L as its angular position from a reference.
- The angular velocity of the platform about the vertical axis, $\dot{\varphi}$, with respect to the chassis, with an angular position of the platform with respect to the chassis φ.

Additionally, the following parameters, generalized coordinates, and velocities are used to describe its kinematics:

- The position vector of the center of the robot with respect to a fixed frame, $\mathbf{p} = \{x, y\}$, and its derivative $\dot{\mathbf{p}} = \{\dot{x}, \dot{y}\}$.
- The angular position of the platform with respect to the fixed frame, θ_p, and its corresponding angular velocity, $\dot{\theta}_p$.
- The angular position of the chassis with respect to the fixed frame, θ_c, and its derivative $\dot{\theta}_c$.
- The radius of the actuated wheels, R.
- The distance between the rotation axis of the actuated wheels and the vertical rotation axis, d_1.
- The distance between the actuated wheels, d_2.

Notice that, with the definitions of these variables, the actuated rotation φ can be expressed as a function of θ_c and θ_p, as shown in Figure 4:

$$\varphi = \theta_p - \theta_c. \tag{1}$$

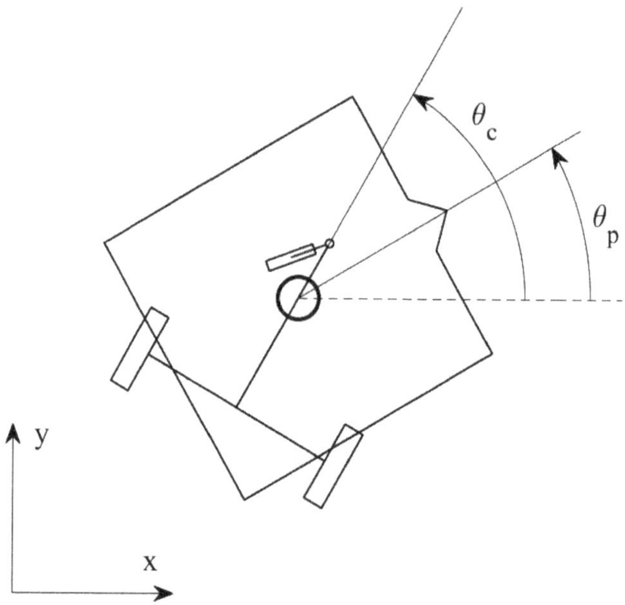

Figure 4. The orientation of the robot as a function of the defined variables.

The wheels of the MOBY platform are standard fixed wheels, which impose the following constraints on the motion of the robot:

$$\dot{x} - (\frac{R\cos(\theta_c)}{2} + \frac{Rd_1\sin(\theta_c)}{d_2})\dot{\alpha}_L - (\frac{R\cos(\theta_c)}{2} - \frac{Rd_1\sin(\theta_c)}{d_2})\dot{\alpha}_R = 0, \tag{2}$$

$$\dot{y} - (\frac{R\sin(\theta_c)}{2} - \frac{Rd_1\cos(\theta_c)}{d_2})\dot{\alpha}_L - (\frac{R\sin(\theta_c)}{2} + \frac{Rd_1\sin(\theta_c)}{d_2})\dot{\alpha}_R = 0. \tag{3}$$

A third constraint transmits the angular velocity of the platform motor from the chassis to the platform,

$$\dot{\theta}_c + \dot{\varphi} - \dot{\theta}_p = 0, \tag{4}$$

enforcing Equation (1). When the relation in Equation (10) is applied to Equation (4), the constraint can be rewritten as

$$\frac{R}{d_2}\dot{\alpha}_R - \frac{R}{d_2}\dot{\alpha}_L + \dot{\varphi} - \dot{\theta}_p = 0 \tag{5}$$

to explicitly show the relation between $\dot{\theta}_p$ and the actuated variables of the system.

To take the dynamic effects of the caster wheel assembly into account, additional variables have to be detailed, which are shown in Figure 5:

- The distance from the center of the platform to the joint of the caster fork, d_4.
- The length of the caster fork from its joint to the axis of the caster wheel, d_3.
- The radius of the caster wheel, R_c.
- The angle of the caster fork with respect to the chassis, β.
- The rotational velocity of the caster wheel, $\dot{\alpha}_c$.

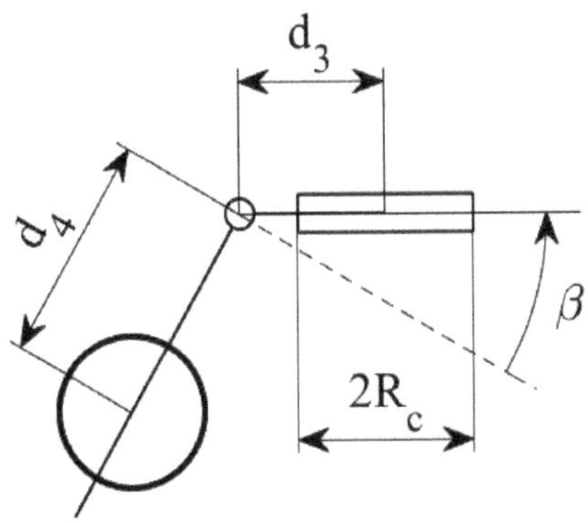

Figure 5. Variables defining the dimensions and configuration of the caster wheel assembly.

These variables also allow for the non-holonomic constraints of the caster wheel assembly to be applied as described in the work of Siegwart & Nourbakhsh [19]. The fourth constraint dictates the velocity of the caster fork as a function of the velocity and configuration of the robot. The fifth and final constraint imposes the no-slip condition on the caster wheel:

$$\cos(\beta+\theta_c)\dot{x}+\sin(\beta+\theta_c)\dot{y}-\frac{R}{d_2}(d_3+d_4\sin\beta)\dot{\alpha}_L+ \\ \frac{R}{d_2}(d_3+d_4\sin\beta)\dot{\alpha}_R-d_3\dot{\beta}=0, \tag{6}$$

and

$$\sin(\beta+\theta_c)\dot{x}-\cos(\beta+\theta_c)\dot{y}+\frac{d_4R\cos\beta}{d_2}\dot{\alpha}_L-\frac{d_4R\cos\beta}{d_2}\dot{\alpha}_R-\dot{\alpha}_cR_c=0. \tag{7}$$

3.1. Forward Kinematics

The Jacobian matrix was used to model the relation between the three motors of the robot and its twist relative to the fixed frame of reference. Consider $W=\{\dot{x},\dot{y},\dot{\theta_p}\}$ the twist of the platform with respect to the fixed frame. Then, the Jacobian matrix $[J]$ satisfies the following equation:

$$\begin{bmatrix}\dot{x}\\ \dot{y}\\ \dot{\theta}_p\end{bmatrix}=\begin{bmatrix} & & \\ & J & \\ & & \end{bmatrix}\begin{bmatrix}\dot{\alpha}_L\\ \dot{\alpha}_R\\ \dot{\varphi}\end{bmatrix}. \tag{8}$$

The wheel constraints in Equations (2) and (3), together with the relations among the orientation of the chassis, platform, and vertical degree of freedom in Equation (5), form the Jacobian matrix:

$$[J]=\begin{bmatrix}\frac{R\cos\theta_c}{2}+\frac{Rd_1\sin\theta_c}{d_2} & \frac{R\cos\theta_c}{2}-\frac{Rd_1\sin\theta_c}{d_2} & 0\\ \frac{R\sin\theta_c}{2}-\frac{Rd_1\cos\theta_c}{d_2} & \frac{R\sin\theta_c}{2}+\frac{Rd_1\cos\theta_c}{d_2} & 0\\ -\frac{R}{d_2} & \frac{R}{d_2} & 1\end{bmatrix}. \tag{9}$$

When the no-slip condition was applied, it was observed that the orientation of the chassis is a function of the position of the wheels:

$$\theta_c=\theta_{c0}+\frac{R(\alpha_R-\alpha_L)}{d_2}, \tag{10}$$

where θ_{c0} is the initial orientation of the chassis, and the initial values of α_R and α_L are set to zero.

3.2. Inverse Kinematics

For the inverse kinematics, the desired relation is

$$\begin{bmatrix} \dot{\alpha}_L \\ \dot{\alpha}_R \\ \dot{\varphi} \end{bmatrix} = \begin{bmatrix} & J^{-1} & \end{bmatrix} \begin{bmatrix} \dot{x} \\ \dot{y} \\ \dot{\theta}_p \end{bmatrix}. \tag{11}$$

In the above expression, $[J^{-1}]$ is the inverse of the previously computed Jacobian matrix, and it can be calculated to obtain the matrix:

$$[J^{-1}] = \begin{bmatrix} \frac{2d_1\cos\theta_c + d_2\sin\theta_c}{2Rd_1} & \frac{2d_1\sin\theta_c - d_2\cos\theta_c}{2Rd_1} & 0 \\ \frac{2d_1\cos\theta_c - d_2\sin\theta_c}{2Rd_1} & \frac{2d_1\sin\theta_c + d_2\cos\theta_c}{2Rd_1} & 0 \\ \frac{\sin\theta_c}{d1} & -\frac{\cos\theta_c}{d1} & 1 \end{bmatrix}. \tag{12}$$

3.3. Passive Degrees of Freedom

To integrate the response of the caster assembly with the velocities of the other degrees of freedom of the robot, the Jacobian matrix was augmented, adding two rows to $[J]$. To produce this augmented Jacobian matrix, Equations (6) and (7) must be expressed in terms of the variables $\dot{\alpha}_L$, $\dot{\alpha}_R$, and $\dot{\varphi}$. From the rewritten expressions, $\dot{\beta}$ and $\dot{\alpha}_c$ were isolated, and the derivatives of their definitions were incorporated into the augmented Jacobian matrix, $[J_{dir}]$.

$$[J_{dir}] = \begin{bmatrix} \frac{R\cos\theta_c}{2} + \frac{Rd_1\sin\theta_c}{d_2} & \frac{R\cos\theta_c}{2} - \frac{Rd_1\sin\theta_c}{d_2} & 0 \\ \frac{R\sin\theta_c}{2} - \frac{Rd_1\cos\theta_c}{d_2} & \frac{R\sin\theta_c}{2} + \frac{Rd_1\cos\theta_c}{d_2} & 0 \\ -\frac{R}{d_2} & \frac{R}{d_2} & 1 \\ -\frac{R(2d_3 - d_2\cos\beta + 2d_1\sin\beta + 2d_4\sin\beta)}{2d_2d_3} & \frac{R(2d_3 + d_2\cos\beta + 2d_1\sin\beta + 2d_4\sin\beta)}{2d_2d_3} & 0 \\ \frac{R(2d_1\cos\beta + 2d_4\cos\beta + d_2\sin\beta)}{2R_cd_2} & -\frac{R(2d_1\cos\beta + 2d_4\cos\beta - d_2\sin\beta)}{2R_cd_2} & 0 \end{bmatrix}. \tag{13}$$

The augmented Jacobian matrix provides the expected values of the velocities $\dot{\beta}$ and $\dot{\alpha}_c$:

$$\begin{bmatrix} \dot{x} \\ \dot{y} \\ \dot{\theta}_p \\ \dot{\beta} \\ \dot{\alpha}_c \end{bmatrix} = \begin{bmatrix} & J_{dir} & \end{bmatrix} \begin{bmatrix} \dot{\alpha}_L \\ \dot{\alpha}_R \\ \dot{\varphi} \end{bmatrix}. \tag{14}$$

Notice that this Jacobian matrix is not square and, thus, is no longer invertible. The augmented inverse Jacobian matrix was obtained by repeating the previous process, this time using $[J^{-1}]$ as a starting point and expressing Equations (6) and (7) in terms of the task space variables: $\dot{x}$, $\dot{y}$, and $\dot{\theta}_p$. The resulting matrix, $[J_{inv}]$, satisfies Equation (15), which produces the response of all the degrees of freedom of the system to a velocity command expressed in the task space:

$$\begin{bmatrix} \dot{\alpha}_L \\ \dot{\alpha}_R \\ \dot{\varphi} \\ \dot{\beta} \\ \dot{\alpha}_c \end{bmatrix} = \begin{bmatrix} & J_{inv} & \end{bmatrix} \begin{bmatrix} \dot{x} \\ \dot{y} \\ \dot{\theta}_p \end{bmatrix}, \tag{15}$$

with

$$[J_{inv}] = \begin{bmatrix} \frac{2d_1 \cos\theta_c + d_2 \sin\theta_c}{2Rd_1} & \frac{2d_1 \sin\theta_c - d_2 \cos\theta_c}{2Rd_1} & 0 \\ \frac{2d_1 \cos\theta_c - d_2 \sin\theta_c}{2Rd_1} & \frac{2d_1 \sin\theta_c + d_2 \cos\theta_c}{2Rd_1} & 0 \\ \frac{\sin\theta_c}{d1} & -\frac{\cos\theta_c}{d1} & 1 \\ \frac{2d_1 \cos(\beta+\theta) + d_4 \cos(\beta+\theta_c) - 2d_3 \sin\theta_c - d_4 \cos(\beta-\theta_c)}{2d_1 d_3} & \frac{2d_1 \sin(\beta+\theta_c) + d_4 \sin(\beta+\theta_c) + 2d_3 \cos\theta_c + d_4 \sin(\beta-\theta_c)}{2d_1 d_3} & 0 \\ \frac{2d_1 \sin(\beta+\theta_c) + d_4 \sin(\beta+\theta_c) - d_4 \sin(\beta-\theta_c)}{2R_c d_1} & -\frac{2d_1 \cos(\beta+\theta_c) + d_4 \cos(\beta+\theta_c) + d_4 \cos(\beta-\theta_c)}{2R_c d_1} & 0 \end{bmatrix}. \quad (16)$$

4. Kinematic Behavior

Most of the non-intuitive behavior of this type of robot is due to the arrangement of the degrees of freedom of the robot and the kinematic constraints imposed on the system. The preferred driving strategy for uneven and rough terrains, which we call the *pulling* strategy, turned out to be unstable, and hence, it needed to be studied in detail. The control of the offset-differential robot is equivalent in terms of its kinematics to that of the look-ahead strategy of differential robots, which are widely used.

4.1. Driving Stability

As mentioned, the main phenomenon of interest was the instability described by Yun & Yamamoto [13] that appears when the machine is commanded to move with its actuated wheels in front, that is, with the wheels pulling the chassis, rather than pushing it. In the cited analysis, the authors derived the state space and the feedback control with input–output linearization for the look-ahead strategy. They showed that the internal dynamics exhibited instability when driving backwards, that is, when the tracking point was behind the driving wheels.

In order to show the instability, they used the equations of the internal dynamics when the reference point is moving and the robot is driving in a straight line, and they applied the Lyapunov function of the angular position of the wheels; see the referenced work for more details.

Yun and Yamamoto found that the robot has two equilibrium points. When applied to the offset differential robot, the first equilibrium point is the configuration in which the center of the robot and the midpoint of the axis of the active wheels are aligned in the direction of motion, with the wheels behind the center of the robot. This state can be called *pure push*, and it is a stable equilibrium point. If the robot is commanded to move in a straight line, it will gradually reach this state.

Unlike the first equilibrium point, the second equilibrium point is unstable. In this state, the wheels are ahead of the center of the robot. This unstable state can be called *pure pull*. With minimal disturbance, the angle between the chassis and the direction of motion will grow until they are perpendicular; at that point, the robot will start pushing and eventually reach the pure pushstate, as shown in Figure 6.

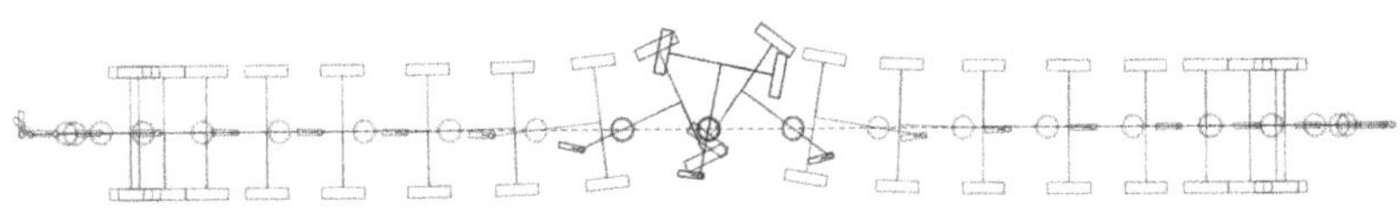

Figure 6. Simulation showing the reversal of the direction of the chassis. From left to right, the robot accelerates backwards, holds its speed, switches from pull to push, and comes to a stop.

4.2. Kinematic Behavior

An easy solution is to avoid driving from this unstable equilibrium point; however, for rough and uneven terrains, such as agricultural ground, the pulling maneuver is advantageous. One easy way to see this is to consider the situation in which the robot has to climb a small obstacle with the caster wheel in front, which will create a jamming effect, as opposed to climbing with the active wheels in front.

Mathematically, the push–pull region of the velocity can be determined using the dot product of the current velocity and the direction of neutral steering. Assuming that the robot has a velocity of magnitude greater than zero, the cosine of γ in Equation (17) will be positive when the velocity is in the push region, negative in the pull region, and zero in the specific case of neutral steering. The particular cases of pure push and pure pull will yield 1 and -1, respectively.

The name *neutral steering* was given to the two particular velocities perpendicular to pure push–pull. These directions are of special interest since they present the tightest possible continuous turn radius achievable without alternating regions. During such a turn, the instant center of rotation of the chassis will be located at the midpoint between the two actuated wheels. This will result in their velocities having the same magnitude but opposite directions:

$$\cos\gamma = \frac{([Rot(\theta_c)]\begin{bmatrix}1\\0\end{bmatrix})^T\begin{bmatrix}\dot{x}\\\dot{y}\end{bmatrix}}{c}, \tag{17}$$

where

$$[Rot(\theta)] = \begin{bmatrix}\cos\theta & -\sin\theta\\ \sin\theta & \cos\theta\end{bmatrix}. \tag{18}$$

The speed at which the chassis approaches the pure push state when moving continuously in the positive x direction can be identified through the relations defined by the inverse kinematics and the derivative of Equation (10). The wheel speeds $\dot{\alpha}_L$ and $\dot{\alpha}_R$, expressed in terms of $\dot{x}$ and $\dot{y}$, are substituted in the velocity of the chassis:

$$\dot{\theta}_c = \frac{R(\dot{\alpha}_R - \dot{\alpha}_L)}{d_2} = -\frac{\sin\theta_c}{d_1}\dot{x}. \tag{19}$$

Assuming a situation in which the robot is reversing in a straight line, a more general case can be obtained, expressing the desired velocity in a frame of reference aligned with the chassis:

$$\dot{\gamma} = -\frac{\sin\gamma}{d_1}c, \tag{20}$$

where γ is the angle between the pure push direction and the direction of the desired velocity, and c is the magnitude of the velocity of the robot. Note that both sides of the expression can be divided by c, yielding

$$\frac{1}{R_{turn}} = -\frac{\sin\gamma}{d_1}, \tag{21}$$

which details the instant turn radius of the robot during the reversal. This points to the trajectory being independent of the magnitude of the velocity.

If the robot is going in a straight line, $\dot{\gamma}$ would ideally be zero. Assuming d_1 is fixed and c is decided via the operator, the only way to influence $\dot{\gamma}$ is by changing the angle between the chassis frame of reference and the desired velocity. Taking advantage of this, the velocity vector may simply be rotated so that it is aligned with the chassis's frame of reference. This would stop the progress of the reversal. However, considering Equation (20), when the robot is in the push region with the same value of c, it is evident that the rate at which the spontaneous turn of the chassis accelerates is the same as that at which it slows down when approaching pure push. Using this knowledge, it becomes clear that there are velocities that can not only halt the progress of the turn but also reverse it. Therefore, it can be inferred that there is a value of γ that yields the opposite to the current value of $\dot{\gamma}$.

Considering only cases in the pull region, $\gamma \in (\frac{\pi}{2}, \frac{3\pi}{2})$, the direction that corrects the attitude of the chassis is $-\gamma$. Thus, to rectify the orientation of the chassis, the velocity is mirrored along the x-axis of the chassis's frame of reference.

Because the correction is prescribed through modifying the task space velocities, it leads to inaccuracies in the trajectory. Ways of mitigating the error between the corrected and planned trajectories are discussed in the following section.

The earlier the tendency to switch the chassis is corrected, the smaller the change in trajectory. If the trend to push is continuously corrected, it will not proliferate, provided that there are no significant unexpected forces involved. This was not the case in the simulation shown in Figure 7, in which γ was allowed to grow freely from the beginning of the motion to the point shown with the red arrow.

This approach can be extended to any trajectory by matching $\dot{\gamma}$ to the instant angular velocity of the chassis. This generalization requires the introduction of an additional variable, γ_{goal}, which is the angle of the velocity with respect to the chassis when the robot is moving in a circular trajectory. It can be found using the relation

$$\gamma_{goal} = \pi - \sin^{-1}(\frac{d_1\omega}{c}), \tag{22}$$

where ω is the angular velocity of the robot with respect to its instant center of rotation (positive is counter-clockwise). The corrected velocity vector is produced by mirroring the input velocity along an axis passing through the center of the robot with the angle γ_{goal} with respect to the input velocity.

$$\begin{bmatrix} \dot{x} \\ \dot{y} \end{bmatrix}_{corrected} = [Rot(\theta_c + \gamma_{goal})] \begin{bmatrix} 1 & 0 \\ 0 & -1 \end{bmatrix} [Rot(\theta_c + \gamma_{goal})]^T \begin{bmatrix} \dot{x} \\ \dot{y} \end{bmatrix}_{input} \tag{23}$$

The chosen approach, to mirror the direction of the velocity without changing its magnitude, returns the chassis to its target orientation at the same rate at which it left it, assuming that the change of region was purely kinematic. If the disturbance that initiated the reversal was dynamic in nature, the maximum torque that the motors can provide may prevent the robot from matching the growth in angular velocity. In that case, the robot will return to the desired configuration as rapidly as the motors allow, attempting to maintain the target speed.

To avoid returning to the pull region unnecessarily, a compensation regionis defined, beyond which the compensation algorithm is not applied. The compensation region is unlike the other regions in the sense that its bounds can be set arbitrarily and change during motion. For instance, if a loss of traction on the wheels or tipping over are concerns, the compensation region may change, depending on the velocity of the robot. An example of a possible compensation region can be seen in Figure 8.

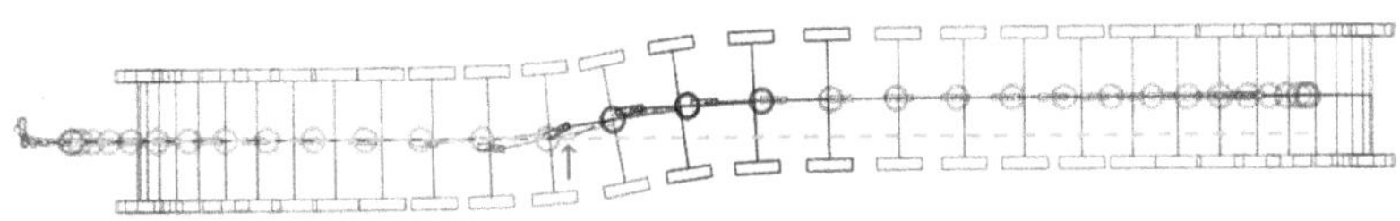

Figure 7. The corrected motion of the chassis when reversing. The red arrow shows the point in the trajectory (the dashed black line) where the correction starts being applied. The dashed green line shows the originally intended trajectory.

To better observe the compensation algorithm in action, the simulation shown in Figure 9 introduced unanticipated forces in the shape of blows to the fork of the caster wheel. In the following simulation, the impact was simulated as a 4000 N force acting on the fork of the caster wheel for 25 ms. The direction of the force was perpendicular to the chassis and roughly towards the center of rotation.

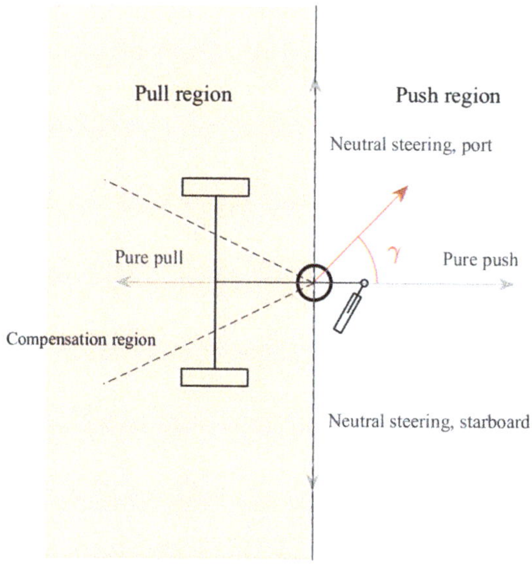

Figure 8. Regions and main directions of the robot. The compensation region is defined by the user and can even include part of the push region. The dark red arrow shows an example of a velocity in the push region.

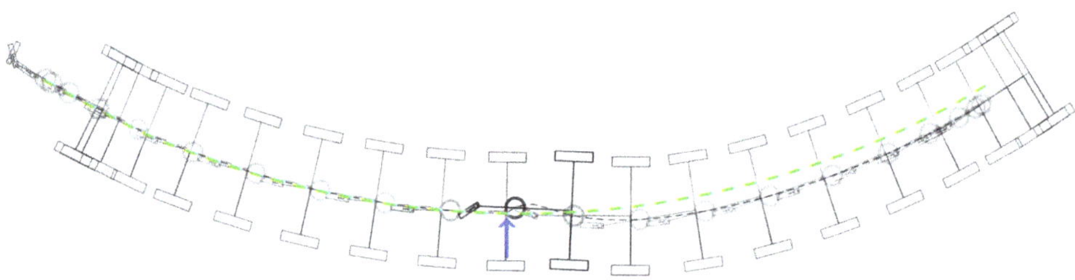

Figure 9. Dynamic simulation of the response of the robot to an impact on the fork of the caster wheel. The blue arrow indicates the position of the center of the robot when the force is applied and its direction. The green dashed line shows the expected trajectory had there been no impact.

As can be seen in Figure 9, the compensation only returns the chassis to the predetermined orientation. This is enough to prevent the spontaneous change of region, but getting the robot to go back to the path requires a recalculation of the trajectory to accommodate the merging back into the original path.

Looking further into Equation (23) and introducing a definition for γ_{err}, which is the difference of γ and γ_{goal}, reveals that the same effect may be achieved by rotating the input velocity vector two times, the opposite of γ_{err}. Note that, while the results of Equations (23) and (24) are the same, the transformations are not equivalent.

$$\begin{bmatrix} \dot{x} \\ \dot{y} \end{bmatrix}_{corrected} = [Rot(-2\gamma_{err})] \begin{bmatrix} \dot{x} \\ \dot{y} \end{bmatrix}_{input} \tag{24}$$

A more general expression can be written as shown in Equation (25), in which k determines the aggressiveness of the compensation, ranging from 0 (no compensation at all) through to 1 (only stopping the progress of the switch but not returning to the desired configuration) and then returning to the desired configuration faster with a higher value of

k. The compensation region should be smaller with a higher value of k to avoid ineffective compensation.

$$\begin{bmatrix} \dot{x} \\ \dot{y} \end{bmatrix}_{corrected} = [Rot(-k\gamma_{err})] \begin{bmatrix} \dot{x} \\ \dot{y} \end{bmatrix}_{input} \tag{25}$$

5. Driving Strategy

The insight gained from the kinematic analysis was used to design maneuvers so that the robot would be in the pull region for extended periods of time. In order to avoid a region change, two rules emerged from the previous sections:

1. The trajectory must be smooth, that is, it must be continuous as well as continuously differentiable. This extends to the initial position of the robot. The vector analogous to the pure pull direction must be tangent to the velocity at the starting point; otherwise, significant compensation will be necessary to remain in the pull region over the long term. If the first part of the path is curved, γ and γ_{goal} should be equal.
2. Turns with a radius smaller than $\mathbf{d_1}$ are possible but should be avoided. While momentarily turning tightly does not guarantee a change in velocity region, sustained rotation with such a small radius will eventually result in a change in the direction of the chassis.

In addition to disturbances due to the terrain, instability manifests inherently due to the fact that commands cannot be given to the motors in a perfectly continuous manner. Every slight change in the desired turn radius will result in a small lateral shift, even if γ_{goal} and γ were equal in the previous instant. The lower the refresh rate and the more sudden the change in turn radius, the more deviation to be expected. For this reason, the trajectory should be recalculated periodically to adapt to changes brought on via compensations during motion. The more important the accuracy is to the particular application, the more often trajectories should be amended to redress the lateral deviation caused by the compensation algorithm. These recalculations may concern only a small part of the path and lead the robot to rejoin the original route or entail the recomputation of the entire motion.

General trajectories with a changing curvature will always exhibit unstable behavior, even in smooth, continuous motion. One easy way of defining a viable trajectory involves concatenating a series of arcs that are tangent to each other at the points where they meet. In this case, the first arc would need to be tangent to the direction of the robot at its starting point.

An interesting characteristic of the motion is that the progress of the switch in chassis orientation is independent of the velocity, as evidenced by Equation (21). This means that, absent any dynamic effects, the trajectories that are consistent with the first two rules can be executed at any speed allowed by the actuators.

Another kinematically viable strategy is to always keep the instant center of rotation on the axis of the actuated wheels of the previous instant. This is also accomplished by building a trajectory of concatenating arcs, as shown in Figure 10. This time, there is a discontinuity in the velocity each time the robot passes from one arc to the next. Instead, what is kept track of is the orientation of the chassis, and the direction of the center and radius of the turn are calculated so that γ_{goal} and γ_{real} remain equal during the motion, absent any disturbances, see detail in Figure 11. This method solves the otherwise inherent need to compensate every time the radius of the turn of the trajectory changes at the cost of decreased maneuverability. Additionally, if the instant radius changes too suddenly, it will result in large on-the-spot wheel velocity variations. Such changes not only lead to spikes in the needed torque but may also cause the friction force between the tires and the ground to exceed the static friction threshold and slide. One final disadvantage of this method is that it only deals with the kinematic causes of the error of γ, meaning that other sources, such as external forces, remain.

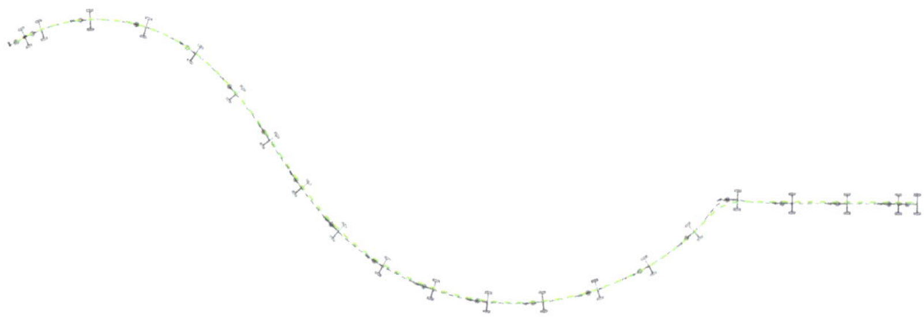

Figure 10. The trajectory determined by concatenating arcs and a straight line. The dashed green line is the pre-defined trajectory. The path begins with a 60° arc with a radius of 5 m; then, it applies a 90° arc with a radius of 8 m and finally a 5-m-long straight line. The angular velocity changes uniformly in the stretches between arcs. The shapes of the chassis and the wheels of the robot show the position and orientation of the robot at evenly spaced instants.

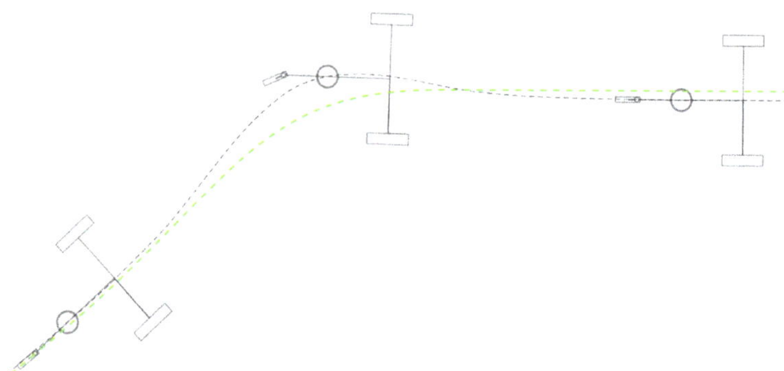

Figure 11. Detail of the previous figure showing how tighter turns are prone to causing an increase in position error due to the discrete nature of the velocity commands.

6. Dynamic Behavior

6.1. Lateral Stability

We denote lateral stability as the property of the robot that avoids the loss of contact on its wheels and rolling over, thus impeding its locomotion. For a study of the stability of mobile robots, see, for instance [20]. For low velocities, when dynamic effects can be neglected, a quasi-static analysis is used.

The presented implementation of the offset-differential robot has three points of contact with the ground, thanks to the caster wheel. Because its velocity in the selected applications (rough and uneven terrain) is low, this is the case we studied in this work. However, checking the stability under dynamic conditions would be an interesting extension of this work.

The tip-over stability margin was computed with a simple static analysis, considering the center of inertia, the angle of the contact plane formed by the three contact points, and the reaction forces at the wheels. In the case of the three-wheeled robot, we can create tip-over planes by joining the contact point of two wheels with the center of inertia of the robot, forming a tetrahedron; see Figure 12.

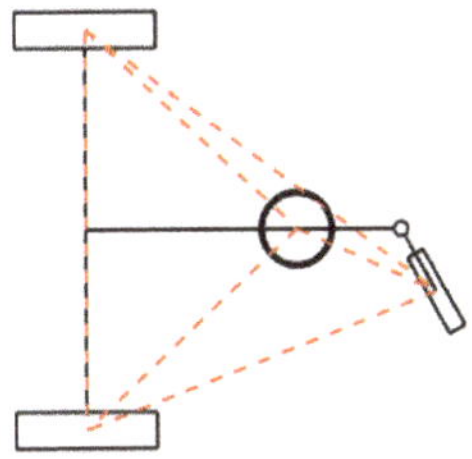

Figure 12. The stability tetrahedron is formed by the contact points of the wheels and the center of inertia of the robot.

In order to ensure static stability, the axis of the weight force, located at the center of inertia, must fall within the tetrahedron in order to determine whether the robot will roll over.

It is easy to see that, for the offset-differential robot, unless the load is perfectly centered at the vertical rotation axis, lateral stability must be checked during the motion planning of the robot, even in smooth, horizontal terrain. The angle of the caster wheel needs to be known in order to perform this check. As we will see later, being able to sense the heading angle of the caster wheel is important for the accurate control of this robot.

Greater stability can be accomplished for this robot using an additional structure that does not change the kinematic behavior. It consists of additional caster wheels connected to the robot via an additional mechanism that allows continuous contact with the surface and load distribution among the wheels. A schematic representation can be found in Figure 13. Joints A and B can be either cylindrical or Cardan joints, depending on the terrain's irregularity. The distance between joints A and B determines the normal reaction on the drive wheels.

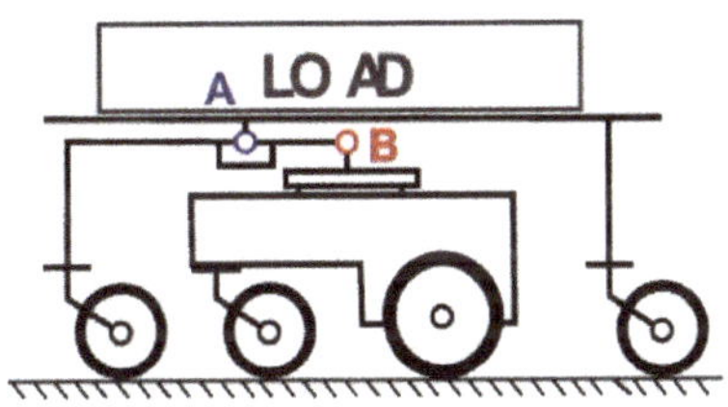

Figure 13. A solution to increase lateral stability without changing the kinematic behavior.

6.2. Dynamic Model

The dynamics of the offset-differential robot were modeled using the Lagrangian formulation. To do so, a few more variables were defined:

- The mass, m_c, and the moment of inertia, I_c, of the chassis (excluding the actuated wheels).
- The mass, m_W, and the moments of inertia, I_W, of the actuated wheels, which were assumed to be the same for both.
  - $I_{W,ax}$ is the moment of inertia of the wheel spinning through its axis.
  - $I_{W,rad}$ is the moment of inertia as it relates to the contribution of the wheel when the whole chassis is rotating. It was calculated as the moment of inertia of the wheel spinning about its radius plus the additional inertia brought about by the distance from the center of rotation of the chassis, which is taken into account through the parallel axis theorem.
- The mass, m_p, and the moment of inertia, I_p, of the platform.
- The mass, m_{cf}, and the moment of inertia, I_{cf}, of the caster fork.
- The mass, m_{cw}, and the moments of inertia, $I_{cw,ax}$ and $I_{cw,rad}$, of the caster wheel.

The distance between the centers of mass of the actuated wheels was considered the same as the distance between the centers of their tires, d_2.

The present case involves the robot moving on flat, even ground, which makes the potential energy term of the Lagrangian formulation null.

The mass matrix of the system, M, was calculated using the definition of the kinetic energy of the bodies, as in [11]. The Coriolis matrix, C, was calculated using the Christoffel symbols of the first kind method [21]. The five constraints applied to the robot were written as Pfaffian constraints [22], such that:

$$A(q)\dot{q} = 0 \tag{26}$$

Here, $A(q)$ is a 5-by-8 constraints Jacobian, and q is a vector containing the degrees of freedom of the system. q can be divided between active, $q_A = (\alpha_L, \alpha_R, \varphi)$, and passive, $q_P = (x, y, \theta_p, \beta, \alpha_c)$, variables. These vectors come together to form q, the 8-by-1 vector that contains all the variables in the order $q = (q_A, q_P)$. This leads to Equation (27):

$$\tau = M(q)\ddot{q} + C(q,\dot{q})\dot{q} + A(q)^T\lambda \tag{27}$$

where λ is a vector containing the Lagrange multipliers.

6.3. Forward Dynamics

Using Equation (27), the forward dynamics is easily solved.

$$\ddot{q} = M^{-1}(\tau_u + \tau_f + F_{ext} - C\dot{q} - A^T\lambda) \tag{28}$$

Note that in the previous expression, the general force term τ was replaced by a term containing the torque supplied by the three motors, τ_u, and a term outlining the dynamic contribution of friction on all the joints affected by it, τ_f. An additional force vector, F_{ext}, was also included in order to introduce external forces on any of the degrees of freedom. This includes torques on the actuated wheels, platform, and caster wheel assembly, as well as forces applied directly in the x and y directions on the center of the robot.

6.4. Inverse Dynamics

As shown in [11], the calculation of the Lagrange multipliers can be conveniently bypassed in the inverse dynamics. Considering the nature of τ_u,

$$\tau_u = \begin{bmatrix} \tau_{u,\alpha_L} \\ \tau_{u,\alpha_R} \\ \tau_{u,\varphi} \\ 0 \\ 0 \\ 0 \\ 0 \\ 0 \end{bmatrix} = \begin{bmatrix} \mathbf{I}_{3\times3} \\ \\ \mathbf{0}_{5\times3} \end{bmatrix} \begin{bmatrix} \tau_{u,\alpha_L} \\ \tau_{u,\alpha_R} \\ \tau_{u,\varphi} \end{bmatrix} = T_u u, \tag{29}$$

the dynamic equilibrium equation can be rewritten as shown in Equation (30).

In the previous equation, T_u is the input tensor that maps the control variables, u, to the coordinates of the system [23]. Note that the external forces in Equation (30) include only those that can be anticipated.

$$u = \begin{bmatrix} \mathbf{I}_{3\times3} & \mathbf{0}_{3\times5} \end{bmatrix} \begin{bmatrix} T_u & -A^T \end{bmatrix} (M\ddot{q} + C\dot{q} - \tau_f - F_{ext}). \tag{30}$$

In the simulations, the friction on the joints of the robot is approximated using a viscous friction model:

$$\tau_f = \begin{bmatrix} c_{\alpha_L} & 0 & 0 & 0 & 0 & 0 & 0 & 0 \\ 0 & c_{\alpha_R} & 0 & 0 & 0 & 0 & 0 & 0 \\ 0 & 0 & c_{\varphi} & 0 & 0 & 0 & 0 & 0 \\ 0 & 0 & 0 & 0 & 0 & 0 & 0 & 0 \\ 0 & 0 & 0 & 0 & 0 & 0 & 0 & 0 \\ 0 & 0 & 0 & 0 & 0 & 0 & 0 & 0 \\ 0 & 0 & 0 & 0 & 0 & 0 & c_{\beta} & 0 \\ 0 & 0 & 0 & 0 & 0 & 0 & 0 & c_{\alpha_c} \end{bmatrix} \dot{q}. \tag{31}$$

7. Tests

To verify the dynamic model, as well as to demonstrate the benefits of applying the compensation, two simulations and two experiments were conducted. Both simulated similar motions, comparing compensated and uncompensated versions of each. In the case of the simulations, an additional test was performed to quantify the impact of the caster wheel on the dynamics of the system.

7.1. Simulations

The dynamic model of the robot was used to understand the impact of the kinematic instability of the robot on the torque needed to comply with the velocity commands. In addition, the modeling of the caster wheel also allowed for the analysis of its own contribution to the system.

Two simulations were carried out to these ends using the mass properties presented in [10] for comparison. Both started with the chassis of the robot anti-parallel to x and an angle of $\beta = 0$. For the first simulation, the instability was allowed to progress freely. The second run applied the compensation algorithm since the beginning of the movement. The raw velocity commands were the same for both, constantly accelerating during the first quarter of the simulations, holding a velocity of $c = 1.2\ \text{m/s}$ and decelerating at the same rate during the final part. The configuration of the robot during the first test can be seen in Figure 6.

For clarity, the simulations were conducted without viscous friction at the joints. The simulated total mass of the robot was around 130 kg, including $m_{cf} = 0.5\ \text{kg}$ of the caster fork and $m_{cw} = 0.7\ \text{kg}$ of the caster wheel. The moment of inertia of the chassis was $I_c = 1.3\ \text{kgm}^2$. In addition, the moments of inertia of the actuated wheels and the components of the caster wheel were also considered for all directions in which they rotated.

In all simulations, the refreshing rate of the engines was 1000 Hz, as was the refreshing rate of the dynamics calculations. The time-integration scheme used to produce the trajectories was the forward Euler method.

Tests were carried out in MATLAB using an i7-10750H CPU and 16 GB of RAM. The chosen time step size was 1 ms.

7.2. Experiments

Two experiments were conducted with the real omnidirectional robot depicted in Figure 3. The robot was ordered to move forward 2 m and then back. By doing so, it was ensured that the return half of the itinerary began in the pull region. During the journey back to its starting point, the robot was "hit" with two impulses of alternating directions. These simulated impacts were short-velocity commands perpendicular to the motion of the robot.

In the first test, the robot was allowed to react naturally. In the second test, the compensation algorithm was implemented so that the robot could recover its original chassis orientation.

The distance between the center of the robot and the wheel axis in Figure 1 in the case of the real robot was $d_1 = 0.145$ m. Commands were given to the real robot once every 4 ms.

7.3. Uncompensated Pull-Region Motion

The robot was ordered to move forward 2 m at a velocity of $c = 0.05$ m/s. Then, the velocity was reversed. After 10 s, an order was given to add a perpendicular component of 0.5 m/s to the velocity during 0.4 s; 22 s into the reverse motion, the same order was given but towards the opposite side.

7.4. Compensated Pull-Region Motion

The previous experiment was repeated, this time with the compensation on. In addition, the duration of the "hits" was doubled to allow the perpendicular velocity to grow more than it had in the previous test.

8. Results

8.1. Simulations

During the switch shown in Figure 6, the robot had to quickly alter the velocity of its motors to maintain the velocity of the center. To do so, high torque needed to be supplied to the actuated wheels, as can be seen in the left graph of Figure 14. Also note that, in addition to the effect of the caster wheel, the torque profile was the same when accelerating and slowing down. This was due to the chassis rotating 180° through the second stage of motion.

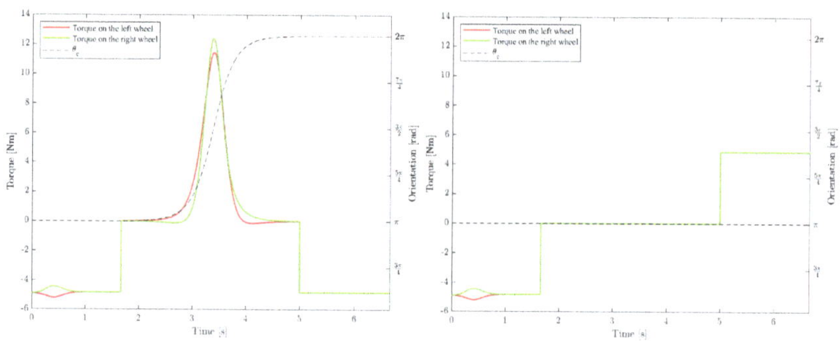

Figure 14. The torque required on the actuated wheels to perform the motion in Figure 6. (**Left**), uncompensated; (**right**), applying the compensation every instant.

When correcting the rotation of the chassis, the torques were those presented in the right graph of Figure 14. The three phases of motion were plainly visible: acceleration, holding $c = 1.2$ m/s, and coming to a stop.

In the first simulation, the highest torque required for any of the wheels was $\tau_{u,\alpha_R} = 12.38$ Nm, whereas the same value for the second simulation was $\tau_{u,\alpha_L} = 5.22$ Nm. In addition, the combined friction with the ground needed for all three wheels to execute the motion, shown in Figure 15, was much greater in the first run, reaching 225.1 N. Had there been no reversal, the maximum necessary static friction would have been 94.5 N.

This points to another issue as a consequence of instability: the fact that the required force between the ground and tires may exceed the maximum stiction.

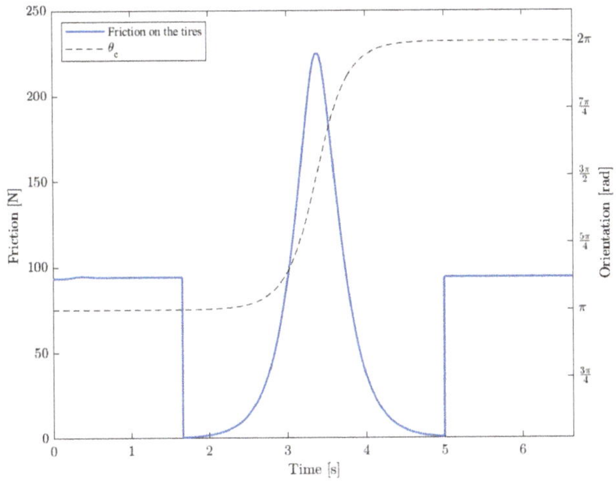

Figure 15. The magnitude of the friction with the ground needed for the robot to perform the motion in Figure 6.

The caster wheel presented instability in some ways analogous to the offset differential drive, the main difference being that it was not active. As such, what dictated its behavior was the movement of the point in the chassis connected to the caster fork with a revolute joint. This motion, through the constraints defined in Equations (6) and (7), informed the response of the constrained degrees of freedom β and α_c.

In the case of the previous simulations, the low velocity of the robot at the beginning of the motion meant that the torques that the motors needed to produce to make up for the increased resistance due to the position of the caster wheel were relatively low, as shown in Figure 16.

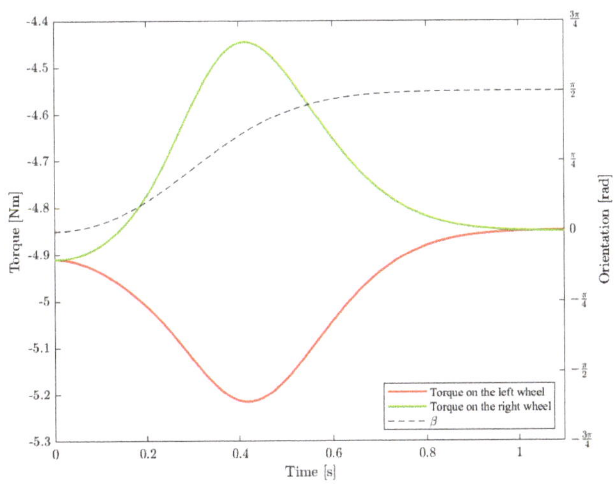

Figure 16. Detail showing the effect of the position of the caster wheel during the acceleration phase in the motion shown in Figure 6. The simulation was the same as the one shown in Figure 14 (right).

Nonetheless, there are cases in which the caster wheel could have a greater impact on the dynamics of the whole system. An example is the common case in which the robot is moving in the pure push direction and then is ordered to move in the pure pull direction. The caster wheel is then in an unstable equilibrium with a value of $\beta = -\frac{\pi}{2}$. Granted that there are no notable disturbances, the robot can reach significant speeds while the caster wheel remains in or close to the unstable equilibrium point.

Despite being far less massive (with a 1.2 kg combined mass), the instability of the caster wheel assembly may have an impact on the torque needed by the wheels comparable to that of the whole chassis. In the situation described in Figure 17, the caster fork stayed at its unstable equilibrium point during the acceleration phase and suddenly came out of it during the constant-velocity stage. In this simulation, the peak torque was $\tau_{u,\alpha_R} = 12.38$ Nm, which was supplied to the right wheel.

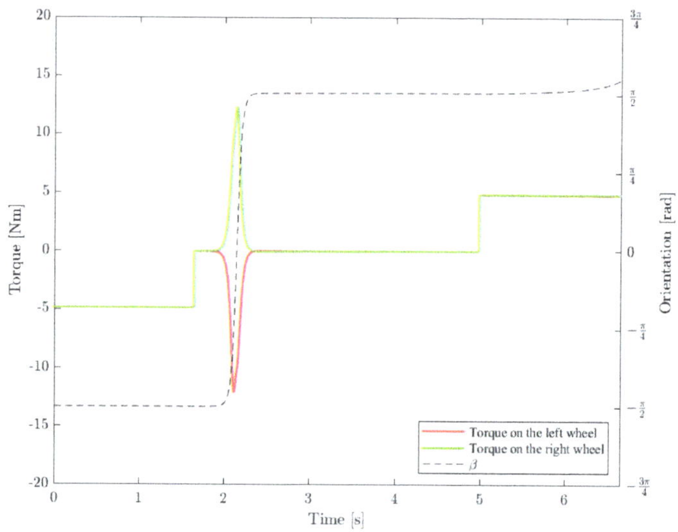

Figure 17. Dynamic effects of the caster wheel when it exited its unstable position during the second phase of compensated motion.

Another important consequence of the dynamics of the caster wheel is its potential to initiate a region switch. As long as the velocity commands are discrete, the response of the motors to the rotation of the caster wheel is always slightly off. If left uncorrected, the deviation induced by the caster wheel has the potential to greatly accelerate the progress of the region switch.

All three simulations lasted 6.6 s. When the time needed to load the dynamic model and initialize the variables was not taken into account, the computation times for the simulations were 1.68 s for the first simulation, 2.05 s for the second, and 2.30 s for the third. Each of these times included the inverse and forward dynamics calculations of the whole simulation.

It is worth noting that there is still room for improvement when it comes to optimizing the simulations so that the results can be produced faster, should the need arise.

8.2. Experiments

As expected, the first disturbance was more than enough to initiate the turn shown in the left graph of Figure 18.

In the second run, the velocities of which are displayed in the right graph of Figure 18, the perpendicular velocities made the magnitude of θ_c increase as the robot was shoved. However, as soon as the disturbance ceased, the chassis quickly recovered its original heading.

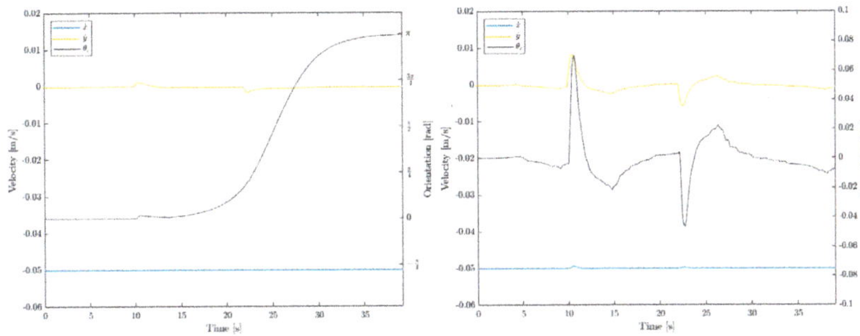

Figure 18. Velocities and chassis angle during the experiments: (**left**), the first experiment; (**right**), the second experiment (with compensation). Note the difference in scale between the plots.

9. Conclusions

In this work, the aim was to fully characterize the dynamics of the offset differential robot, especially its behavior in the unstable driving zone. While this region has been studied, and some of its advantages have been identified for non-omnidirectional, differential-drive robots, the use in the omnidirectional robots that present this feature has not been studied so far, to our knowledge.

With the results of the simulations, it became apparent that the reversal of the robot's region of velocity is a phenomenon of kinematic origin that has dynamic consequences. The simulations also showed that the orientation of the caster wheel can be problematic in some instances. For this reason, for relatively high-speed applications, the caster wheel should be instrumented, and its orientation should be taken into account when trajectories are planned.

There are clear advantages to the use of the unstable pull region in the operation of offset differential drive robots, from simple increases in speed in certain situations to reductions in the risk of getting stuck while navigating difficult terrain. In the context of agricultural applications, advancing with the actuated wheels in front makes soft and irregular ground easier to traverse, not only on account of their larger diameter but also because they use the terrain to propel themselves and to compensate for the moment that may make the robot stuck on some terrain.

As shown in the simulations, the cost of allowing the robot to choose its velocity region freely can have detrimental effects on its odometry and the accuracy of its movement. Often, this would mean having to choose not only between precision and velocity but also safety since a loss of traction in the wheels could endanger the robot and its environment.

As a result of the instability inherent to the design, constant corrections are necessary to remain in the pull region. These corrections can be successfully applied using the methods described in this paper. These methods can be generalized to a broad range of trajectories, encompassing most situations. However, trajectories involving fast, tight turns and great variations in turning speed are susceptible to a noticeable position error. In these circumstances, there is still a decision to be made between speed and accuracy, but the risks of losing traction or not having the adequate capacity to supply the necessary torque to the wheels are similar to those faced with the robot in its pure pull (or equivalent) direction. If the robot is left unmonitored, there is the possibility of a region switch being caused at an unexpected point, greatly increasing the chance of failing to perform as required.

Finally, the dynamic effect of the caster wheel was considered, opening the door to torque control for the offset differential drive robot structure. The inclusion of the caster wheel assembly in the dynamic model proved especially relevant to understanding the way it can contribute to the convergence into the stable equilibrium of the robot by accelerating it. For an accurate calculation of torques, the orientation of the caster wheel may need to be known, for instance via an encoder.

The dynamic model can also serve as the basis for more research into issues that offset-differential machines may run into, like a loss of traction, tipping over when turning, driving up slopes, and getting wheels stuck in agricultural settings.

Future Research

The current model is to be used to create more accurate control strategies for offset-differential robots. With this goal, the model will be further refined by incorporating the inertia of the rotary actuation and transmission systems. In existing market solutions, the centers of inertia for both the actuation–transmission part and the upper platform parts of the robot are not completely aligned with the vertical axis of rotation. This results in additional dynamic effects causing alterations in the robot's position during rapid movements. Studying these dynamic effects in detail is also important for implementing path planning and control strategies to mitigate their impact.

The dynamic effects of the orientation of the caster wheel on the computed torques make identifying this orientation necessary in order to obtain accurate torque commands. The intention is to equip the robot with a caster wheel featuring an encoder to constantly monitor its orientation in order to obtain better torque characterization and also a continuous computation of lateral stability.

The need for more precise and accurate mobile robots is also related to their applications in mobile manipulation. Exploration is underway for a joint solution involving a mobile platform and a robotic arm designed to function as a unified system with some shared axes. The goal is to have an omnidirectional manipulation system with the needed accuracy for interaction with the environment.

Author Contributions: Conceptualization, A.P.G. and C.D.-M.; methodology, A.P.G. and J.B.T.; software, J.B.T.; validation, C.D.-M. and J.B.T.; formal analysis, J.B.T.; writing—original draft, J.B.T.; writing—review and editing, A.P.G.; supervision, A.P.G. and C.D.-M. All authors have read and agreed to the published version of the manuscript.

Funding: This research was supported in part by award TED2021-131877B-I00 and the AgroMOBY project (PDR 2014-2022), award 56 30141 2021, from the Operació 01.02.01 de Transferència Tecnològica del Programa de desenvolupament rural de Catalunya 2014–2022. The contents are solely the authors' responsibility.

Data Availability Statement: Data are contained within the article.

Conflicts of Interest: The authors declare no conflicts of interest.

References

1. Prados Sesmero, C.; Buonocore, L.R.; Di Castro, M. Omnidirectional Robotic Platform for Surveillance of Particle Accelerator Environments with Limited Space Areas. *Appl. Sci.* **2021**, *11*, 6631. [CrossRef]
2. Qian, J.; Zi, B.; Wang, D.; Ma, Y.; Zhang, D. The Design and Development of an Omni-Directional Mobile Robot Oriented to an Intelligent Manufacturing System. *Sensors* **2017**, *17*, 2073. [CrossRef] [PubMed]
3. Jacobs, T. Omnidirectional Robot Undercarriages with Standard Wheels—A Survey. In Proceedings of the 25th IEEE International Conference on Mechatronics and Machine Vision in Practice (M2VIP), Stuttgart, Germany, 20–22 November 2018.
4. Taheri, H.; Zhao, C. Omnidirectional mobile robots, mechanisms and navigation approaches. *Mech. Mach. Theory* **2020**, *153*, 103958. [CrossRef]
5. Ramirez-Serrano, A.; Kuzyk, R. Modified Mecanum Wheels for Traversing Rough Terrains. In Proceedings of the 2010 Sixth International Conference on Autonomic and Autonomous Systems, Cancun, Mexico, 7–13 March 2010; pp. 97–103. [CrossRef]
6. Chung, W.; Moon, C.; Jung, C.; Jin, J. Design of the Dual Offset Active Caster Wheel for Holonomic Omni-directional Mobile Robots. *Int. J. Adv. Robot. Syst.* **2010**, *7*, 105–110.
7. Nasu, S.; Wada, M. Mechanical Design of an Active-caster Robotic Drive with Dual-Wheel and Differential Mechanism. In Proceedings of the Annual Conference of the IEEE Industrial Electronics Society—IECON2015, Yokohama, Japan, 9–12 November 2015.
8. You, Y.; Fan, Z.E.A. Design and Implementation of Mobile Manipulator System. In Proceedings of the International Conference on CYBER Technology in Automation, Control, and Intelligent Systems, Suzhou, China, 29 July–2 August 2019.
9. Jung, M.J.; Shim, H.-S.; Kim, H.S.; Kim, J.H. The miniature omni-directional mobile robot OmniKity-I (OK-I). In Proceedings of the IEEE International Conference on Robotics and Automation 1999 (ICRA'99), Detroit, MI, USA, 10–15 May 1999.

10. Giro Perez, P. Dynamic Modelling, Parameter Identification, and Motion Control of an Omnidirectional Tire-Wheeled Robot. Master's Thesis, Universitat Politecnica de Catalunya (UPC), Barcelona, Spain, 2021.
11. Giró, P.; Celaya, E.; Ros, L. The Otbot project: Dynamic modelling, parameter identification, and motion control of an omnidirectional tire-wheeled robot. *arXiv* **2023**, arXiv:RO/2311.10834.
12. Canuto Gil, J.; Domènech, C. Patent: Omnidirectional Platform and Omnidirectional Conveyor. U.S. Patent WO2019020861A2, 17 May 2018.
13. Yun, X.; Yamamoto, Y. Stability analysis of the internal dynamics of a wheeled mobile robot. *J. Robot. Syst.* **1997**, *14*, 697–709. [CrossRef]
14. Shojaei, K.; Shahri, A.M.; Tarakameh, A.; Tabibian, B. Adaptive trajectory tracking control of a differential drive wheeled mobile robot. *Robotica* **2010**, *1*, 11–23. [CrossRef]
15. Shojaei, K.; Shahri, A.M.; Tabibian, B. Design and Implementation of an Inverse Dynamics Controller for Uncertain Nonholonomic Robotic Systems. *J. Intell. Robot. Syst.* **2013**, *71*, 65–83. [CrossRef]
16. Mondal, K.; Wallace, B.; Rodriguez, A.A. Stability Versus Maneuverability of Non-holonomic Differential Drive Mobile Robot: Focus on Aggressive Position Control Applications. In Proceedings of the 2020 IEEE Conference on Control Technology and Applications (CCTA), Montreal, QC, Canada, 24–26 August 2020. [CrossRef]
17. Borkar, K.K.; Aljrees, T.; Pandey, S.K.; Kumar, A.; Singh, M.K.; Sinha, A.; Singh, K.U.; Sharma, V. Stability Analysis and Navigational Techniques of Wheeled Mobile Robot: A Review. *Processes* **2023**, *11*, 3302. [CrossRef]
18. Fue, K.G.; Porter, W.M.; Barnes, E.M.; Rains, G.C. An Extensive Review of Mobile Agricultural Robotics for Field Operations: Focus on Cotton Harvesting. *AgriEngineering* **2020**, *2*, 150–174. [CrossRef]
19. Siegwart, R.; Nourbakhsh, I. *Introduction to Autonomous Mobile Robots*; The MIT Press: Cambridge, MA, USA, 2011.
20. Kim, I.; Jeon, W.; Yang, H. Design of a transformable mobile robot for enhancing mobility. *Int. J. Adv. Robot. Syst.* **2017**, *14*, 1729881416687135.
21. Echeandia, S.; Wensing, P.M. Numerical Methods to Compute the Coriolis Matrix and Christoffel Symbols for Rigid-Body Systems. *J. Comput. Nonlinear Dyn.* **2021**, *16*, 091004. [CrossRef]
22. Murray, R.M.; Sastry, S.; Li, Z. *A Mathematical Introduction to Robotic Manipulation*; CRC Press: Boca Raton, FL, USA, 1994.
23. Lynch, K.; Bloch, A.; Drakunov, S.; Reyhanoglu, M.; Zenkov, D. *Control of Nonholonomic and Underactuated Systems*; CRC Press: Boca Raton, FL, USA, 2011; pp. 1–36. [CrossRef]

robotics

MDPI

Article

Motion Planning of Differentially Flat Planar Underactuated Robots

Michele Tonan, Matteo Bottin, Alberto Doria * and Giulio Rosati

Department of Industrial Engineering, University of Padova, 35131 Padova, Italy; michele.tonan@phd.unipd.it (M.T.)
* Correspondence: alberto.doria@unipd.it; Tel.: +39-049-827-6803

Abstract: Differential flat underactuated robots have fewer actuators than degrees of freedom (DOFs). This characteristic makes it possible to design light and cost-effective robots with great dexterity. The primary challenge associated with these robots lies in effectively controlling the passive joint, in particular, when collisions with obstacles in the workspace have to be avoided. Most of the previous research focused on point-to-point motions without any control on the actual robot trajectory. In this work, a new method is presented to plan trajectories that include one or more via points. In this way, the underactuated robot can avoid the obstacles in the workspace, similarly to traditional fully actuated robots. First, a trajectory planning strategy is analytically described; then, numerical results are presented. The numerical results show the effects of the via points and of the order of the polynomials adopted to define the motion laws. In addition, experimental tests performed on a two-DOF underactuated robot are presented, and their results validate the proposed method.

Keywords: robot; underactuated; trajectory planning; via point

Citation: Tonan, M.; Bottin, M.; Doria, A.; Rosati, G. Motion Planning of Differentially Flat Planar Underactuated Robots. *Robotics* **2024**, *13*, 57. https://doi.org/10.3390/robotics13040057

Academic Editor: Raffaele Di Gregorio

Received: 29 February 2024
Revised: 21 March 2024
Accepted: 22 March 2024
Published: 30 March 2024

1. Introduction

Most mechanical systems used in engineering applications have one controlled actuator for each degree of freedom (DOF) or have no actuator and are driven by external forces that cannot be controlled. The first class includes machine tools and robots, and the second class includes vibrating systems excited by base motion, unbalance, wind, and other physical phenomena. Underactuated mechanical systems are less common and have more DOFs than actuators [1,2]. In recent years, the interest in underactuated systems has increased in the field of robotics [3–5] and legged locomotion [6,7].

Underactuated robots are a promising solution for applications requiring an increase in dexterity and, at the same time, decreases in the cost, encumbrance, and weight of the robot [8,9]. In this class of robots, one or more joints of the kinematic chain are not driven by motors but are equipped with springs. From the mathematical point of view, underactuated robots are characterized by the presence of nonholonomic second-order constraints that represent the dynamics of the passive joints [10,11]. The main problem of underactuated robots is control, since the passive joints have to be indirectly controlled by the motors of the commanded joints.

A great deal of research has been carried out on the planning of point-to-point motions exploiting differential flatness properties [12–15]. In order to achieve differential flatness, the last links of the robot must have a particular mass distribution with the center of mass (COM) of some links lying on joint axes [12,16]. This property simplifies the last rows of the mass matrix and leads to the cancellation of some gravity, Coriolis, and centrifugal torques, and the nonholonomic constraint becomes integrable [10]. When a mechanical system—in particular, a robot—is differentially flat, a set of variables can be defined to express the state variables, which are called flat variables [8,12]. The number of flat variables is equal to the number of actuated DOFs. The relationship between state variables and flat variables is called diffeomorphism [17]. Actually, the diffeomorphism transforms a system of second-order differential equations into a reduced set of higher-order equations.

In robotic applications, often there are obstacles in the workspace. The possibility of avoiding an obstacle can be strongly increased by not only specifying the initial and final configurations but also introducing one or more via points that modify the trajectory of the end effector near the obstacle [18,19]. However, to the best of the authors' knowledge, the trajectory planning for differentially flat underactuated robots has never been performed including via points. In fact, many aspects of the control of differentially flat robots have been studied over the years, but only point-to-point motions have been considered [13,20], leaving a research gap to be filled. It is worth noticing that trajectory planning with via points for underactuated robots has been studied before (e.g., for mobile robots [21,22]), but not in the case of differentially flat robots. For this reason, this paper deals with motion planning of differentially flat underactuated robots with one or more via points.

The main contribution of the paper is the development and validation of a general trajectory planning algorithm in the space of flat variables. It is worth noting that for this class of robots, trajectory planning is not a purely kinematic problem, but it is influenced by dynamics, since the last joint is not directly driven.

The paper is organized as follows. In Section 2, the conditions of differential flatness are stated, and the equations of motion of the planar underactuated robot are presented and their special features are discussed. Section 3 discusses the main point of the research and deals with path planning considering one or more via points. In Section 4, numerical results are presented and discussed. An experimental validation of the method is presented in Section 5. Finally, conclusions and possible future developments are presented in Section 6.

2. Mathematical Model of the Underactuated Robot

In the framework of this research, the analyzed robots are differentially flat since they comply with the requirements of the theory of differential flatness [8] as follows:

- The last link center of mass (n) lies on the axis of the n-th joint;
- The overall center of mass of the links n and $n-1$ lies on the axis of the $(n-1)$-th joint.

Such requirements remain valid for the last j links (i.e., $n, n-1, \ldots, n-j+1$), where $j-1$ is the number of passive joints. As a result, the center of mass of the last j links lies on the $(n-j+1)$-th joint axis, and the final j links are considered "fully balanced".

To achieve controllability of the passive joints, a torsional spring is placed on the last $j-1$ joints.

This work focuses solely on a single passive joint; therefore, $j=2$. Therefore, only the joints n and $n-1$ are fully balanced.

The scheme of the robot is depicted in Figure 1, in which for each i-th link, q_i is the joint angle, a_i is the link length, a_{Gi} is the position of the center of mass with respect to the joint axis, and m_i and I_{Gi} are the mass and the moment of inertia about the center of mass, respectively. In the fully balanced links, a_{ci} is the distance of the balancing mass m_{ci} from joint i. At the passive joints, the torsional spring has stiffness k_i. A viscous damper with coefficient c_i is included to introduce dissipative phenomena in the model. Neither elastic nor dissipative phenomena are included in the actuated joints because it is assumed that the control system of the actuated joints compensates for such effects.

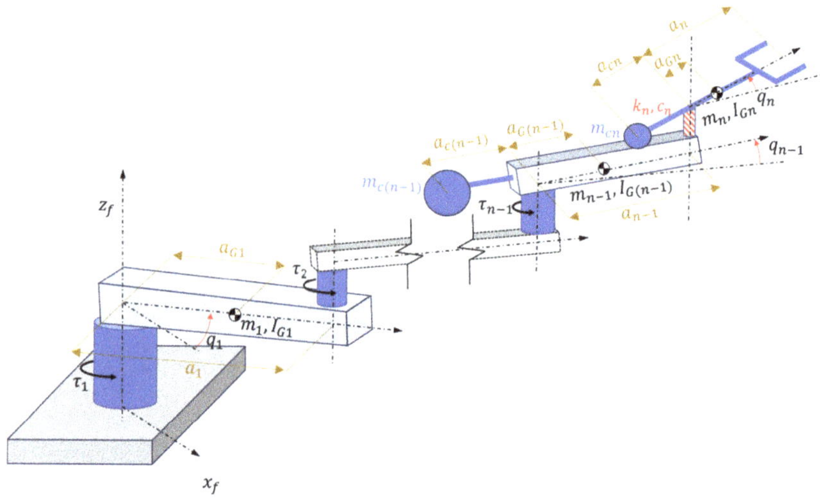

Figure 1. Scheme of a robot with n degrees of freedom (DOFs) equipped with one passive joint.

The equations of motion of the planar robot can be obtained using Lagrange's method. The system of equations of motion can be expressed in matrix form as follows [15]:

$$\boldsymbol{M}_n(\boldsymbol{q})\ddot{\boldsymbol{q}} + \boldsymbol{C}_n\dot{\boldsymbol{q}} + \boldsymbol{K}_n\boldsymbol{q} + \boldsymbol{b}(\boldsymbol{q},\dot{\boldsymbol{q}}) + \boldsymbol{g}(\boldsymbol{q}) = \boldsymbol{\tau}$$

$$\boldsymbol{M}_n(\boldsymbol{q})\begin{bmatrix}\ddot{q}_1\\ \ddot{q}_2\\ \vdots\\ \ddot{q}_{n-2}\\ \ddot{q}_{n-1}\\ \ddot{q}_n\end{bmatrix} + \boldsymbol{C}_n\begin{bmatrix}\dot{q}_1\\ \dot{q}_2\\ \vdots\\ \dot{q}_{n-2}\\ \dot{q}_{n-1}\\ \dot{q}_n\end{bmatrix} + \boldsymbol{K}_n\begin{bmatrix}q_1\\ q_2\\ \vdots\\ q_{n-2}\\ q_{n-1}\\ q_n\end{bmatrix} + \begin{bmatrix}b_1(\boldsymbol{q},\dot{\boldsymbol{q}})\\ b_2(\boldsymbol{q},\dot{\boldsymbol{q}})\\ \vdots\\ b_{n-2}(\boldsymbol{q},\dot{\boldsymbol{q}})\\ 0\\ 0\end{bmatrix} + \begin{bmatrix}g_1(\boldsymbol{q})\\ g_2(\boldsymbol{q})\\ \vdots\\ g_{n-2}(\boldsymbol{q})\\ 0\\ 0\end{bmatrix} = \begin{bmatrix}\tau_1\\ \tau_2\\ \vdots\\ \tau_{n-2}\\ \tau_{n-1}\\ 0\end{bmatrix} \quad (1)$$

where $\boldsymbol{M}_n(\boldsymbol{q})$ is the mass matrix; matrices $\boldsymbol{C}_n$ and $\boldsymbol{K}_n$ contain the joint damping and stiffness terms, respectively; vector $\boldsymbol{b}(\boldsymbol{q},\dot{\boldsymbol{q}})$ contains the Coriolis and centrifugal terms; vector $\boldsymbol{g}(\boldsymbol{q})$ contains the gravitational terms; and vector $\boldsymbol{\tau}$ contains the motor torques. The following three features of the system of Equation (1) are worth noting:

- There are neither gravitational nor Coriolis-centrifugal torques on the last joints because the last two links are fully balanced (the last two elements of vectors $\boldsymbol{b}(\boldsymbol{q},\dot{\boldsymbol{q}})$ and $\boldsymbol{g}(\boldsymbol{q})$ are null);
- The last element of $\boldsymbol{\tau}$ is null because there is no motor on the last joint (i.e., a passive joint);
- Matrices $\boldsymbol{C}_n$ and $\boldsymbol{K}_n$ are entirely null except the last bottom-right element, in which there are the torsional stiffness (k_n) and damping coefficient (c_n) of the passive joint; this happens because neither elastic nor dissipative phenomena are included in the actuated joints.

Matrix $\boldsymbol{M}_n(\boldsymbol{q})$ is configuration-dependent. However, since the robot satisfies the conditions of differential flatness, the last $[n-1, n]$ rows and columns of the matrix are constant:

$$\boldsymbol{M}_n(\boldsymbol{q}) = \begin{bmatrix} & & & I^*_{n-1} & I^*_n\\ & & & I^*_{n-1} & I^*_n\\ & \boldsymbol{I}^*_{(n-2)\times(n-2)}(\boldsymbol{q}) & & \vdots & \vdots\\ & & & I^*_{n-1} & I^*_n\\ I^*_{n-1} & \cdots & I^*_{n-1} & I^*_{n-1} & I^*_n\\ I^*_n & \cdots & I^*_n & I^*_n & I^*_n \end{bmatrix} \quad (2)$$

Addtionally, only the submatrix $\boldsymbol{I}^*_{(n-2)\times(n-2)}(\boldsymbol{q})$ depends on the robot configuration. I^*_{n-1} and I^*_n are the moments of inertia reduced to joints $n-1$ and n, respectively.

The theory of differential flatness is then used to define a reduced set of variables $\boldsymbol{y}$ (called flat variables), equal to the number of actuators [23]. For a planar robot equipped with only one passive joint, the flat variables are defined as follows [15]:

$$\begin{aligned} y_1 &= \sum_{i=1}^{n} q_i \\ y_i &= q_{i-1} \quad \text{with} \quad i = 2, 3, \ldots, n-1 \end{aligned} \tag{3}$$

The flat variables are used to manipulate Equation (1), performing a diffeomorphism that maps the joint values into the flat variables, i.e., $\boldsymbol{q} \in \mathbb{R}^n \mapsto \boldsymbol{y} \in \mathbb{R}^{n-1}$.

The last row of Equation (1) becomes

$$I^*_n(\ddot{q}_1 + \ddot{q}_2 + \cdots + \ddot{q}_{n-1} + \ddot{q}_n) + c_n\dot{q}_n + k_n q_n = 0 \tag{4}$$

which can be rewritten through the flat variable:

$$I^*_n\ddot{y}_1 + c_n\dot{q}_n + k_n q_n = 0 \tag{5}$$

This equation is linear since the inertia term I^*_n is constant. As a result, the Laplace transform can be calculated [15]:

$$I^*_n s Y_1(s) + (c_n s + k_n)Q_n(s) = 0 \tag{6}$$

From this equation, the last joint rotation can be related to the flat variable:

$$Q_n(s) = -\frac{I^*_n}{c_n s + k_n} s^2 Y_1(s) \tag{7}$$

Collecting the torsion spring stiffness k_n, the right term of the equation can be rearranged as

$$Q_n(s) = -\frac{I^*_n}{k_n}\left(1 + \frac{c_n}{k_n}s\right)^{-1} s^2 Y_1(s) \tag{8}$$

If low-friction bearings are used, the torsion spring stiffness is much larger than the damping coefficient c_n. Hence, using the first-order Taylor expansion, the term $(1 + c_n/k_n\, s)^{-1}$ can be simplified, yielding the following:

$$Q_n(s) = -\frac{I^*_n}{k_n}\left(1 - \frac{c_n}{k_n}s\right)s^2 Y_1(s) = -\frac{I^*_n}{k_n}s^2 Y_1(s) + \frac{I^*_n c_n}{k_n^2}s^3 Y_1(s) \tag{9}$$

Finally, the passive joint position can be expressed as a function of time using the inverse Laplace transform, in which the initial conditions are null:

$$q_n(t) = -\frac{I^*_n}{k_n}\ddot{y}_1(t) + \frac{I^*_n c_n}{k_n^2} y_1^{(3)}(t) \tag{10}$$

It is worth noting that Equation (10) is more complex than the one usually found in the literature. Actually, many papers (e.g., [24,25]) simplify the system by neglecting damping (thus, $c_n = 0$) so that q_n is related only to the second derivative of y_1 rather than to a combination of $\ddot{y}_1$ and $y_1^{(3)}$. For this reason, the following sections will deal with trajectory planning for robots with both damped and undamped passive joint.

The dynamic model of Equation (1) contains joint accelerations. Hence, Equation (10) must be derived twice to obtain the passive joint acceleration, which depends on the flat variable:

$$\ddot{q}_n(t) = -\frac{I^*_n}{k_n} y_1^{(4)}(t) + \frac{I^*_n c_n}{k_n^2} y_1^{(5)}(t) \tag{11}$$

Joint rotations $q_i(t)$ should be twice continuously differentiable (C^2) to avoid vibratory phenomena. As a result, Equation (11) yields that $y_1(t)$ must be continuously differentiable five times (C^5). The other flat variables ($y_2, \ldots, y_{n-1}$) are equal to the actuated joint values; hence, to avoid vibratory phenomena, these flat variables must be twice continuously differentiable (C^2).

3. Trajectory Planning

The typical trajectory planning of differentially flat underactuated robots considers only point-to-point movements and is performed in the space of flat variables. These movements provide fixed initial and final conditions for joint positions and their derivatives. The conditions for joint positions are the joint angles required to complete the task, apart from the passive joints, whose angles are set to zero to ensure the static equilibrium of the torsional spring. The conditions for the joint variable derivatives are null values at the beginning and at the end of the motion.

In trajectory planning with via points, the trajectories of fully actuated robots are usually planned in the joint space so that the via points are reached by the robot in specific configurations (i.e., joint variables), while the derivatives of joint variables are typically different from zero. Even if joint velocities and accelerations can be imposed, the continuity of joint variable derivatives is more important than the specific values. As a result, the conditions in the via points are limited to the joint positions and the continuity of joint variable derivatives. (please note that trajectory planning can also be performed in the Cartesian space by choosing proper Cartesian via points; however, Cartesian trajectory planning can be converted into joint space via robot inverse kinematics).

Considering an n-DOF robot equipped with only one passive joint, trajectory planning with via points is again performed in the space of flat variables. Since there are $n-1$ flat variables, the definition of the values of flat variables at the via point does not completely define robot configuration. This problem is solved by including at the via point a further condition that comes from the last equation of motion (Equation (10)) and links the joint variable of the passive joint with the derivatives of the first flat variable.

Trajectory planning can be performed through very different motion laws [26]. In this paper, polynomial laws are used; such laws ensure the continuity of the derivatives and the polynomial degree can be changed according to the conditions.

The planning of the i-th flat variable (with $i > 1$) can be performed via any function that can satisfy the condition on the double continuity (C^2). The damping of the passive joint has no effect since the flat variable depends only on the joint variable of the actuated joint (Equation (3)). Hence, this paper focuses on the first flat variable $y_1(t)$.

The trajectory of $y_1(t)$ is planned using polynomial functions between each pair of points. Hereby, the j-th polynomial of the trajectory is named $y_{1j}(\lambda)$, with $\lambda = (t - t_{j,0})/T_j \in [0,1]$, $t_{j,0}$ is the initial time of the polynomial, and T_j is the polynomial duration. The final polynomial is named $y_{1f}(\lambda)$. The polynomials of degree p are defined in the following way:

$$y_{1j}(\tau) = \sum_{k=0}^{p} a_{jk}\lambda^k \tag{12}$$

The i-th derivative of the j-th polynomial with respect to real time t can be calculated as follows:

$$y_{1j}^{(i)}(\lambda) = \frac{d^i y_{1j}(\lambda)}{dt^i} = \frac{d\lambda^i}{dt^i}\frac{d^i y_{1j}(\lambda)}{d\lambda^i} = \frac{1}{(t_j - t_{j-1})^i}\frac{d^i y_{1j}(\lambda)}{d\lambda^i} = \frac{1}{T_j^i}\frac{d^i y_{1j}(\lambda)}{d\lambda^i} \tag{13}$$

3.1. Undamped Robot ($c_n = 0$)

If damping in the passive joint of the underactuated robot can be neglected, Equation (11) shows that y_1 must be continuously differentiable four times (C^4). If there are n_{via} via points, $n_{via} + 1$ polynomials are needed.

The conditions at the initial point, the generic via point, and the final point are as follows:

- Initial point: Initial values of flat variable y_1 are based on the initial values of joint variables, and flat variable derivatives are set to zero:

$$\begin{aligned} y_{11}(0) &= \sum_{i=1}^{n} q_{i,0} \\ \dot{y}_{11}(0) = \ddot{y}_{11}(0) &= y_{11}^{(3)}(0) = y_{11}^{(4)}(0) = 0 \end{aligned} \tag{14}$$

- Via point: The value of flat variable y_1 is based on joint values at the via point, with continuity of the flat variable derivatives and Equation (10) with $c_n = 0$:

$$\begin{gathered} y_{1j}(1) = \sum_{i=1}^{n} q_{i,via} \\ y_{1j}(1) = y_{1,j+1}(0)\,,\ \dot{y}_{1j}(1) = \dot{y}_{1,j+1}(0)\,,\ \ddot{y}_{1j}(1) = \ddot{y}_{1,j+1}(0) \\ y_{1j}^{(3)}(1) = y_{1,j+1}^{(3)}(0)\,,\ y_{1j}^{(4)}(1) = y_{1,j+1}^{(4)}(0) \end{gathered} \tag{15}$$

$$\ddot{y}_{1j}(1) = -\frac{q_{n,via}(t)k_n}{I_n^*} \tag{16}$$

- Final point: The final joint values of flat variable y_1 are based on the final values of joint variables, and flat variable derivatives are set to zero:

$$\begin{aligned} y_{1f}(1) &= \sum_{i=1}^{n} q_{i,f} \\ \dot{y}_{1f}(1) = \ddot{y}_{1f}(1) &= y_{1f}^{(3)}(1) = y_{1f}^{(4)}(1) = 0 \end{aligned} \tag{17}$$

The conditions of Equations (15) and (16) are repeated for each via point that the robot has to reach, while the conditions of Equations (14) and (17) are applied to the first and last polynomial, respectively. If n_{via} via points are used, the total number of conditions n_{cond} on the trajectory is

$$n_{cond} = 10 + 7n_{via} \tag{18}$$

Such conditions can be enforced by using polynomials of specific degrees. In particular, since, for a polynomial of p-th degree, there are $p+1$ coefficients, the degrees of the polynomials can be calculated as follows:

$$(p_1 + 1) + (p_2 + 1) + (n_{via} - 1)(p_{via} + 1) = n_{cond} \tag{19}$$

where p_1 and p_2 are the degrees of the first and last polynomials, respectively, and p_{via} is the degree of the polynomials connecting via points (if needed). For a single via point, it yields

$$p_1 + p_2 = 15 \tag{20}$$

As a result, for a single via point the degree of the two polynomials cannot be the same. In this work, two polynomials of 8-th degree and 7-th degree are chosen. Such polynomial differentiability class is greater than C^4; thus, the conditions are met.

If multiple via points are employed ($n_{via} \geq 2$), $p_1 = 8$, and $p_2 = 7$, Equations (18) and (19) yield

$$p_{via} = \frac{7n_{via} - 7}{n_{via} - 1} - 1 = 6 \tag{21}$$

It is worth noticing that Equation (19) allows for multiple polynomial combinations. However, the proposed solution with $p_1 = 8$, $p_2 = 7$, and $p_{via} = 6$ ensures constancy in the degree of the polynomials even with many via points.

3.2. Damped Robot ($c_n \neq 0$)

If the passive joint presents some damping ($c_n \neq 0$), the first flat variable must be continuously derivable five times (C^5). Also, in this case, $n_{via} + 1$ polynomials are needed.

The conditions are similar to the one stated in Section 3.1, with some notable exceptions:

- Initial point: Initial values of flat variable y_1 are based on the initial values of joint variables, and flat variable derivatives are set to zero up to the fifth derivative:

$$\begin{gathered} y_{11}(0) = \sum_{i=1}^{n} q_{i,0} \\ \dot{y}_{11}(0) = \ddot{y}_{11}(0) = y_{11}^{(3)}(0) = y_{11}^{(4)}(0) = y_{11}^{(5)}(0) = 0 \end{gathered} \tag{22}$$

- Via point: The value of flat variable y_1 is based on joint values at the via point, with continuity of the flat variable derivatives and Equation (10):

$$\begin{gathered} y_{1j}(1) = \sum_{i=1}^{n} q_{i,via} \\ y_{1j}(1) = y_{1,j+1}(0)\,,\ \dot{y}_{1j}(1) = \dot{y}_{1,j+1}(0)\,,\ \ddot{y}_{1j}(1) = \ddot{y}_{1,j+1}(0) \\ y_{1j}^{(3)}(1) = y_{1,j+1}^{(3)}(0)\,,\ y_{1j}^{(4)}(1) = y_{1,j+1}^{(4)}(0)\,,\ y_{1j}^{(5)}(1) = y_{1,j+1}^{(5)}(0) \end{gathered} \tag{23}$$

$$q_{n,via} = -\frac{I_n^*}{k_n}\frac{\ddot{y}_{1j}(1)}{T_j^2} + \frac{I_n^* c_n}{k_n^2}\frac{y_{1j}^{(3)}(1)}{T_j^3} \tag{24}$$

- Final point: The final joint values of flat variable y_1 are based on the final values of joint variables, and flat variable derivatives are set to zero up to the fifth derivative:

$$\begin{gathered} y_{1f}(1) = \sum_{i=1}^{n} q_{i,f} \\ \dot{y}_{1f}(1) = \ddot{y}_{1f}(1) = y_{1f}^{(3)}(1) = y_{1f}^{(4)}(1) = y_{1f}^{(5)}(1) = 0 \end{gathered} \tag{25}$$

Following Section 3.1, the total number of conditions n_{cond} for a damped system passing through n_{via} via points is

$$n_{cond} = 12 + 8n_{via} \tag{26}$$

since, in the case of multiple via point, the conditions of Equations (23) and (24) are repeated for each via point that the robot must reach, while the conditions of Equations (22) and (25) are applied to the first and last polynomial, respectively.

Equation (19) holds true for a damped system as well. Thus, for a single via point, it yields

$$p_1 + p_2 = 18 \tag{27}$$

In this work, two polynomials of 9-th degree are chosen. The differentiability class of these polynomials is larger than C^5; thus, the conditions are met.

If multiple via points are employed ($n_{via} \geq 2$) and $p_1 = p_2 = 9$, Equations (26) and (19) yield

$$p_{via} = \frac{8n_{via} - 8}{n_{via} - 1} - 1 = 7 \tag{28}$$

The condition stated in Equation (24) is difficult to manipulate for a general function since it is a differential equation. However, since y_{1j} is a polynomial function, the condition can be manipulated by introducing the polynomial derivatives. For the first via point, ($j = 1$) y_{11} is a 9-th degree polynomial function; thus, its second and third derivatives are

$$\frac{d^2 y_{11}(\lambda)}{d\lambda^2} = \frac{1}{T_1^2}\left(72a_{19}\lambda^7 + 56a_{18}\lambda^6 + 42a_{17}\lambda^5 + 30a_{16}\lambda^4 + 20a_{15}\lambda^3 + 12a_{14}\lambda^2 + 6a_{13}\lambda + 2a_{12}\right) \tag{29}$$

$$\frac{d^3y_{11}(\lambda)}{d\lambda^3} = \frac{1}{T_1^3}\left(504a_{19}\lambda^6 + 336a_{18}\lambda^5 + 210a_{17}\lambda^4 + 120a_{16}\lambda^3 + 60a_{15}\lambda^2 + 24a_{14}\lambda + 6a_{13}\right) \tag{30}$$

which, combined with the conditions of Equation (22) in Equation (24), yield

$$\begin{aligned} &a_{19}\left(504\frac{c_n}{T_1} - 72k_n\right) + a_{18}\left(336\frac{c_n}{T_1} - 56k_n\right) + \\ &+ a_{17}\left(210\frac{c_n}{T_1} - 42k_n\right) + a_{16}\left(120\frac{c_n}{T_1} - 30k_n\right) = \frac{q_{n,via}k_n^2T_1^2}{I_n^*} \end{aligned} \tag{31}$$

For the following via points, y_{1j} are 7-th degree polynomials; thus, the derivatives are

$$\frac{d^2y_{1j}(\lambda)}{d\lambda^2} = \frac{1}{T_j^2}\left(42a_{j7}\lambda^5 + 30a_{j6}\lambda^4 + 20a_{j5}\lambda^3 + 12a_{j4}\lambda^2 + 6a_{j3}\lambda + 2a_{j2}\right) \tag{32}$$

$$\frac{d^3y_{1j}(\lambda)}{d\lambda^3} = \frac{1}{T_j^3}\left(210a_{j7}\lambda^4 + 120a_{j6}\lambda^3 + 60a_{j5}\lambda^2 + 24a_{j4}\lambda + 6a_{j3}\right) \tag{33}$$

which, introduced in Equation (24), yield

$$\begin{aligned} &a_{j7}\left(210\frac{c_n}{T_j} - 42k_n\right) + a_{j6}\left(120\frac{c_n}{T_j} - 30k_n\right) + a_{j5}\left(60\frac{c_n}{T_j} - 20k_n\right) + \\ &+ a_{j4}\left(24\frac{c_n}{T_j} - 12k_n\right) + a_{j3}\left(6\frac{c_n}{T_j} - 6k_n\right) + a_{j2}(30k_n) = \frac{q_{n,via}k_n^2T_j^2}{I_n^*} \end{aligned} \tag{34}$$

Equations (31) and (34), although limited to the polynomial trajectories, can be easily implemented in a linear system to obtain the polynomial coefficients.

4. Numerical Results

To test the proposed approach, a two-DOF robot was simulated in Matlab. The robot lies on the horizontal plane; thus, $\boldsymbol{g}(\boldsymbol{q}) = \boldsymbol{0}$. Link 2 is symmetric and complies with the requirements of the differential flatness theory.

The geometrical and inertial parameters of the robot are summarized in Table 1.

Table 1. Geometrical and inertial parameters of the 2-DOF robot.

Parameters	Units	Link 1	Link 2
m_i	kg	3.0×10^{-2}	1.2×10^{-2}
m_{ci}	kg	0	1.2×10^{-2}
I_{Gi}	$kg \cdot m^2$	5.8×10^{-5}	4.9×10^{-5}
a_i	m	1.3×10^{-1}	8.5×10^{-2}
a_{Gi}	m	7.1×10^{-2}	0
a_{ci}	m	0	8.5×10^{-2}

Simulations aim to show that the introduction of proper via points allows the robot to avoid obstacles in the workspace. The simulations comprise eight tests.

The first six tests are summarized in Table 2. They differ in the number of obstacles and in the damping value ($c_n = 0$ or $c_n \neq 0$). The number of via points is set equal to the number of obstacles. The damping coefficient c_n can take different values in the trajectory planning and in the dynamic model:

- $c_n = 0$ both for planning and in the model. This is the case of an ideal robot;

- $c_n = 0$ for planning and $c_n \neq 0$ in the model. This test shows the effect of neglecting the damping during trajectory planning, whereas the actual robot has relevant damping phenomena in the passive joint;
- $c_n \neq 0$ both for planning and in the model. This test is the most realistic case. Because of the presence of damping, the polynomial orders increase [15].

Table 2. Numerical tests.

Test	n_{via}	c_n in Planning	c_n in Model	Order of Polynomials
1	1	0	0	8-7
2	1	0	$\neq 0$	8-7
3	1	$\neq 0$	$\neq 0$	9-9
4	2	0	0	8-6-7
5	2	0	$\neq 0$	8-6-7
6	2	$\neq 0$	$\neq 0$	9-7-9

The motion parameters are listed in Table 3, considering that the initial joint values are zero for both joints. It is worth noting that for an underactuated system, the trajectory depends on robot dynamics, which is influenced by the total motion duration (t_f) and the real time when the robot passes through the via point (t_{via}). At the moment, this is a limitation of the proposed trajectory planning and will be studied in future works.

Table 3. Numerical tests motion parameters. Each couple of $\boldsymbol{q}_{via}$ and each value of t_{via} are related to one via point.

Test	$\boldsymbol{q}_{via}$ (°)	$\boldsymbol{q}_f$ (°)	t_{via} (s)	t_f (s)
1, 2, 3	$[95, -87]$	$[180, 0]$	0.17	0.64
4, 5, 6	$[104, -92]$, $[80, 88]$	$[180, 0]$	0.27, 0.52	0.7

Figures 2–4 show the simulated trajectory with one obstacle. The obstacle is a cylinder of radius 35 mm; the center of the base of the cylinder is placed at the point with coordinates $[75, 182]$ mm in the base reference system. For completeness, in Figures 2 and 4, the trajectories without via points and with the same final configuration $\boldsymbol{q}_f$ and t_f are shown. It is clear how the robot, without a via point, is not able to avoid the collision with the obstacle. The Cartesian trajectories of the end effector (Figure 2) are similar, although Test 3 (Figure 2c) shows an entangled path in the middle of the trajectory. The flat variables (Figure 3) are very similar in magnitude; this result is expected since the Cartesian trajectories are similar and the Cartesian trajectory is the result of the joint trajectories, which, in turn, are the result of the flat variables. For the flat variable derivatives, Figure 3 shows the continuity of the derivatives up to the fourth order. Moreover, the regularity of the fourth derivative of the flat variable in Test 3 confirms the continuity of the fifth derivative. The same behavior cannot be found for Tests 1 and 2.

Tests 1 and 3 show no oscillations of the end effector around the final point of the trajectory (Figure 4a,c). This is an expected result since the model used for trajectory planning is the same one adopted for simulation. Conversely, the dynamic model of Test 2 is different from the one assumed in the planning phase. The result is that the robot mathematically does not exactly pass on the via point since all conditions of Section 3.2 are not met. This phenomenon is not highlighted in Figure 2b since the difference is very small; however, this difference may increase for different values of the via points $\boldsymbol{q}_{via}$ and t_{via}. Finally, a wide natural oscillation of q_2 can be found after the end of the trajectory (Figure 4b).

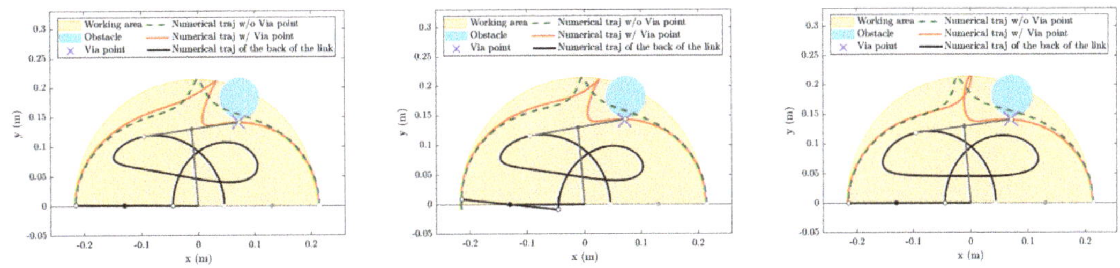

(**a**) Test 1 (**b**) Test 2 (**c**) Test 3

Figure 2. Simulated trajectories. with one via point. The white points are the two extremities of the underactuated link, and the filled points represent the passive joint.

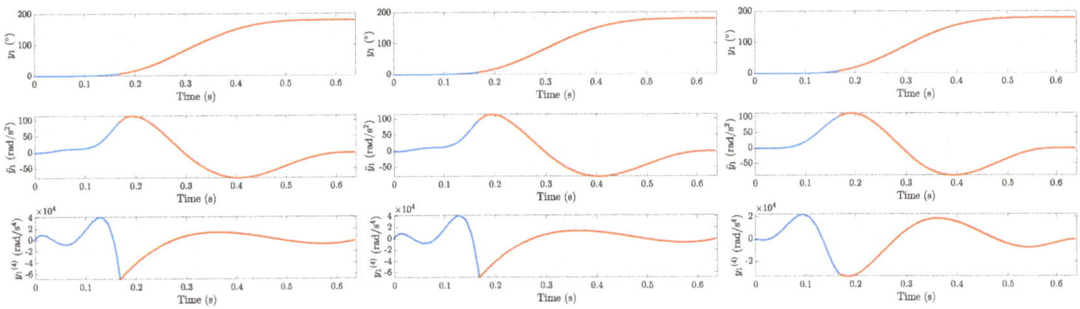

(**a**) Test 1 (**b**) Test 2 (**c**) Test 3

Figure 3. Simulated flat variables with one via point. The different colors represent the different polynomial functions joining the subsequent pairs of points.

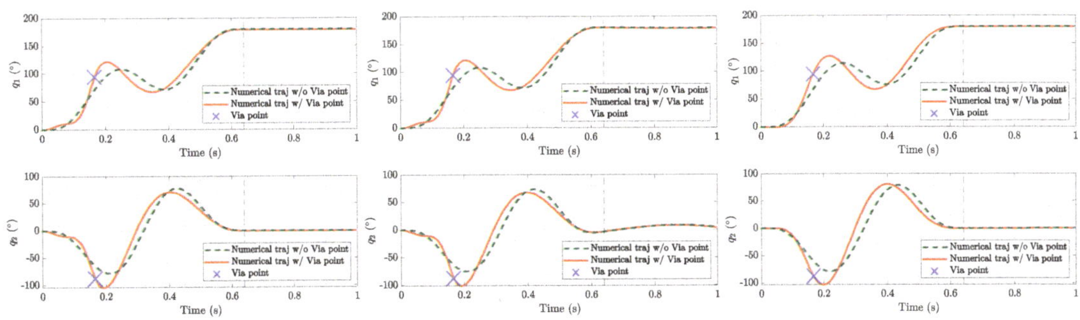

(**a**) Test 1 (**b**) Test 2 (**c**) Test 3

Figure 4. Simulated joint values with one via point.

The results of the simulations with two obstacles in the workspace are reported in Figures 5–7. The obstacles are the cylinder of Tests 1, 2, and 3 (in the same position) and a cube of 50 mm per side. The centroid of the cube is placed at the point with coordinates $[-80, 178]$ mm in the base reference system, with the sides parallel to the base reference system axes. In this case, two via points are considered. Test 5 shows that when there is a relevant damping value, but trajectory planning is carried out neglecting damping, oscillations appear at the end of the motion, and the via points positions are not reached.

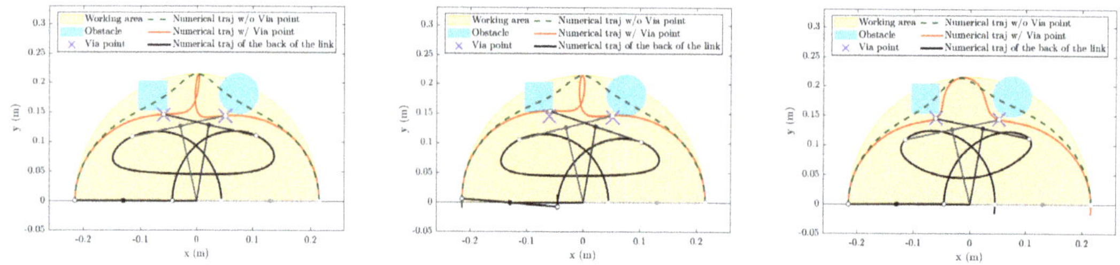

(**a**) Test 4 (**b**) Test 5 (**c**) Test 6

Figure 5. Simulated trajectories with two via points. The white points are the two extremities of the underactuated link, and the filled points represent the passive joint.

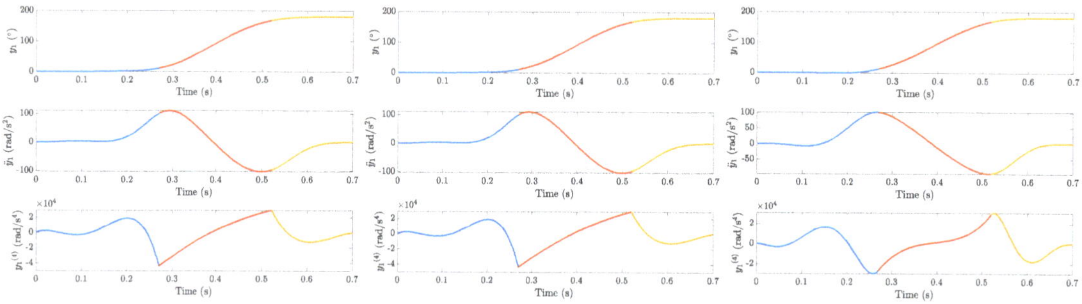

(**a**) Test 4 (**b**) Test 5 (**c**) Test 6

Figure 6. Simulated flat variables with two via points. The different colors represent the different polynomial functions joining the subsequent pairs of points.

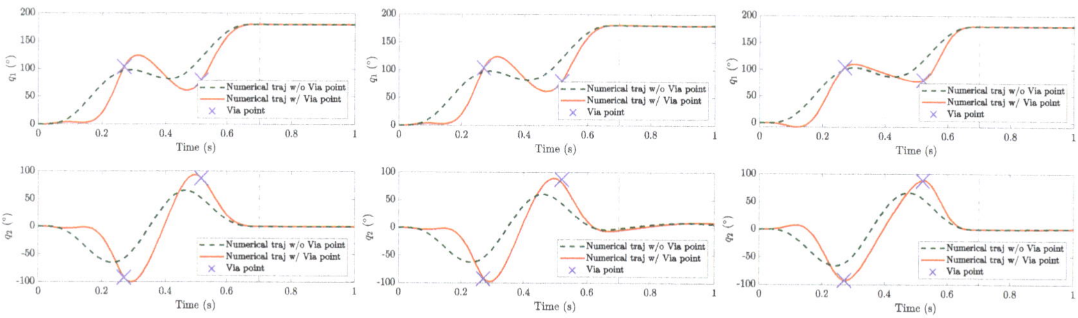

(**a**) Test 4 (**b**) Test 5 (**c**) Test 6

Figure 7. Simulated joint values with two via points.

At the Cartesian trajectories (Figure 5), two results can be note. Firstly, the trajectory shapes are rather different between Tests 4, 5, and 6; this is due to the fast movements, which results in different shapes of the polynomials. In particular, the trajectory of the damped robot (Figure 5c) does not present the entanglement of Tests 4 and 5. Secondly, the trajectory of Test 5 collides with both obstacles, although the trajectory of Test 4 does not; this aspect may be crucial in very tight environments.

Finally, it is worth noting that the designer must be aware of the trajectory of the back of link 2 (black lines of Figures 2 and 5) since this point may collide with the obstacle, especially for high displacements of q_2.

The last two tests highlight how the trajectory changes when different via points are chosen. In Figure 8, two different trajectories (orange and green lines) are shown, which are obtained with $c_n \neq 0$ and $t_f = 0.64$ s (the same as Test 3) but different via points and t_{via}. In particular, the orange trajectory is obtained with a via point placed at $\boldsymbol{q}_{via} = [80, -75]^\circ$ and $t_{via} = 0.11$ s; the green trajectory is obtained with a via point placed at $\boldsymbol{q}_{via} = [55, 70]^\circ$ and $t_{via} = 0.32$ s. The results show that the obstacle can be avoided with very different via points and t_{via}. An optimization of via point placement will be the focus of future work.

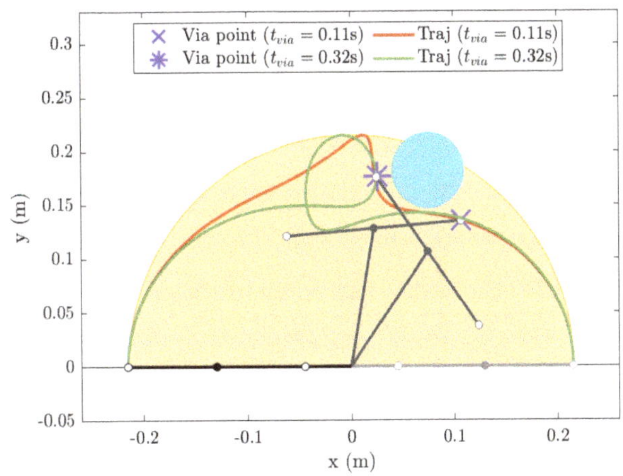

Figure 8. Two different trajectories with two different via points but the same obstacle within the workspace.

5. Experimental Validation

The proposed approach was adopted for trajectory planning of a prototype two-DOF planar underactuated robot with the same inertial properties as the one simulated in Section 4. The task of the robot was to move from one point to another, avoiding two obstacles. Although no damper was installed on the passive joint, damping phenomena due to friction were present. Friction torques are usually proportional to velocity [27]; hence, when the robot performed fast trajectories, damping could not be neglected. The experimental validation is equivalent to Test 6 of Section 4.

The first link was directly connected to the motor, which was a brushed DC motor (Portescap 35NT2R82 426SP by Mclennan Servo Supplies Ltd., Surrey, UK). The rotation of this joint was acquired by means of an incremental encoder Baumer BHK 16.05A2000-I8-5 (by Baumer Italia S.r.l., Milan, Italy) mounted on the motor. The joint was controlled using a Teensy 4.0 and employing a feed-forward control strategy with PD to compensate for uncertainties. The passive joint torsional spring stiffness ($k_2 = 0.0026$ Nm/rad) and damping ($c_2 = 2.4704 \times 10^{-5}$ Nms/rad) were identified by means of experimental modal analysis [28]. The trajectory of link 2 was recorded using a high frame-rate camera for industrial use (Dalsa Genie Nano GM30-M2050 by Teledyne DALSA, Waterloo, ON, Canada) that detects the white markers fixed to the second link. For this camera, at the maximum resolution (2064 × 1544 pixels), the maximum frame rate is 187 fps. However, the frame rate was increased to 280 fps by setting a region of interest (ROI) of dimensions 1600 × 800 pixels.

The test setup is shown in Figure 9a. To replicate the simulation, a paper cylinder and a paper cube were placed in the work area in the same position as in the simulation. The two obstacles are shown in Figure 9b. The movement was performed with the via points of Test 6 but with a different set of t_{via} ($t_{via} = 0.175$ s for the first via point, and $t_{via} = 0.4185$ s for the second via point).

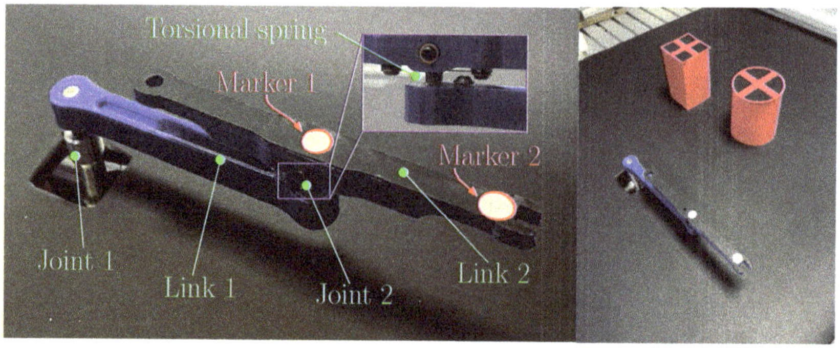

(**a**) (**b**)

Figure 9. (**a**) Prototype of the two-DOF robot used for the experimental validation. (**b**) The two obstacles positioned in the working area.

Before the recording, the camera was calibrated on the plane containing the markers to remove any distortions and to ensure a correct mm/pixel ratio. The camera feed was post-processed after the robot movement, and the marker positions were measured. The coordinates of marker 2 (Figure 9a) were compared to the end effector trajectory of the numerical simulation (i.e., the point placed at a distance a_2 with respect to joint 2). At the same time, since no position sensors were located on joint 2, the joint trajectory $q_2(t)$ was obtained by measuring the absolute angle of the line connecting marker 1 and 2 and by subtracting the joint 1 trajectory $q_1(t)$.

Both Cartesian and joint trajectories (Figure 10a and Figure 10b, respectively) showed very good agreement between experimental and numerical results. The overshoot that can be seen at the end of the Cartesian trajectory is present both in the numerical and experimental trajectories and is due to robot dynamics. The robot was able to reach the via points according to the planning, and no noticeable oscillations of the passive joint were present at the end of the movement. It is worth noting that the movement was very fast ($t_f = 0.7$ s); thus, the inertial forces play a relevant role in the system dynamics. The fact that the robot is differentially flat ensures that the centrifugal and Coriolis forces of the last links are zero; thus, the dynamic model becomes very accurate.

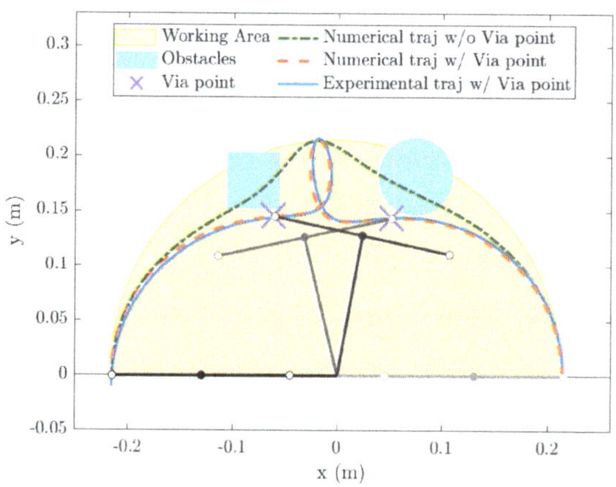

(**a**)

Figure 10. *Cont.*

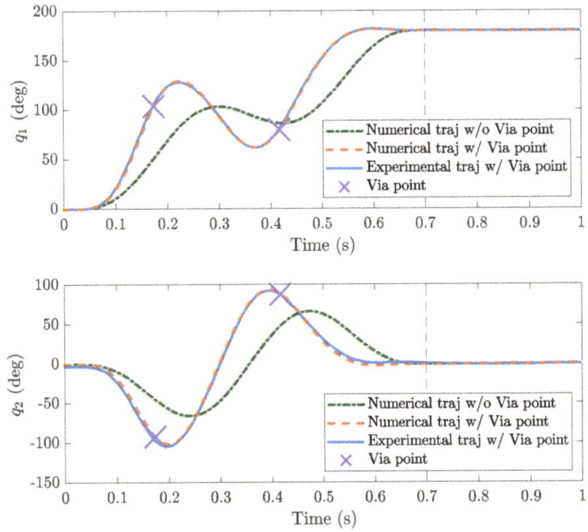

(**b**)

Figure 10. Experimental results with a two-DOF robot: Cartesian trajectories (**a**) and joint values (**b**).

The result was obtained without having any feedback on the position of the last joint; in fact, both the Cartesian and joint trajectories were post-processed from the video captured by the camera and were not used in the control system. Conversely, the robot was driven by calculating the trajectory of the first joint after the planning of the flat variable and the employment of the dynamic model.

6. Conclusions

The main contribution of this paper is the development and validation of a method for motion planning in the space of flat variables considering the presence of via points. It is an extension of the approach adopted for point-to-point motion. A general formulation with n_{via} via points and the flat variables interpolated by high-order polynomial functions has been presented. Since the number of flat variables is lower than the number of DOFs, the indetermination of the configuration of the robot at the via point is solved, adding another condition that comes from the equations of motion.

A series of numerical simulations and an experimental test on a prototype were carried out to assess the validity of the proposed method. The results show that the introduction of one or more via points strongly increases the possibility of avoiding obstacles in the workspace. The numerical results show that the trajectories of the robot in the Cartesian space depend on the position of the via points, the total duration of the motion, and the duration of the various motion sections between the via points.

Since, in actual industrial applications, these parameters can be varied within assigned ranges, future research will deal with the optimization of these parameters. The goal of the optimization can be the minimization of the distance from the obstacles or the minimization of the torques demanded by the actuated joints for an assigned distance of the robot path from the obstacles. Further developments will deal with the interpolation of the flat variables by means of different families of functions as well.

Author Contributions: Conceptualization, M.T., A.D., M.B. and G.R.; methodology, M.T., A.D., M.B. and G.R.; software, M.T. and M.B.; validation, M.T., M.B. and A.D.; formal analysis, M.T., A.D., M.B. and G.R.; investigation, M.T., M.B. and G.R.; writing—original draft preparation, M.T. and M.B.; writing—review and editing, A.D. and G.R.; supervision, A.D. and G.R. All authors have read and agreed to the published version of the manuscript.

Funding: This research was funded by the University of Padova grant number BOTT_BIRD23_01.

Data Availability Statement: The raw data supporting the conclusions of this article will be made available by the authors on request.

Conflicts of Interest: The authors declare no conflicts of interest.

References

1. Deng, M.; Kubota, S. Nonlinear Control System Design of an Underactuated Robot Based on Operator Theory and Isomorphism Scheme. *Axioms* **2021**, *10*, 62. [CrossRef]
2. Zhang, T.; Zhang, D.; Zhang, W. Task-based configuration synthesis of an underactuated resilient robot. *Robotics* **2023**, *12*, 121. [CrossRef]
3. Firouzeh, A.; Salehian, S.S.M.; Billard, A.; Paik, J. An under actuated robotic arm with adjustable stiffness shape memory polymer joints. In Proceedings of the 2015 IEEE International Conference on Robotics and Automation (ICRA), Seattle, WA, USA, 26–30 May 2015; pp. 2536–2543.
4. Qin, G.; Ji, A.; Cheng, Y.; Zhao, W.; Pan, H.; Shi, S.; Song, Y. Design and motion control of an under-actuated snake arm maintainer. *Robotica* **2022**, *40*, 1763–1782. [CrossRef]
5. Quaglia, G.; Tagliavini, L.; Colucci, G.; Vorfi, A.; Botta, A.; Baglieri, L. Design and Prototyping of an Interchangeable and Underactuated Tool for Automatic Harvesting. *Robotics* **2022**, *11*, 145. [CrossRef]
6. Gupta, S.; Kumar, A. A brief review of dynamics and control of underactuated biped robots. *Adv. Robot.* **2017**, *31*, 607–623. [CrossRef]
7. He, B.; Wang, S.; Liu, Y. Underactuated robotics: A review. *Int. J. Adv. Robot. Syst.* **2019**, *16*, 1729881419862164. [CrossRef]
8. Franch, J.; Agrawal, S.K.; Sangwan, V. Differential Flatness of a Class of *n*-DOF Planar Manipulators Driven by 1 or 2 Actuators. *IEEE Trans. Autom. Control* **2010**, *55*, 548–554. [CrossRef]
9. Tonan, M.; Doria, A.; Bottin, M.; Rosati, G. Influence of Joint Stiffness and Motion Time on the Trajectories of Underactuated Robots. *Appl. Sci.* **2023**, *13*, 6939. [CrossRef]
10. Oriolo, G.; Nakamura, Y. Control of mechanical systems with second-order nonholonomic constraints: Underactuated manipulators. In Proceedings of the 30th IEEE Conference on Decision and Control, Brighton, UK, 11–13 December 1991; Volume 3, pp. 2398–2403.
11. De Luca, A.; Oriolo, G. Trajectory planning and control for planar robots with passive last joint. *Int. J. Robot. Res.* **2002**, *21*, 575–590. [CrossRef]
12. Agrawal, S.; Sangwan, V. Differentially flat designs of underactuated open-chain planar robots. *IEEE Trans. Robot.* **2008**, *24*, 1445–1451. [CrossRef]
13. Sangwan, V.; Kuebler, H.; Agrawal, S.K. Differentially flat design of under-actuated planar robots: Experimental results. In Proceedings of the 2008 IEEE International Conference on Robotics and Automation, Pasadena, CA, USA, 19–23 May 2008; pp. 2423–2428.
14. Bottin, M.; Rosati, G. Comparison of Under-Actuated and Fully Actuated Serial Robotic Arms: A Case Study. *J. Mech. Robot.* **2022**, *14*, 034503. [CrossRef]
15. Tonan, M.; Doria, A.; Bottin, M.; Rosati, G. Oscillation-free point-to-point motions of planar differentially flat under-actuated robots: A Laplace transform method. *Robotica* **2024**, 1–19. [CrossRef]
16. Franch, J.; Reyes, À.; Agrawal, S.K. Differential flatness of a class of n-DOF planar manipulators driven by an arbitrary number of actuators. In Proceedings of the 2013 European Control Conference (ECC), Zurich, Switzerland, 17–19 July 2013; pp. 161–166.
17. Sangwan, V.; Agrawal, S.K. Robustness of a flatness based controller against parametric uncertainties for a class of under-actuated planar manipulators. In Proceedings of the 2017 American Control Conference (ACC), Seattle, WA, USA, 24–26 May 2017; pp. 3735–3740.
18. Chettibi, T. Smooth point-to-point trajectory planning for robot manipulators by using radial basis functions. *Robotica* **2019**, *37*, 539–559. [CrossRef]
19. Gasparetto, A.; Boscariol, P.; Lanzutti, A.; Vidoni, R. Path Planning and Trajectory Planning Algorithms: A General Overview. In *Motion and Operation Planning of Robotic Systems: Background and Practical Approaches*; Springer International Publishing: Cham, Switzerland, 2015; pp. 3–27.
20. Sangwan, V.; Agrawal, S. Effects of viscous damping on differential flatness-based control for a class of under-actuated planar manipulators. *IEEE Control Syst. Lett.* **2018**, *2*, 67–72. [CrossRef]
21. Wang, C.; Savkin, A.V.; Garratt, M. Collision free navigation of flying robots among moving obstacles. In Proceedings of the 2016 35th Chinese Control Conference (CCC), Chengdu, China, 27–29 July 2016; pp. 4545–4549.
22. Wang, J.; Wang, J.; Han, Q.L. Receding-Horizon Trajectory Planning for Under-Actuated Autonomous Vehicles Based on Collaborative Neurodynamic Optimization. *IEEE/CAA J. Autom. Sin.* **2022**, *9*, 1909–1923. [CrossRef]
23. Agrawal, S.K.; Sangwan, V. Design of under-actuated open-chain planar robots for repetitive cyclic motions. In Proceedings of the International Design Engineering Technical Conferences and Computers and Information in Engineering Conference, Philadelphia, PA, USA, 10–13 September 2006; Volume 42568, pp. 1057–1066.

24. Chen, W.; Xiong, C.; Chen, W.; Yue, S. Mechanical adaptability analysis of underactuated mechanisms. *Robot. Comput. Integr. Manuf.* **2018**, *49*, 436–447. [CrossRef]
25. Narikiyo, T.; Sahashi, J.; Misao, K. Control of a class of underactuated mechanical systems. *Nonlinear Anal. Hybrid Syst.* **2008**, *2*, 231–241. [CrossRef]
26. Biagiotti, L.; Melchiorri, C. *Trajectory Planning for Automatic Machines and Robots*; Springer Science & Business Media: Berlin/Heidelberg, Germany, 2008.
27. Trinh, M.; Schwiedernoch, R.; Gründel, L.; Storms, S.; Brecher, C. Friction Modeling for Structured Learning of Robot Dynamics. In *Proceedings of the Congress of the German Academic Association for Production Technology*; Springer: Berlin/Heidelberg, Germany, 2022; pp. 396–406.
28. Tonan, M.; Bottin, M.; Doria, A.; Rosati, G. A Modal Approach for the Identification of Joint and Link Compliance of an Industrial Manipulator. *Mech. Mach. Sci.* **2022**, *122*, 628–636.

robotics

MDPI

Article

DDPG-Based Adaptive Sliding Mode Control with Extended State Observer for Multibody Robot Systems

Hamza Khan [1,†], Sheraz Ali Khan [2], Min Cheol Lee [1], Usman Ghafoor [1,3], Fouzia Gillani [4,†] and Umer Hameed Shah [5,*]

[1] School of Mechanical Engineering, Pusan National University, Busan 46241, Republic of Korea; hamzakhan.0496@gmail.com (H.K.); mclee@pusan.ac.kr (M.C.L.); usman@pusan.ac.kr (U.G.)
[2] Department of Mechatronics Engineering, University of Engineering and Technology, Peshawar 25000, Pakistan; sherazalik@uetpeshawar.edu.pk
[3] Department of Mechanical Engineering, Institute of Space Technology, Islamabad 44000, Pakistan
[4] Department of Mechanical Engineering & Technology, Government College University, Faisalabad 37000, Pakistan; fouziagillani@gcuf.edu.pk
[5] Department of Mechanical Engineering and Artificial Intelligence Research Center, College of Engineering and Information Technology, Ajman University, Ajman P.O. Box 346, United Arab Emirates
* Correspondence: m.shah@ajman.ac.ae
† These authors contributed equally to this work.

Abstract: This research introduces a robust control design for multibody robot systems, incorporating sliding mode control (SMC) for robustness against uncertainties and disturbances. SMC achieves this through directing system states toward a predefined sliding surface for finite-time stability. However, the challenge arises in selecting controller parameters, specifically the switching gain, as it depends on the upper bounds of perturbations, including nonlinearities, uncertainties, and disturbances, impacting the system. Consequently, gain selection becomes challenging when system dynamics are unknown. To address this issue, an extended state observer (ESO) is integrated with SMC, resulting in SMCESO, which treats system dynamics and disturbances as perturbations and estimates them to compensate for their effects on the system response, ensuring robust performance. To further enhance system performance, deep deterministic policy gradient (DDPG) is employed to fine-tune SMCESO, utilizing both actual and estimated states as input states for the DDPG agent and reward selection. This training process enhances both tracking and estimation performance. Furthermore, the proposed method is compared with the optimal-PID, SMC, and H_∞ in the presence of external disturbances and parameter variation. MATLAB/Simulink simulations confirm that overall, the SMCESO provides robust performance, especially with parameter variations, where other controllers struggle to converge the tracking error to zero.

Keywords: multibody dynamics; sliding mode control; extended state observer; DDPG

Citation: Khan, H.; Khan, S.A.; Lee, M.C.; Ghafoor, U.; Gillani, F.; Shah, U.H. DDPG-Based Adaptive Sliding Mode Control with Extended State Observer for Multibody Robot Systems. *Robotics* **2023**, *12*, 161. https://doi.org/10.3390/robotics12060161

Academic Editor: Raffaele Di Gregorio

Received: 23 October 2023
Revised: 21 November 2023
Accepted: 22 November 2023
Published: 26 November 2023

1. Introduction

The expanding capabilities of multibody robot systems in autonomous operation and their versatility in performing a wide range of tasks have gathered significant attention from both researchers and industries, emphasizing the persistent need for precision and reliability in their operations. As a result, multibody robot systems require robust control algorithms. However, controlling multibody robot dynamics can be a challenging task, especially when the robot dynamics are unknown. In this effort, different robust control algorithms have been proposed in which sliding mode control (SMC) has been of great interest due to outstanding robustness against parametric uncertainties and external disturbances [1,2]. Subsequent developments resulted in different types of SMC, including integral SMC (ISMC) [3], super twisting SMC (STSMC) [4], terminal SMC (TSMC) [5], SMC with a nonlinear disturbance observer known as sliding perturbation observer (SMC-SPO) [6], and SMC with extended state observer (SMCESO) [7].

This research is conducted for the robust control of multibody industrial robot systems with aims of enhancing the trajectory tracking results. Therefore, we consider the nonlinear control SMC with ESO (SMCESO) for the robot control. ESO considers the system dynamics and external disturbances as perturbations to the system. Therefore, with ESO, the system is only affected by the perturbation estimation error because of the compensation provided by the ESO. Another advantage of the ESO is that it requires no system dynamics information and only uses partial state feedback (position) for estimating the states and the perturbation. Subsequently, the robustness of SMCESO now depends on the quality of estimation of the ESO, which is dependent on the selection of control parameters. However, tuning the parameters manually becomes a challenging task. Therefore, optimal parameter selection can be achieved through adapting the parameters for different sliding conditions.

Various methods for adaptive SMC have been explored, including model-free adaptation, intelligent adaptation, and observer-based adaptation. A. J. Humaidi et al. introduced particle swarm optimization-based adaptive STSMC [8]. The adaptation is carried out based on the Lyapunov theory to guarantee global stability. Y. Wang and H. Wang introduced model-free adaptive SMC, initially estimating unknown dynamics through the time delay estimation method [9,10]. Nevertheless, this approach exhibited undesirable chattering in the control input during experiments, which is deemed unacceptable in the present research. On the other hand, R-D. Xi et al. presented adaptive SMC with a disturbance observer for robust robot manipulator control [11]. Observer-based adaptive SMC stands out for its ability to ensure robustness through minimizing the impact of lumped disturbances, a feature similarly emphasized by C. Jing et al. [12]. Conclusively, this study states that implementing a disturbance observer can lead to finite-time stability and specific tracking performance quality. Furthermore, H. Zao et al. introduced fuzzy SMC for robot manipulator trajectory tracking [13]. H. Khan et al. proposed extremum seeking (ES)-based adaptive SMCSPO for industrial robots [14]. A unique cost function is used which consists of estimation error, and error dynamics to guarantee accurate states and perturbation estimation. H. Razmi et al. proposed neural network-based adaptive SMC [15], and Z. Chen et al. presented radial basis function neural network (NN)-based adaptive SMC [16], both demonstrating commendable performance. However, it is worth noting that the systems under consideration in these studies were relatively smaller than the industrial robot in our current research. Furthermore, a model-free reinforcement learning algorithm known as deep deterministic policy gradient (DDPG) has been observed to provide optimal SMC parameters, enhancing performance through learning and adapting to different sliding patterns [17–19].

Considering the diverse literature, initially, the model-free extremum seeking algorithm was a consideration. However, in the current study, the need to tune multiple (four different) parameters simultaneously led to the exploration of learning-based algorithms such as NN and DDPG for adapting controller parameters. Notably, NN is well suited for simpler systems, while DDPG is preferred for complex, high-dimensional systems with unknown dynamics. DDPG is a model-free, online, and off-policy reinforcement learning algorithm. It employs an actor–critic architecture, where the actor focuses on learning the optimal policy, while the critic approximates the Q-function [20]. The Q-function is responsible for estimating the expected cumulative long-term rewards for state–action pairs. The critic achieves this by minimizing the temporal difference error, which represents the disparity between the predicted Q-value and the actual Q-value derived from environmental feedback. This process equips the critic to evaluate action quality in various states, guiding the actor in selecting actions that maximize expected rewards. Ensuring the convergence of the temporal difference error is a pivotal aspect of effective DDPG agent training.

The primary contribution of this study is the optimal tuning of SMCESO using the DDPG algorithm for a heavy-duty industrial robot manipulator with six degrees of freedom (DOF). Robust performance can be achieved through minimizing estimation errors, ensuring accurate perturbation estimation and compensation. To accomplish this, the

DDPG input states incorporate tracking error, estimation error, current joint angle, and estimated joint angle. A reward has been designed, integrating an overall error tolerance of 0.01 rad for both tracking and estimation errors, yielding positive rewards if error is below the threshold. Conversely, if errors exceed this threshold, negative rewards are assigned. Through this approach, the DDPG agent learns a control pattern based on actual and estimated results, ultimately achieving optimal estimation and robust control performance. The proposed algorithm was implemented and compared with optimal proportional—integral–derivative (PID) control and SMC, and H_∞ control in an extensive MATLAB/Simulink simulation environment. The results demonstrated that SMCESO outperforms all the three controllers, particularly in the presence of variable system parameters, as it effectively reduces the effect of the actual perturbations on system performance.

The reminder manuscript Is organized as follows: Section 2 describes the general multibody dynamics and formulates the SMC. Section 3 presents the ESO and the DDPG algorithm. Section 4 then presents the simulation environment and the results of the proposed algorithm, whereas Section 5 provides the conclusions.

2. Preliminaries

2.1. Multibody Dynamics Description

Consider the second-order multibody dynamics [14] as follows:

$$\ddot{x}_j = f_j(x) + \Delta f_j(x) + \sum_{i=1}^{n} \left[(b_{ji}(x) + \Delta b_{ji}(x))u_i \right] + d_j(t)\ j = 1, \ldots, n \tag{1}$$

where $x \triangleq [x_1 \ldots\ x_n]^T$ are the state vectors representing the position, and $f_j(x)$ and $\Delta f_j(x)$ are the linear dynamic and dynamics uncertainties, respectively. Similarly, the control gain matrix and their uncertainties are represented by $b_{ji}(x)$ and $\Delta b_{ji}(x)$, respectively. u_i and d_j are the control input and external disturbance, respectively. Combining the system nonlinearities, dynamics uncertainties, and disturbances as the perturbation (ψ) can be written as

$$\psi_j(x,t) = \Delta f_j(x) + \sum_{i=1}^{n} \left[\Delta b_{ji}(x) u_i \right] + d_j(t) \tag{2}$$

whereas it is assumed that the perturbation is bounded by an unknown continuous function, i.e., $\left|\psi_j(x,t)\right| \leq \Gamma > 0$, and, in addition, that it is smooth with the bounded derivative $\left|\dot{\psi}_j(x,t)\right| \leq \bar{\Gamma}$.

2.2. Sliding Mode Control

The main concept of SMC is to design a sliding surface σ in the state space (position x_1, and velocity x_2) [21], which is given as

$$\sigma = \dot{e} + ce \tag{3}$$

where $e = x_d - x$ is the tracking error, and $c > 0$ is a positive constant. Now, in order to drive the system dynamics, the state variable should converge to zero: i.e., $\lim_{t \to \infty} \dot{e}, e = 0$ asymptotically in the presence of perturbation. Therefore, SMC tends to bring system states on the sliding surface by means of control force u. Subsequently, SMC has two phases: The first is the reaching phase, during which the system states are not on the sliding surface and require a switching control u_{sw} to reach the sliding surface. The second phase is the sliding phase, in which the system states have reached the sliding surface and now require continuous control, generally known as equivalent control u_{eq}, to remain on the sliding surface, where the overall control input becomes $u = u_{eq} + u_{sw}$. To compute the control input, the derivative of the sliding surface is defined as follows:

$$\dot{\sigma} = \ddot{e} + c\dot{e} = -K_{smc} \cdot \mathrm{sat}(\sigma) \tag{4}$$

where K_{smc} represents the switching control gain, and 'sat' is the saturation function with a boundary layer thickness ε_c, given as

$$\text{sat}(\sigma) = \begin{cases} \frac{\sigma}{|\sigma|} \; if \; |\sigma| > \varepsilon_c \\ \frac{\sigma}{\varepsilon_c} \; if \; |\sigma| \leq \varepsilon_c \end{cases} \tag{5}$$

Assuming unknown system dynamics, $\ddot{x} = u$ is presumed. Substituting this condition with the dynamics error $\ddot{e} = \ddot{x}_d - \ddot{x}$ in (4) results in the following control input.

$$u = -K_{smc} \cdot \text{sat}(\sigma) + \ddot{x}_d - c\dot{e} \tag{6}$$

Here, $K_{smc} \cdot \text{sat}(\sigma)$ is denotes the switching control (u_{sw}), and the negative sign embodies the error convention. The remaining terms are considered equivalent control (u_{eq}). Subsequently, taking the derivative of the sliding surface, with the system disturbed by perturbation (such as (10) in the subsequent section) yields (7):

$$\dot{\sigma} = \ddot{x} - \ddot{x}_d + c\dot{e} = u + \psi(x,t) - \ddot{x}_d + c\dot{e} \tag{7}$$

Substituting the control law from (6) and solving results in (8):

$$\dot{\sigma} = -K_{smc} \cdot \text{sat}(\sigma) + \psi(x,t) \tag{8}$$

Equation (8) shows that in SMC, the sliding surface is affected by the perturbation. Once the system states have reached the sliding phase, the relationship between the sliding surface and the perturbation is given as the following transfer function [6].

$$\frac{\sigma}{\psi(x,t)} = \frac{1}{p + \frac{K_{smc}}{\varepsilon_c}} \tag{9}$$

where p is the s-domain variable. Increasing the boundary layer will decrease the breaking frequency, making the system less sensitive to the higher frequency perturbations. However, at $\sigma \approx 0$, increasing the boundary layer thickness reduces controller performance, leading to higher tracking error. If the sliding surface is tightly bounded, with a very small boundary layer, chattering occurs.

2.3. Problem Forumlation

Calculating the dynamics of a multibody robot system is a challenging task, further compounded by the presence of inaccurate dynamics, which introduce uncertainties. Therefore, for the later study, considering the complete dynamic model as perturbation and $b = 1$, the resulting dynamics is as follows.

$$\ddot{x} = \psi(x,t) + u \tag{10}$$

Subsequently, to ensure the sliding condition outside the boundary later, the sliding dynamics can be written as

$$\dot{\sigma} = c\dot{e} + \psi(x,t) + u, \; \sigma(0) = \sigma_0 \tag{11}$$

For the asymptotic stability of (11) about the equilibrium point, $\dot{V} < 0$ for $\sigma \neq 0$ must be satisfied [22]. The derivative of V is computed as

$$\dot{V} \leq \sigma\dot{\sigma} = \sigma\left[c\dot{e} + \psi(x,t) + u\right] \tag{12}$$

Taking $\eta = c\dot{e} + u$ in (12) will result in

$$\dot{V} \leq \sigma\dot{\sigma} = \sigma[\psi(x,t) + \eta] = \sigma \cdot \psi(x,t) + \sigma \cdot \eta \tag{13}$$

$$\dot{V} \leq |\sigma|\,\Gamma + \sigma \cdot \eta \tag{14}$$

Selecting $\eta = -K_{smc} \cdot \mathrm{sat}(\sigma)$, and with $K_{smc} > 0$, (14) becomes

$$\dot{V} \leq |\sigma|\,\Gamma - |\sigma| \cdot K_{smc} = -|\sigma|(K_{smc} - \Gamma) \tag{15}$$

Consequently, the overall control input becomes

$$u = K_{smc} \cdot \mathrm{sat}(\sigma) + c\dot{e} \tag{16}$$

Equation (15) further emphasizes that for stability, $K_{smc} > \Gamma$ to satisfy the Lyapunov condition. However, obtaining information about Γ can be a complex and tedious task.

3. Proposed Algorithm

There are two concerns: First, based on (9), as the perturbation affects the sliding dynamics, the correct dynamics are unknown. Therefore, a perturbation observer has been used to estimate and compensate the actual perturbation effects. For this purpose, an extended state observer (ESO) has been implemented, which offers the advantage of not requiring the system dynamics information. Secondly, we optimally tune the control gain for SMC and ESO to stabilize the system in finite time, ensuring that both tracking and estimation error converge to zero. Subsequently, deep deterministic policy gradient (DDPG) has been employed for control gain tuning.

3.1. Extended State Observer

ESO provides real-time estimations of unmeasured system states and perturbations, which is the combination of modelled and unmodelled dynamics and external disturbances, enhancing control system performance and robustness. This means ESO considers the system's linear and nonlinear dynamics as the perturbation and estimates them [23]. Consequently, only the control input u in (1) is known. Furthermore, ESO does not require system dynamics information and uses only partial state feedback (position) for estimation. In addition to the system states (position x_1 and velocity x_2), an extended state x_3 is introduced, such as

$$x_3 = \psi(x,t) = f(x) + \Delta f(x) + \sum\nolimits_{i=1}^{n} [\Delta b_i(x)u] + d(t) \leq \Gamma \tag{17}$$

Subsequently, the system dynamics in (1) can be simplified as

$$\begin{aligned} \dot{x}_1 &= x_2 \\ \dot{x}_2 &= x_3 + u \\ x_3 &= \Gamma \end{aligned} \tag{18}$$

With the new system information, the mathematical model of nonlinear ESO [7] is then written as

$$\begin{aligned} \dot{\hat{x}}_1 &= \hat{x}_2 + l_1 \cdot \rho\left(\tilde{x}_1\right) \\ \dot{\hat{x}}_2 &= \hat{x}_3 + u + l_2 \cdot \rho\left(\tilde{x}_1\right) \\ \dot{\hat{x}}_3 &= l_3 \cdot \rho\left(\tilde{x}_1\right) \end{aligned} \tag{19}$$

where the components with " $\wedge$ ", and " $\sim$ " represent the estimated states and the error between the actual and the estimated value, e.g., $\tilde{x}_1 = x_1 - \hat{x}_1$. ρ is the saturation function, which is selected as

$$\rho\left(\tilde{x}_1\right) = \begin{cases} \tilde{x}_1 / |\tilde{x}_1| & \text{if } \left|\tilde{x}_1\right| > \varepsilon_o \\ \tilde{x}_1 / \varepsilon_o & \text{if } \left|\tilde{x}_1\right| \leq \varepsilon_o \end{cases} \tag{20}$$

ε_o is the boundary layer of the ESO such that the estimation error should be $\left|\tilde{x}_1\right| \leq \varepsilon_o$. The estimation errors are calculated as

$$\begin{aligned} \dot{\tilde{x}}_1 &= \tilde{x}_2 - l_1 \cdot \rho\left(\tilde{x}_1\right) \\ \dot{\tilde{x}}_2 &= \tilde{x}_3 - l_2 \cdot \rho\left(\tilde{x}_1\right) \\ \dot{\tilde{x}}_3 &= \Gamma - l_3 \cdot \rho\left(\tilde{x}_1\right) \end{aligned} \tag{21}$$

As the estimated error should be bounded by a boundary later, therefore, (21) can be rewritten as follows.

$$\begin{aligned} \dot{\tilde{x}}_1 &= \tilde{x}_2 - l_1 \cdot \tilde{x}_1/\varepsilon_o \\ \dot{\tilde{x}}_2 &= \tilde{x}_3 - l_2 \cdot \tilde{x}_1/\varepsilon_o \\ \dot{\tilde{x}}_3 &= \Gamma - l_3 \cdot \tilde{x}_1/\varepsilon_o \end{aligned} \tag{22}$$

Subsequently, the state space of the error dynamics can be written as

$$\dot{\tilde{x}} = A\tilde{x} + B\Gamma \tag{23}$$

where

$$A = \begin{bmatrix} -l_1/\varepsilon_o & 1 & 0 \\ -l_2/\varepsilon_o & 0 & 1 \\ -l_3/\varepsilon_o & 0 & 0 \end{bmatrix}, \; and \; E = \begin{bmatrix} 0 \\ 0 \\ 1 \end{bmatrix} \tag{24}$$

The characteristic equation of A can be calculated as follows:

$$|\lambda I - A| = \begin{vmatrix} \lambda + l_1/\varepsilon_o & -1 & 0 \\ l_2/\varepsilon_o & \lambda & -1 \\ l_3/\varepsilon_o & 0 & \lambda \end{vmatrix} = \lambda^3 + (l_1/\varepsilon_o)\lambda^2 + (l_2/\varepsilon_o)\lambda + l_3/\varepsilon_o \tag{25}$$

The error dynamics are stable if the gains l_1, l_2, and l_3 are positive. Therefore, these gains are selected using the pole placement method as follows:

$$(s+\lambda)^3 = s^3 + 3 \cdot s \cdot \lambda^2 + 3 \cdot \lambda \cdot s^2 + \lambda^3 \tag{26}$$

Comparing the coefficients of (25) and (26) results in the following selection of gains:

$$l_1 = 3 \cdot \lambda \cdot \varepsilon_o, \; l_2 = 3 \cdot \lambda^2 \cdot \varepsilon_o, \; \text{and } l_3 = \lambda^3 \cdot \varepsilon_o \tag{27}$$

3.2. *Extended State Observer-Based Sliding Mode Control (SMCESO)*

For enhanced system performance, the final control input u_o for the system with estimated perturbation $\hat{\psi}(x,t)$ from ESO and switching control from SMC can be written as

$$u_o = u - \hat{x}_3 = u - \hat{\psi}(x,t) \tag{28}$$

where u is from (16). Consequently, the system dynamics from (10) can be rewritten as follows.

$$\ddot{x} = u_o - \hat{\psi}(x,t) + \psi(x,t) = u_o + \tilde{\psi}(x,t) \tag{29}$$

where $\tilde{\psi}(x,t) = \psi(x,t) - \hat{\psi}(x,t)$ is the perturbation estimation error. Now, it is evident that with ESO, the system is only affected by the perturbation estimation error as compared to the actual perturbation. This follows $\left|\tilde{\psi}(x,t)\right| \ll |\psi(x,t)|$, ensuring that ESO-based SMC is more stable than the individual SMC. Subsequently, the Lyapunov function in (15) will become

$$\dot{V} \leq -|\sigma|\left(K'_{smc} - \tilde{\psi}(x,t)\right) \tag{30}$$

The stability of SMCESO with the Lyapunov function $\sigma\dot{\sigma} \leq 0$ can be calculated as

$$\sigma\dot{\sigma} \leq |\sigma|(\ddot{e} + c\dot{e}) \leq |\sigma|(\ddot{x} - \ddot{x}_d + c\dot{e}) \leq 0 \tag{31}$$

With the system dynamics and combined control input from (6) and (28), according to (7), this will result in the following condition:

$$\sigma\dot{\sigma} \leq |\sigma|(-K'_{smc} \cdot \text{sat}(\sigma) + \ddot{x}_d - c\dot{e} - \hat{\psi}(x,t) + \psi(x,t) - \ddot{x}_d + c\dot{e}) \leq 0 \tag{32}$$

Simplifying (32) yields (33):

$$\sigma\dot{\sigma} \leq |\sigma|\left(-K'_{smc} \cdot \text{sat}(\sigma) + \tilde{\psi}(x,t)\right) \leq 0 \tag{33}$$

Subsequently, to keep the system stable, the control gain should follow the following condition.

$$K'_{smc} > \left|\tilde{\psi}(x,t)\right| \tag{34}$$

Now, the new control gain K'_{smc} is small in comparison to conventional gain K_{smc}, with $K'_{smc} < K_{smc}$. The reduced gain will result in smoother switching control, eliminating any chattering for improved performance. Furthermore, the control parameters K'_{smc}, c, ε_c, and λ are then optimally tuned using DDPG to reduce manual tuning efforts.

3.3. DDPG-Based SMCESO

Deep deterministic policy gradient (DDPG) is a reinforcement learning algorithm designed for solving continuous action space problems. It combines elements of deep neural networks and the deterministic policy gradient theorem to achieve remarkable performance in control tasks. DDPG employs an actor–critic architecture, with the actor network modeling the policy and the critic network estimating the state–action value function. A key innovation in DDPG is the use of target networks to stabilize training, with periodic updates to slowly track the learned networks. This approach, coupled with experience replay, enables stable and efficient learning, making DDPG a prominent choice for complex, high-dimensional control problems.

Similar to other reinforcement learning algorithms, the DDPG algorithm operates within the framework of a Markov decision process (MDP) [24], denoted by $(\mathcal{S}, \mathcal{A}, \mathcal{P}, \mathcal{R})$, where $\mathcal{S}$ and $\mathcal{A}$ represent the environment's state space and the agent's action space, respectively. $\mathcal{P}$ signifies the probability of state transitions. During agent training, the reward function $\mathcal{R}$ serves as the training target. In core, while training the agent, the system's state $s \epsilon \mathcal{S}$ is observed, and the associated reward $r \epsilon \mathcal{R}$ is acquired. Subsequently, the optimal policy $\pi^a(a|s)$ is determined through maximizing the state–action value function.

$$Q(s,a) = E[R_c | S_t = s,\ A_t = a] \tag{35}$$

where R_c represents the cumulative reward, and $R_c = \sum_{k=0}^{\infty} \gamma^k r_{k+1}$ with $0 \leq \gamma \leq 1$ is the discount factor that reflects the importance of the reward value at future moments. To enhance the controller performance, the DDPG has to study the regulation strategy μ (actor network) and calculate the probability of each action. Consequently, the controller parameters are updated in real time to maximize the total reward [25,26].

$$\begin{cases} \max\limits_{\mu}\left[\sum_{k=0}^{\infty} \gamma^k r_k(x_1(k), \hat{x}_1(k))\right] \\ st: \ \theta(k) = \mu(K'_{smc},\ c, \varepsilon_c, \lambda) \\ \qquad \theta_{\min} \leq \theta(k) \leq \theta_{\max} \end{cases} \tag{36}$$

$\theta(k)$ is the set of action parameters, with minimum limit $\theta_{\min}$ and maximum limit $\theta_{\max}$. The structure of DDPG is presented in Figure 1. The selection of a suitable state space is crucial for ensuring the convergence of reinforcement learning. In the context of the present challenges, the chosen state space should inherently pertain to the robot's position and its estimated dynamics. As a result, for the sake of computational efficiency and enhanced learning, the state space is straightforwardly defined as $S = [x(k), \hat{x}(k)]$, and the state vector is defined as $s_k = \left[x_1,\ \hat{x}_1, e, \widetilde{x}_1\right]$.

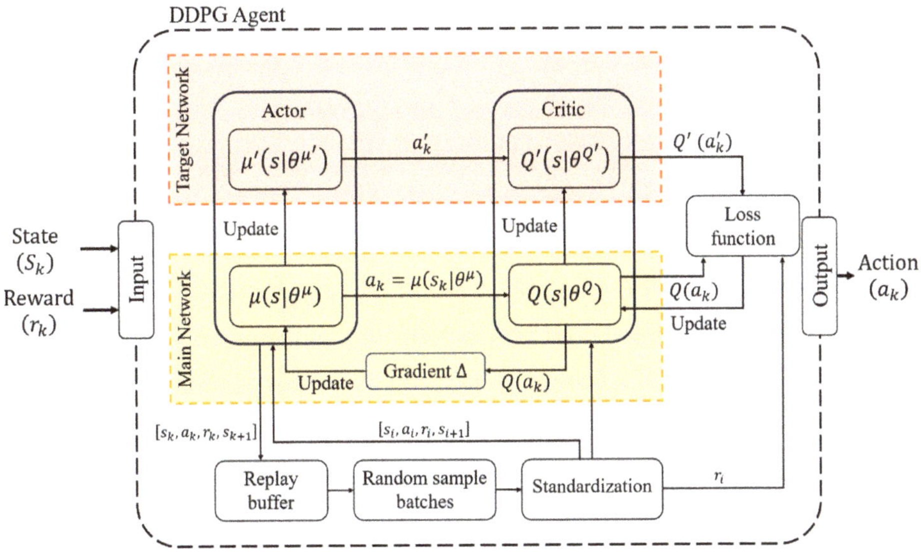

Figure 1. Structure of DDPG.

The actor–critic value network for the robot system is established, which is a double-layer structure including the target network and main network. The replay buffer stores data in the form of $[s_k, a_k, r_k, s_{k+1}]$, which is used for network training. Both the main networks and target networks share the same structure but differ in their parameters. The actor network is denoted by $a_k = \mu(s_k|\theta^{\mu})$, with θ^{μ} as the network parameter. The critic network is denoted as $Q(s_k,\ a_k|\theta^{Q})$, with the network parameter as θ^{Q}. When training, small batches of sample information $[s_i, a_i, r_i, s_{i+1}]$ are randomly selected from the replay buffer for learning. In brief, the training process involves the four networks to ensure that actions generated by the actor network can be used as input for the critic network to maximize the state–action value function in (35). The training process is provided in Algorithm 1.

Algorithm 1: Training DDPG Agent

Initialize the networks $\mu(s_k|\theta^\mu)$, and $Q(s_k,\ a_k|\theta^Q)$ randomly.
Initialize the target network $\mu'\left(s_k\middle|\theta^{\mu'}\right)$, and $Q'\left(s_k,\ a_k\middle|\theta^{Q'}\right)$ with weights.
Initialize the replay buffer.
While $ep\ \le ep_{max}$
  Randomly initialize the process $\mathcal{N}$ for action exploration.
  Receive the states s_k
  while $k < k_{max}$
    $a_k = \mu(s_k|\theta^\mu) + \mathcal{N}$.
    Execute the environment to update the reward r_k, and s_{k+1}.
    Store $[s_k, a_k, r_k, s_{k+1}]$ in replay buffer R.
    Sample a random minibatch of m transitions transitions $[s_i, a_i, r_i, s_{i+1}]$ from R.
    Set target $y_i = r_i + \gamma \cdot Q'\left(s_{i+1}, \mu'\left(s_{i+1}\middle|\theta^{\mu'}\right)\middle|\theta^{\mu'}\right)$.
    Update the critic by minimizing the loss function $J = \frac{1}{m}\Sigma\left(y_i - Q(s_i, a_i|\theta^Q)\right)^2$.
    Update the actor using the sampled policy gradient
$\nabla_{\theta^\mu} J \approx \frac{1}{m}\Sigma\,\nabla_a Q(s,a|\theta^Q)|_{s=s_i,\ a=\mu_s} \nabla_{\theta^\mu}\mu(s|\theta^\mu)$.
    Update the target network with soft update
$\theta^{Q'} \leftarrow \tau\theta^Q + (1-\tau)\theta^{Q'},\ \theta^{\mu'} \leftarrow \tau\theta^\mu + (1-\tau)\theta^{\mu'}$.
    If isdone == 1
      Reset.
    End if
  end while k
  if $r_{average} \ge r_{stopping}$
    Stop training.
  End
end

The DDPG-based SMCESO block diagram is presented in Figure 2. For robust performance, the tracking error should be eliminated. Subsequently, the estimation should be accurate, i.e., $\tilde{x} \to 0$. Consequently, the true perturbation will be estimated and well compensated. Therefore, the reward function for the current study is designed as follows.

$$
\begin{aligned}
r &= R_1 + R_2 + R_3 \\
R_1 &= \begin{cases} -1 & e \ge e_{tol} \\ 5 & e < e_{tol} \end{cases} \\
R_2 &= \begin{cases} -1 & \tilde{x}_1 \ge \tilde{x}_{1,tol} \\ 5 & \tilde{x}_1 < \tilde{x}_{1,tol} \end{cases} \\
R_3 &= \begin{cases} -100 & x_1 \ge x_{1,stop} \\ 0 & x_1 < x_{1,stop} \end{cases}
\end{aligned}
\tag{37}
$$

where e_{tol} is the error tolerance for accepting good performance of tracking control. Similarly, R_2 is for the good performance of ESO with estimation error tolerance as $\tilde{x}_{1,tol}$. R_3 is for the stopping condition (isdone in Algorithm 1), meaning the robot is not stable exceeding the movement limits $x_{1,stop}$.

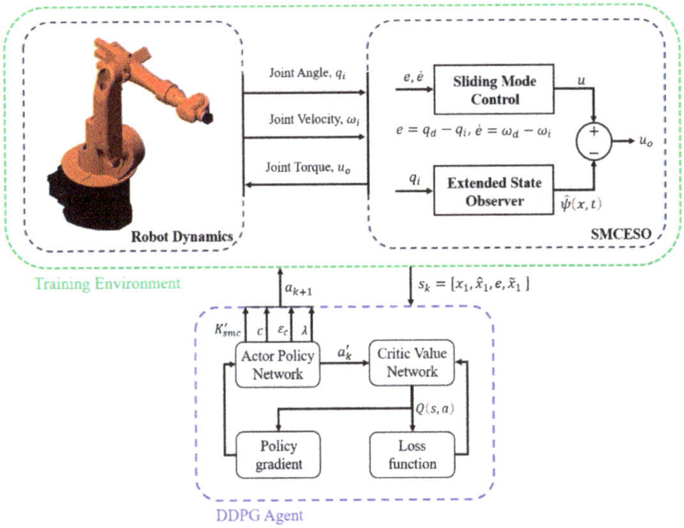

Figure 2. Block diagram of DDPG-based SMCESO.

4. Simulations and Discussion

This section provides details about the simulation system and the environment. It also includes the presentation of results and their subsequent discussion.

4.1. System and Environment Descrption

For the DDPG implementation, a simulation environment is created in MATLAB/Simulink, featuring an object pick-and-place task using the Simscape Multibody model of the KUKA KR 16 S industrial robot arm, as presented in Figure 2. The KR 16 is a six-degrees-of-freedom (DOF) high-speed, heavy-duty industrial robot arm with a substantial payload capacity. Demonstrating robust performance with such robot will validate the efficiency of the proposed method. Consequently, the robot must exhibit robust performance and a minimal tracking error in the presence of nonlinear dynamics. The sampling time for the DDPG algorithm is set to 0.5 s, while the control algorithm operates with a sampling time of 5 ms. The computations are carried out on a computer equipped with an Intel i7 processor and an RTX 3090 ti GPU.

4.2. Simulations

Simulations are conducted in two phases. First is the implementation of the proposed algorithm on a simple linear system to explain the basics or the workings of the ESO. Second is the implementation on the multibody dynamics of the robot arm, with a sine wave as the desired position. For simulation, the DDPG hyperparameters are presented in Table 1

4.2.1. Simple System Implementation

For a simple linear system, consider the following second-order dynamics.

$$\ddot{x} = u_o - b\dot{x} - kx + d(t),\ d(t) = a \cdot \sin(t) \tag{38}$$

$$\psi(x,t) = -10\dot{x} - 50x + d(t) \tag{39}$$

where a is the magnitude of disturbance $(d(t))$, $b = 10$ is the damping coefficient, and $k = 50$ is the stiffness. The performance of DDPG-based SMCESO has been compared with SMC, proportional–integral–derivative (PID) control optimally tuned using the Control

System Tuner toolbox in Simulink, and H_∞ control [27]. The control gains are provided in Table 2, and the trajectory tracking error is shown in Figure 3.

Table 1. DDPG parameters.

Reinforcement	Parameters	
	Parameter	**Value**
Critic	Learn rate	1×10^{-3}
	Gradient Threshold	1
Actor	Learn rate	1×10^{-4}
	Gradient threshold	1
Agent	Sample time	0.5
	Target smooth factor	1×10^{-3}
	Discount factor	1
	Minibatch size	64
	Experience buffer length	1×10^{6}
	Noise variance	0.3
	Noise variance decay rate	1×10^{-5}
Training	Maximum episode	2000
	Maximum steps	20
	Average reward window length	10

Table 2. Control gains.

Control Algorithm	Gains
PID	$K_p = 200$, $K_i = 1000$, and $K_d = 20$.
SMC	$K_{smc} = 300$, $c = 35$, and $\epsilon_c = 0.5$.
SMCESO	$K'_{smc} = 50$, $c = 30$, $\epsilon_c = 0.5$, $\lambda = 137.31$, and $\epsilon_o = 1$.
H_∞	Sensitivity Function $W_s = s + 40/4s + 0.36$

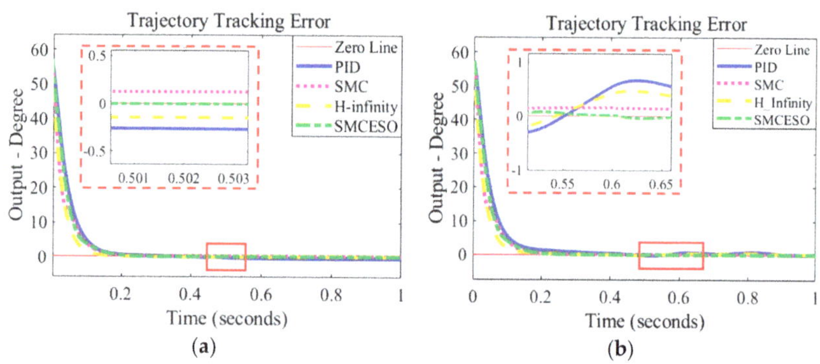

Figure 3. Controller performance evaluation: (**a**) with disturbance; (**b**) with parameter variation.

The error results of the step response in Figure 3a reveal that when a disturbance (a = 10) is present, all three controllers except for SMCESO demonstrate good performance with high control gains but fail to fully converge the error to zero. In contrast, SMCESO effectively estimates and compensates for the perturbation, as depicted in Figure 4 (on the next page), leading to error convergence toward zero. Moreover, as anticipated in Section 3.2, the new control gain K'_{smc} is notably smaller than the conventional gain K_{smc} (in Table 1), which is tuned using the DDPG algorithm. Additionally, the algorithms underwent testing with parameter changes, where the stiffness was chosen as $k = 50 \pm 8$. These variations were introduced using the Simulink random number block with a variance of 20. The tracking errors for variable stiffness are presented in Figure 3b, illustrating that

PID exhibits the maximum deviation, while H_∞ outperforms PID. However, SMC now surpasses H_∞ due to a model mismatch between the actual system and the dynamics used for controller synthesis. Finally, SMCESO outperforms all three controllers through maintaining the error very close to zero. This validates that SMCESO effectively estimates system uncertainties and compensates for their effects on the system response, resulting in robust performance.

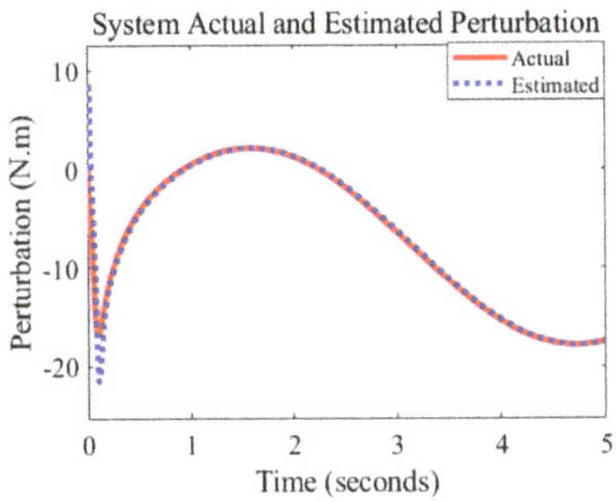

Figure 4. Actual and estimated perturbation comparison.

4.2.2. Adaptive SMCESO with Multibody Robot

With a multibody robot system, the DDPG agent has been trained to fine-tune the controller parameters. For controller evaluation, Joint 2 (q_2) of the robot manipulator has been considered as it holds the maximum weight of the robot against gravity. Therefore, the robot arm is fully extended, and only q_2 is moving. The desired trajectory is defined as $q_{2,d} = \sin(w{\cdot}t)$, with initial frequency $w_o = 1$, which resets after every episode as $w = 1 + rand[-0.5, 0.5]$. Furthermore, the total simulation time is 10seconds, with an ideal reward $r_{max} = 210$. The training stops when the average award reaches $r_c \geq 199$, considering the average reward window length. The DDPG agent took 343 episodes for the training. The episode reward and the cumulative reward are presented in the following Figure 5, and subsequently, the tuned parameters are shown in Figure 6 and the trajectory tracking error and joint torques are in Figure 7.

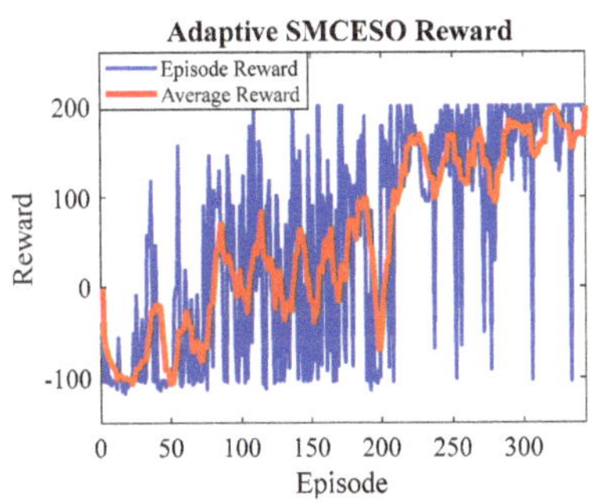

Figure 5. SMCESO training reward.

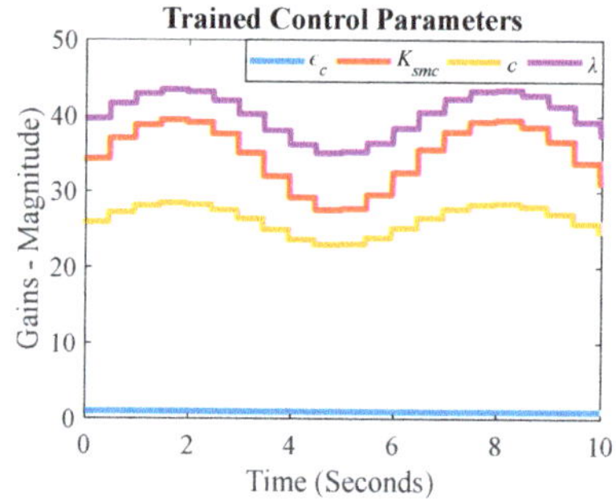

Figure 6. SMCESO fine-tuned gains.

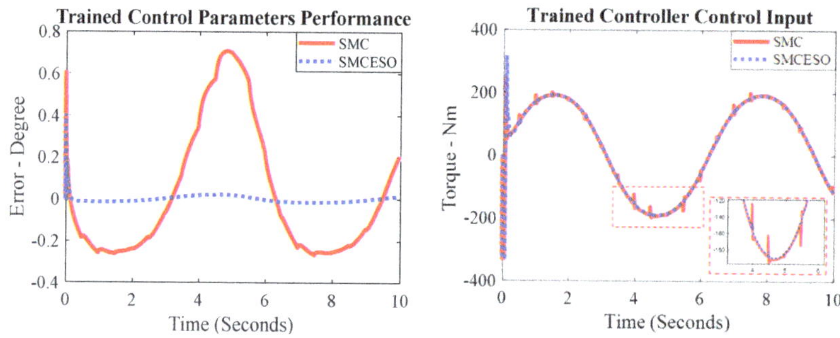

Figure 7. Fine-tuned controller tracking performance.

The joints were equipped with electromechanical motor dynamics with the motor parameters given in Table 3. Consequently, both control algorithms (SMC and SMCESO) can achieve joint tracking errors with the range ± 1 *degree*. However, it is evident from the control input that SMC has sudden spikes throughout the simulation. Reducing gains can eliminate these spikes but will reduce the control performance, resulting in larger errors. Similarly, to reduce the error of SMC, higher gains (more than double those of SMCESO) are required. This, in turn, increases the spikes and occasionally introduces chattering in the response. In contrast, SMCESO shows very smooth performance and keeps the error within the range of ± 0.1 *degree*. This validates the robustness of SMC integrated with ESO, which overcomes the perturbation effects of the system with a total mass $m > 55$ Kg on joint 2. Overall, the initial jump in the control input is primarily attributed to motor dynamics such as friction, which stabilizes once the robot starts moving. Moreover, for a deeper understanding of achieving robust performance, observing the estimated states in Figure 8.

Table 3. Motor dynamics parameters.

Parameter	Value
Inductance, L	0.573×10^{-3} H
Resistance, R	$0.978\ \Omega$
Torque constant, k_t	33.5×10^{-3} N·m/A
Voltage constant, k_e	33.5×10^{-3} V·s/rad
Gear Ratio	100

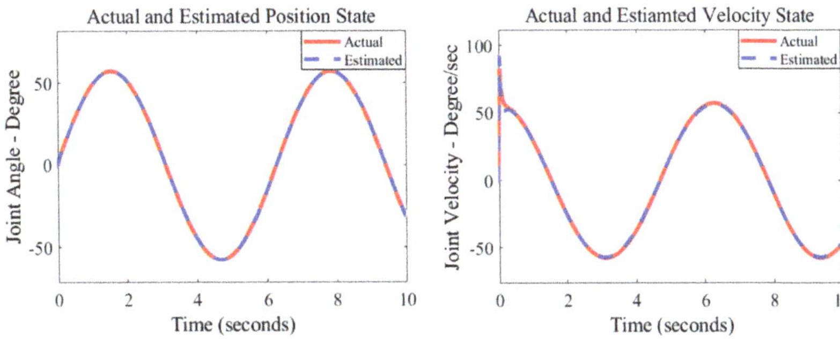

Figure 8. Actual and estimated states of the system.

The position and velocity results show that the state observer is performing very well, with estimations showing nearly zero error. This suggests that the system may have highly effective perturbation estimation and compensation capabilities to enhance tracking performance. Moreover, the Simscape multibody toolbox allows obtaining the dynamics components of the robot system, including the mass matrix $M(q)$, velocity product torque $C(q,\dot{q})\cdot\dot{q}$ with $C(q,\dot{q})$ Coriolis terms, and gravitational torque $G(q)$. This can be achieved through first creating the rigid body tree and then utilizing the Manipulator Algorithm library from Robotics System Toolbox. Subsequently, similar to (10), the expected perturbation is presumed as

$$\psi(x,t) = C(q,\dot{q})\cdot\dot{q} + G(q) \tag{40}$$

The assumed and estimated perturbations are presented in Figure 9, below.

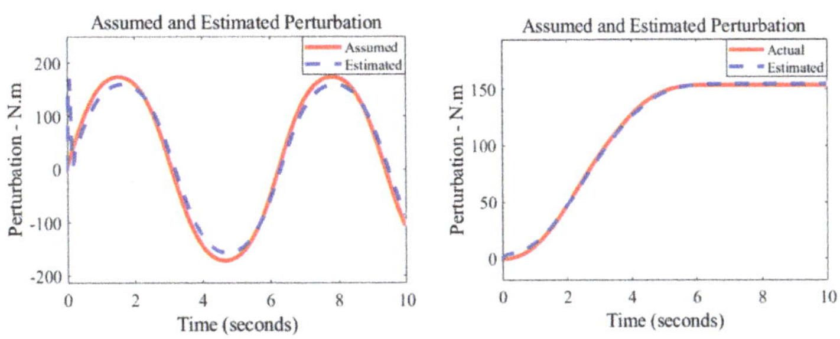

Figure 9. Perturbation results.

The estimated perturbation closely aligns with the assumed perturbation. With the desired trajectory being a sine wave, the velocity is continuously changing, leading to some perturbation estimation error, as expected due to the motor dynamics, which are not factored into the perturbation calculation. However, this error can be compensated by the SMC in Equation (16), further validating the theory in Equation (29) that, with ESO, the system dynamics are primarily influenced by the perturbation estimation error. From a magnitude perspective, it is evident that the perturbation estimation error is considerably smaller than the actual perturbation, making the system achieve robust performance. Furthermore, when the robot comes to a stop, the estimated perturbation converges to match the assumed perturbation, confirming the accurate working of the ESO.

5. Conclusions

In this study, an approach to control and stabilize multibody robotic systems with inherent dynamics and uncertainties is presented. The approach leverages extended state observer (ESO) and sliding mode control (SMC) (SMCESO), combined with the optimization capabilities of deep deterministic policy gradients (DDPGs). One of the advantages of ESO is that it requires only partial state feedback (position) to estimate the perturbation, which includes the system dynamics and external disturbances. Initially, the proposed algorithm is implemented on a simple second-order system with introduced sinusoidal disturbance. Subsequently, the control parameters were fine-tuned using a DDPG agent, which was trained based on system tracking error, joint angle, estimated joint angle, and estimation error. This training allowed the DDPG-based SMCESO to outperform the optimally tuned PID control (via a control tuner toolbox), conventional SMC (tuned through DDPG), and H_∞ control in terms of robustness, significantly enhancing system stability and performance. Even in the presence of disturbances, the SMCESO consistently converges to zero error due to its perturbation rejection capabilities. It was also demonstrated that with ESO, the system dynamics are primarily affected by the perturbation estimation error, which was validated through simulations showing close alignment between estimated and actual perturbations, leaving only minor estimation errors to be handled by the SMC control input. As a result, the multibody robot system's overall performance is highly robust.

Author Contributions: Conceptualization, H.K. and M.C.L.; Data curation, S.A.K.; Formal analysis, F.G.; Funding acquisition, U.H.S.; Investigation, S.A.K. and F.G.; Methodology, H.K.; Project administration, M.C.L.; Resources, M.C.L.; Software, H.K.; Validation, U.G. and U.H.S.; Writing—original draft, H.K.; Writing—review and editing, U.G. and U.H.S. All authors have read and agreed to the published version of the manuscript.

Funding: This research received no external funding.

Data Availability Statement: Data are contained within the article.

Acknowledgments: This work was supported by the Deanship of Graduate Studies and Research (DGSR) Program, Ajman University, United Arab Emirates.

Conflicts of Interest: The authors declare no conflict of interest.

References

1. Shtessel, Y.; Edwards, C.; Fridman, L.; Levant, A. Introduction: Intuitive Theory of Sliding Mode Control. In *Sliding Mode Control and Observation*; Control Engineering; Birkhäuser: New York, NY, USA, 2014; pp. 1–42.
2. Afifa, R.; Ali, S.; Pervaiz, M.; Iqbal, J. Adaptive Backstepping Integral Sliding Mode Control of a MIMO Separately Excited Dc Motor. *Robotics* **2023**, *12*, 105. [CrossRef]
3. Khan, H.; Abbasi, S.J.; Lee, M.C. DPSO and Inverse Jacobian-based Real-time Inverse Kinematics with Trajectory Tracking using Integral SMC for Teleoperation. *IEEE Access* **2020**, *8*, 159622–159638. [CrossRef]
4. Hollweg, G.V.; de Oliveira Evald, P.J.; Milbradt, D.M.; Tambara, R.V.; Gründling, H.A. Design of continuous-time model reference adaptive and super-twisting sliding mode controller. *Math. Comput. Simul.* **2022**, *201*, 215–238. [CrossRef]
5. Mobayen, S.; Bayat, F.; ud Din, S.; Vu, M.T. Barrier function-based adaptive nonsingular terminal sliding mode control technique for a class of disturbed nonlinear systems. *ISA Trans.* **2023**, *134*, 481–496. [CrossRef] [PubMed]
6. Khan, H.; Abbasi, S.J.; Lee, M.C. Robust Position Control of Assistive Robot for Paraplegics. *Int. J. Control Autom. Syst.* **2021**, *19*, 3741–3752. [CrossRef]
7. Abbasi, S.J.; Khan, H.; Lee, J.W.; Salman, M.; Lee, M.C. Robust Control Design for Accurate Trajectory Tracking of Multi-Degree-of-Freedom Robot Manipulator in Virtual Simulator. *IEEE Access* **2022**, *10*, 17155–17168. [CrossRef]
8. Humaidi, A.J.; Hasan, A.F. Particle Swarm Optimization-Based Adaptive Super-Twisting Sliding Mode Control Design for 2-Degree-of-Freedom Helicopter. *Meas. Control* **2019**, *52*, 1403–1419. [CrossRef]
9. Wang, Y.; Zhu, K.; Yan, F.; Chen, B. Adaptive Super-Twisting Nonsingular Fast Terminal Sliding Mode Control for Cable-Driven Manipulators using Time-Delay Estimation. *Adv. Eng. Softw.* **2019**, *128*, 113–124. [CrossRef]
10. Wang, H.; Fang, L.; Song, T.; Xu, J.; Shen, H. Model-free Adaptive Sliding Mode Control with Adjustable Funnel Boundary for Robot Manipulators with Uncertainties. *Rev. Sci. Instrum.* **2021**, *92*, 065101. [CrossRef]
11. Xi, R.-D.; Xiao, X.; Ma, T.-N.; Yang, Z.-X. Adaptive Sliding Mode Disturbance Observer-Based Robust Control for Robot Manipulators Towards Assembly Assistance. *IEEE Robot. Autom. Lett.* **2022**, *7*, 6139–6146. [CrossRef]
12. Jing, C.; Xu, H.; Niu, X. Adaptive Sliding Mode Disturbance Rejection Control with Prescribed Performance for Robotic Manipulators. *ISA Trans.* **2019**, *91*, 41–51. [CrossRef]

13. Zhao, H.; Tao, B.; Ma, R.; Chen, B. Manipulator trajectory tracking based on adaptive fuzzy sliding mode control. *Concurr. Comput. Pract. Exp.* **2023**, *35*, e7620. [CrossRef]
14. Khan, H.; Lee, M.C. Extremum Seeking-Based Adaptive Sliding Mode Control with Sliding Perturbation Observer for Robot Manipulators. In Proceedings of the IEEE International Conference on Robotics and Automation (ICRA), London, UK, 29 May–2 June 2023; pp. 5284–5290.
15. Razmi, H.; Afshinfar, S. Neural Network-Based Adaptive Sliding Mode Control Design for Position and Attitude Control of a Quadrotor UAV. *Aerosp. Sci. Technol.* **2019**, *91*, 12–27. [CrossRef]
16. Chen, Z.; Huang, F.; Chen, W.; Zhang, W.; Sun, W.; Chen, J.; Zhu, S.; Gu, J. RBFNN-Based Adaptive Sliding Mode Control Design for Delayed Nonlinear Multilateral Telerobotic System with Cooperative Manipulation. *IEEE Trans. Ind. Inform.* **2020**, *16*, 1236–1247. [CrossRef]
17. Wang, D.; Shen, Y.; Sha, Q.; Li, G.; Kong, X.; Chen, G.; He, B. Adaptive DDPG Design-Based Sliding-Mode Control for Autonomous Underwater Vehicles at Different Speeds. In Proceedings of the IEEE Underwater Technology (UT), Kaohsiung, Taiwan, 16–19 April 2019; pp. 1–5.
18. Mosharafian, S.; Afzali, S.; Bao, Y.; Velni, J.M. A Deep Reinforcement Learning-Based Sliding Mode Control Design for Partially Known Nonlinear Systems. In Proceedings of the European Control Conference (ECC), London, UK, 12–15 July 2022; pp. 2241–2246.
19. Lei, C.; Zhu, Q. U-Model-Based Adaptive Sliding Mode Control using a Deep Deterministic Policy Gradient. *Math. Probl. Eng.* **2022**, *2022*, 8980664. [CrossRef]
20. Pantoja-Garcia, L.; Parra-Vega, V.; Garcia-Rodriguez, R.; Vázquez-García, C.E. A Novel Actor—Critic Motor Reinforcement Learning for Continuum Soft Robots. *Robotics* **2023**, *12*, 141. [CrossRef]
21. Abbasi, S.J.; Lee, S. Enhanced Trajectory Tracking via Disturbance-Observer-Based Modified Sliding Mode Control. *Appl. Sci.* **2023**, *13*, 8027. [CrossRef]
22. Raoufi, M.; Habibi, H.; Yazdani, A.; Wang, H. Robust Prescribed Trajectory Tracking Control of a Robot Manipulator Using Adaptive Finite-Time Sliding Mode and Extreme Learning Machine Method. *Robotics* **2022**, *11*, 111. [CrossRef]
23. Saleki, A.; Fateh, M.M. Model-free control of electrically driven robot manipulators using an extended state observer. *Comput. Electr. Eng.* **2020**, *87*, 106768. [CrossRef]
24. Zheng, Y.; Tao, J.; Sun, Q.; Zeng, X.; Sun, H.; Sun, M.; Chen, Z. DDPG-Based Active Disturbance Rejection 3D Path-Following Control for Powered Parafoil Under Wind Disturbances. *Nonlinear Dyn.* **2023**, *111*, 1–17. [CrossRef]
25. Sun, M.; Zhang, W.; Zhang, Y.; Luan, T.; Yuan, X.; Li, X. An Anti-Rolling Control Method of Rudder Fin System Based on ADRC Decoupling and DDPG Parameter Adjustment. *Ocean. Eng.* **2023**, *278*, 114306. [CrossRef]
26. Yang, J.; Peng, W.; Sun, C. A Learning Control Method of Automated Vehicle Platoon at Straight Path with DDPG-Based PID. *Electronics* **2021**, *10*, 2580. [CrossRef]
27. Dey, N.; Mondal, U.; Mondal, D. Design of a H-Infinity Robust Controller for a DC Servo Motor System. In Proceedings of the 2016 International Conference on Intelligent Control Power and Instrumentation (ICICPI), Kolkata, India, 21–23 October 2016; pp. 27–31.

robotics

MDPI

Article

An Experimental Investigation of the Dynamic Performances of a High Speed 4-DOF 5R Parallel Robot Using Inverse Dynamics Control

Paolo Righettini [1,*], Roberto Strada [1], Filippo Cortinovis [1], Federico Tabaldi [1], Jasmine Santinelli [1] and Andrea Ginammi [2]

[1] Department of Engineering and Applied Sciences, University of Bergamo, 24044 Dalmine, Italy; roberto.strada@unibg.it (R.S.); filippo.cortinovis@unibg.it (F.C.); federico.tabaldi@unibg.it (F.T.); jasmine.santinelli@unibg.it (J.S.)

[2] Mechatronics and Dynamic Devices s.r.l., 24044 Dalmine, Italy; andrea.ginammi@mdquadro.com

* Correspondence: paolo.righettini@unibg.it

Abstract: High-speed pick-and-place industrial applications often use parallel kinematic robots due to their high stiffness and dynamic performance; furthermore, the latter not only depends on the mechanical characteristics of the robots but also on the control algorithm. The literature shows several theoretical contributions to such controllers, mainly tested at the simulation level or on simple proof-of-concept laboratory equipment that execute low-speed and simple trajectories. This paper presents an experimental investigation of the dynamic performance of an industrial high-speed 4-DOF 5R parallel robot designed for pick-and-place applications on moving objects. The inverse dynamics control in the task space is used as a control algorithm. The results show the contribution of all the components of the control algorithm to the motor torque, and the inverse dynamics controller performances are discussed also in comparison to those achievable with simpler PD or PID controllers in a joint space. Moreover, the paper shows the controller synthesis from a modern mechatronic point of view, and the effectiveness of the proposed solution for the tracking of complex high-speed trajectories in an industrial application.

Keywords: parallel robot; dynamic modelling; inverse dynamics control; high speed pick-and-place; pick-and-place with moving points; mechatronic design

Citation: Righettini, P.; Strada, R.; Cortinovis, F.; Tabaldi, F.; Santinelli, J.; Ginammi, A. An Experimental Investigation of the Dynamic Performances of a High Speed 4-DOF 5R Parallel Robot Using Inverse Dynamics Control. *Robotics* **2024**, *13*, 54. https://doi.org/10.3390/robotics13030054

Academic Editor: Raffaele Di Gregorio

Received: 15 February 2024
Revised: 17 March 2024
Accepted: 18 March 2024
Published: 20 March 2024

1. Introduction

Pick-and-place operations are among the most common in industrial applications. Within the production process, which may include, e.g., assembly, packaging, bin picking, and inspection, manipulators are commonly used in modern manufacturing environments. Given the wide range of activities that pick-and-place manipulators can perform, research related to the efficiency of these applications is of evident importance.

The productivity of this type of line, where several robots work synergistically, strictly depends on each individual manipulator's cycle time. The reduction in the cycle time is closely related to the maximum achievable speed, which depends on many factors, including manipulator structure, efficient trajectory programming, and an effective control strategy.

In many applications of this kind, where high speed and precision are required, parallel kinematics manipulators offer many advantages due to their features. Compared with their serial counterparts, their closed-loop architecture makes them generally stiffer. In addition, they also offer better dynamic performance due to the positioning of the motors on the base, which reduces significantly the total moving mass.

As mentioned earlier, making the most of the structural advantages of these manipulators requires appropriate control systems. The existing research on control systems for

PKMs (parallel kinematics machines) can be categorized into the following two distinct groups: error-based controllers and model-based controllers. Alternatively, there are also hybrid versions that combine aspects of both types. Error-based controllers, such as PID and its variations, are decentralized. They only consider joint errors and do not take into account the manipulator's dynamic model, unlike model-based controllers. The computed torque controller family falls under this second category. The use of a descriptive model of the system and the knowledge of its characteristic parameters can provide a number of performance advantages, notwithstanding difficulties related to the model's uncertainties. Both types of controllers have several advantages and disadvantages; for this reason, several research studies have been conducted to compare the performance of the two families of control systems [1], focusing on different kinds of parallel kinematic robots as test rigs. Many contributions not only compare model-based controllers with those most widely used to control PKMs such as PD, but offer solutions to improve the efficiency of both.

An effective way to improve performance is the use of variable gains, which has been proposed in several studies. This idea has resulted in numerous contributions towards enhancing performance. A recent example presented in [2] deals with the motion control of a 6-DOF Altinay Stewart–Gough platform, comparing experimentally the following two control techniques: PD and computed torque control (CTC), also known as inverse dynamics control (IDC). Non-linear gains (NPD and NCTC) and a non-linear observer were added for velocity estimation in both controllers to enhance the performance. The experimental results show that the non-linearity of the gains contributes more to improving performance than the controller structure. The performance improvement of the classical CTC using a non-linear PD component was presented earlier, applied to a planar parallel manipulator, in [3]. In [4], a sliding mode control structure was enhanced through the use of fuzzy logic aimed at a reduction in the control effort and in the chattering phenomena; moreover, the developed controller was tested on a simulated Stewart–Gough platform. Another recent article that focuses on a Stewart-like PKM is [5], in which a robot is controlled using an LQI controller whose gains are tuned in real-time using an artificial neural network. The proposed control strategy is first developed co-simulating the mechanical system and the regulator, and a prototype is subsequently used to experimentally characterize the controller. In [6], a novel 4-DOF 3T1R parallel robot is controlled using a robust control based on a grey-box dynamic model of the manipulator. Some of the dynamic contributions are modelled analytically; others, such as the Coriolis, centrifugal, and gravitational actions, are approximated using a neural network to reduce the modelling effort. A robust sliding mode control approach is then developed and experimentally tested both on a multisine trajectory and on a simple pick-and-place motion. In [7], a four-limb parallel manipulator with Schoenflies motion, which is designed for pick-and-place applications, is presented. The paper proposes an experimental validation of an inverse dynamics control applied to a simple pick-and-place trajectory. Starting from the obtained results, the same authors propose a more advanced control system in [8]. In this case, a PD controller with the addition of an offline precomputed torque is presented. The controller gains are modified in real-time by exploiting fuzzy logic and a bat algorithm. The authors justify the proposal's use of fuzzy logic due to the need to use model information to achieve good performance and, simultaneously, the difficulty when building an accurate model. Therefore, a simplified model is used for the bat algorithm, which does not consider the rods' inertia. The evaluation of the proposed system is only simulated on the system model using the same trajectory of the previous work. In addition to the variable gains, the article exploits the dynamic model of the manipulator by adding a feedforward contribution to the PD controller. In [9], the same strategy is applied on a 3PRRR prototype, a planar kinematically redundant parallel manipulator, on which the authors test a PID controller with a feedforward component dependent on the robot's model. Using a camera to detect the end-effector position in space, the paper finally proposes a hybrid joint-task space computed torque control strategy, experimentally proven to be the most effective. Visual servoing is used also in [10] to improve the performance of controllers applied to a parallel

6-RSS robot. Three controllers were implemented and compared: a joint space controller, which does not rely on the camera feedback; a task space controller with visual feedback; and a visual servoing dynamic sliding mode controller. The best results in terms of tracking errors were obtained with the sliding mode controller, while the joint space regulator led to the worst performance. Focused only on performance comparison, ref. [11] presents the experimental validation of a CT controller applied to a 3-DOF translational parallel manipulator. The results show that, compared to a PD controller, the model-based controller offers better positioning performances.

Among the parallel kinematics manipulators covered in the various surveys, several papers focus on 5R robots.

In 2022, Rodriguez [12] shows the application of a CTC to a five-bar manipulator. The 2-DOF system is a prototype with small motors. The considered trajectory develops in 70 s, requiring the addition of an integrative component to compensate for errors due to low motion dynamics. In the same year, Coutinho [13] shows an experimental comparison of two controllers, a PD and a CTC, and the contribution of a Sliding Mode (SM) component. With experimental tests performed on a 2-DOF prototype, both the PD + SM and CT + SM hybrid controllers were shown to reduce the tracking errors, with the one based on the dynamic model being proved more effective. The combination of sliding mode control with classical CTC is also evaluated in [14]. The comparison between the CT + SM and the generic CTC is performed on the simulated system, showing good performances for both controllers. Another contribution proposing the simulation-based comparison of different types of controllers is [15], where SMC is compared with PD and CTC, concluding that SMC is more robust to uncertainties. In [16], the application of a PD controller with precomputed torque feedforward is proposed, where non-linear gains are calibrated in real-time by exploiting neural networks. Compared with a classical CTC, the presented controller offers better performances; however, the results were not validated experimentally, but rather with simulations on the manipulator model. Another research applied to the five-bar mechanism, focused instead on PID controllers, is presented in [17]. Numerical simulations of a finite-time non-linear PID regulation controller, applied to the model of a five-bar mechanism were conducted. The simulations' results confirm the usefulness of the proposed approach.

Recalling the interest of PKM robots for industrial pick-and-place applications, the analysis of articles concerning the use of different controllers on parallel kinematics manipulators reveals some of the following main shortcomings:

- The literature does not show research conducted on an industrial 5R robot for the inverse dynamics control evaluation. Some authors use simple mechanical systems, such as in [12] or [13], where experimental tests are conducted on 2-DOF prototypes that move slowly and hence with negligible dynamic contributions to the joint torques;
- In almost all articles, except for [7], the trajectories used to evaluate the performance of control systems are not meaningful to the pick-and-place application. Specifically, the trajectories with low dynamics, or not representative of a typical pick-and-place motion, are used;
- None of the reviewed papers quantifies the contributions of the various components of validated control systems;
- The inverse dynamics controller is almost always applied only in the joint space.

The present paper deals with an experimental evaluation of the dynamic performances of a 5R 4-DOF robot controlled by an inverse dynamics controller. The design procedure commonly used for robotic or mechatronic systems gives a high confidence level for the evaluation of the mass parameters of the robot. Based on this knowledge already available at the system design level, and on the dynamic model of the robot, the authors chose the inverse dynamics control structure, whose parameters directly come from the design procedure. Moreover, the robot performances obtained using the inverse dynamics controller is compared with the ones of error-based PD/PID controllers, which, being in widespread use, can be considered useful benchmarks. In light of the previous analysis of the literature, the main contributions of the present paper can be summarized as follows:

- The experimental activities are conducted on a robot designed for high-speed pick-and-place operations in industrial applications; furthermore, the performances of the system are investigated using a pick-and-place trajectory [14] that fully exploits the manipulator's characteristics;
- The pick-and-place trajectory used for the experimental activities does not lie just on a plane, but it also has out-of-plane sections; moreover, it features both highly dynamic and quasi-static portions;
- The ID controller is applied in the task space and not, as is commonly done, in the joint space;
- The different contributions that constitute the global controller signal are evaluated and discussed.

The article is structured as follows: Section 2 presents the dynamic and the kinematic model of the robot object of the investigation (Section 2.1), the control schemes (Section 2.2), the reference pick-and-place task (Section 2.3), and the software and hardware equipment (Section 2.4) used for the experimental activities. Section 3 discusses the collected experimental results and compares the investigated controllers' performances. Finally, some conclusions are drawn in Section 4.

2. Methods

2.1. Dynamic Model of the Manipulator

The considered robot is a 4-DOF parallel kinematic manipulator designed for fast pick-and-place applications. As shown in Figure 1, the robot's mechanical structure is composed of the following two main subsystems: a 5R planar linkage, which determines the position of point p_{5R} in the horizontal plane, and a ball-screw-spline (BSS) mechanism mounted in p_{5R} that enables the roto-translation of the end-effector along and around the vertical axis. The manipulator presents significant masses and inertias distributed among its main constitutive elements, marking a difference with other notable PKM architectures, such as the Delta robot, whose main masses are concentrated in a few key components. To counterbalance the presence of considerable masses, the mechanical design favoured high structural stiffness, in contrast, e.g., with the PKM investigated in [18].

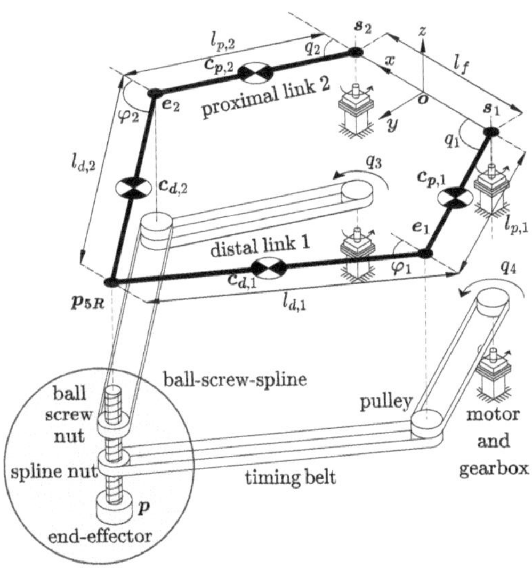

Figure 1. Schematic representation of the 4-DOF manipulator with its main functional subelements. All the notable points and angles are also indicated.

All the parameters needed to describe the geometry, the mass distribution and the actuators of the device are determined during its mechatronic design phase, and can therefore be also used for the synthesis of the control system. The geometrical and mass parameters are in particular listed, described, and quantified in Table 1, with symbols and nomenclature consistent with those indicated in Figure 1. The numerical values of each quantity were provided by the manufacturer, Mechatronics and Dynamic Devices s.r.l., Dalmine, Italy [19]. The properties of the actuation systems are reported, on the other hand, in Table 2, the selection of the transmission and of the power drive systems having been performed according to the procedure detailed in [20].

Table 1. Mass and geometrical properties of the robot

Description	Symbol	Value
Length of the proximal links	l_p	250 mm
Length of the distal links	l_d	250 mm
Frame length	l_f	180 mm
Mass of the proximal links	$m_{p,1}, m_{p,2}$	2.9 kg
Mass of the distal links	$m_{d,1}, m_{d,2}$	2.9 kg
Barycentric inertia of the proximal links	$J_{p,1}, J_{p,2}$	5.22×10^{-2} kg m^2
Barycentric inertia of the distal links	$J_{d,1}, J_{d,2}$	5.22×10^{-2} kg m^2
Mass of the BSS and of the end-effector	m_{ee}	0.36 kg
BSS pitch	p_{bss}	20 mm
Rotational inertia of the end effector	J_{ee}	6.40×10^{-6} kg m^2
Ball screw nut's moment of inertia	J_3	1.20×10^{-6} kg m^2
Spline nut's moment of inertia	J_4	1.20×10^{-6} kg m^2

Table 2. Main parameters of the four servoaxes.

Description	Symbol	Axis 1	Axis 2	Axis 3	Axis 4
Rated torque	$\tau_{pds,rated}$, [N m]	0.7	0.7	0.36	0.36
Peak torque	$\tau_{pds,max}$, [N m]	1.4	1.4	0.72	0.72
Peak velocity	$\omega_{pds,max}$, $[\mathrm{rad\,s^{-1}}]$	500	500	500	500
Motor inertia	J_m, $[\mathrm{kg\,mm^2}]$	17	17	2.4	2.4
Transmission efficiency	η_t	~1	~1	~1	~1
Transmission inertia	J_t, $[\mathrm{kg\,mm^2}]$	24.85	24.85	20.5	20.5
Reduction ratio	i_t	64	64	40	40

The planar linkage is actuated by two servoaxes constituted each by a brushless motor and a planetary gearbox attached to the robot's fixed frame. The BSS is actuated by another pair of similarly constituted servoaxes, through two timing belt transmission systems housed inside the hollow links. The pulleys constituting the transmissions have all the same radius; as a result, the BSS and the planar 5R mechanism are kinematically decoupled. In fact, the configuration of the 5R linkage is a function only of generalized coordinates $\boldsymbol{q}_{5R} = \begin{bmatrix} q_1 & q_2 \end{bmatrix}^\top$, which, as shown in Figure 1, represent the proximal link rotations. Conversely, the configuration of the BSS is described by the joint coordinates $\boldsymbol{q}_{bss} = \begin{bmatrix} q_3 & q_4 \end{bmatrix}^\top$, which ultimately correspond to the rotations of the ball screw and spline nuts that constrain the ball-screw-spline. As a result, the in-plane position $\boldsymbol{p}_{5R}$ of the end-effector is determined exclusively by $\boldsymbol{q}_{5R}$, whereas its vertical and angular positions z_{ee} and φ_{ee} are functions of $\boldsymbol{q}_{bss}$ only.

For motion planning purposes, it is useful to study the forward and inverse kinematics of the device up to the accelerations. A dynamic model of the robot is on the other hand useful for the development of centralized regulators.

The kinematics of the two mechanical subsystems is well-known and easily composed. The authors have already treated, in greater detail, the kinematics of the manipulator in [21]. An in-depth discussion of the singularities for the five-bar linkage can be found in [22,23], even though it should be remarked that the case study presented in this work does not involve the crossing of any singularity. Finally a more detailed discussion of the kinematics and dynamics of the 5R robot can be found in [14]. The dynamics of the entire mechanism may be written as follows:

$$\left(\begin{bmatrix} \boldsymbol{M}_{5R} & 0 \\ 0 & \boldsymbol{M}_{bss} \end{bmatrix} + \boldsymbol{M}_{act}\right)\begin{bmatrix} \ddot{\boldsymbol{q}}_{5R} \\ \ddot{\boldsymbol{q}}_{bss} \end{bmatrix} + \begin{bmatrix} \boldsymbol{C}_{5R} & 0 \\ 0 & 0 \end{bmatrix}\begin{bmatrix} \dot{\boldsymbol{q}}_{5R} \\ \dot{\boldsymbol{q}}_{bss} \end{bmatrix} + \begin{bmatrix} 0 \\ \nabla U_{bss} \end{bmatrix} = \boldsymbol{\tau}, \tag{1}$$

or more compactly as follows:

$$\boldsymbol{M}(\boldsymbol{q})\ddot{\boldsymbol{q}} + \boldsymbol{C}(\boldsymbol{q},\dot{\boldsymbol{q}})\dot{\boldsymbol{q}} + \nabla U = \boldsymbol{\tau}\,. \tag{2}$$

In Equation (1) $\boldsymbol{M}_{5R}$ and $\boldsymbol{M}_{bss}$ are the mass matrices associated with the planar linkage and to the BSS, respectively, $\boldsymbol{C}_{5R}$ is the 5R 's Coriolis and centrifugal actions matrix; ∇U_{bss} is the gradient of the gravitational potential associated with the ball-screw-spline, while $\boldsymbol{\tau}$ are the torques applied at the output shaft of the four gearboxes. It should be noted that the planar linkage operates in the horizontal plane, and hence its potential energy is constant; also it may be remarked that the BSS is a linear time-invariant system and, therefore, not subject to Coriolis and centrifugal actions. Finally, the following matrix:

$$\boldsymbol{M}_{act} = \mathrm{diag}\left(\left[i_{t1}^2(J_{m1} + J_{t1}) \quad i_{t2}^2(J_{m2} + J_{t2}) \quad i_{t2}^2(J_{m2} + J_{t2}) \quad i_{t4}^2(J_{m4} + J_{t4})\right]\right) \tag{3}$$

accounts for the effects of the moments of inertia J_{mj} and J_{tj} (with $j = 1, \dots, 4$) associated with the four motor shafts and gearboxes, considering as appropriate the reduction ratios i_{tj}. The matrix $\boldsymbol{M}_{5R}$ can be written as follows:

$$\boldsymbol{M}_{5R} = \sum_{k=1}^{2}\left(m_{p,k}\boldsymbol{D}_{cp,k}^{\top}\boldsymbol{D}_{cp,k} + m_{d,k}\boldsymbol{D}_{cd,k}^{\top}\boldsymbol{D}_{cd,k} + J_{d,k}\boldsymbol{D}_{\varphi d,k}^{\top}\boldsymbol{D}_{\varphi d,k}\right) + m_{ee}\boldsymbol{D}_{5R}^{\top}\boldsymbol{D}_{5R} + \boldsymbol{J}_p \tag{4}$$

with the following:

$$\boldsymbol{J}_p = \begin{bmatrix} J_{p,1} & 0 \\ 0 & J_{p,2} \end{bmatrix}. \tag{5}$$

In Equation (4) $\boldsymbol{D}_{cp,k}$ and $\boldsymbol{D}_{cd,k}$ represent the Jacobian matrices associated with the centers of mass of the proximal and distal links; $\boldsymbol{D}_{\varphi d,k}$ indicate the Jacobian matrices of distal angles $\varphi_{d,k}$, while $\boldsymbol{D}_{5R}$ is the Jacobian of $\boldsymbol{p}_{5R}$. The mass and inertial parameters can all be found in Table 1 with their description and quantification. The matrix $\boldsymbol{C}_{5R}$ has been written as follows:

$$\boldsymbol{C}_{5R} = \sum_{k=1}^{2}\left(m_{p,k}\boldsymbol{D}_{cp,k}^{\top}\dot{\boldsymbol{D}}_{cp,k} + m_{d,k}\boldsymbol{D}_{cd,k}^{\top}\dot{\boldsymbol{D}}_{cd,k} + J_{d,k}\boldsymbol{D}_{\varphi d,k}^{\top}\dot{\boldsymbol{D}}_{\varphi d,k}\right) + m_{ee}\boldsymbol{D}_{5R}^{\top}\dot{\boldsymbol{D}}_{5R}\,, \tag{6}$$

exploiting the fact that the 5R linkage undergoes purely planar motions. In Equation (6), it can be seen that the time derivatives of the Jacobian matrices also appear. Fully working out these kinematic quantities is outside the scope of this work; moreover, they all have been determined exactly using as appropriate analytical and geometrical methods.

The kinematics of the BSS is invariant with respect to the joint configuration, and is thus fully described by a constant Jacobian matrix. Let p_{bss} be the pitch of the ball-screw-spline; the following kinematic relationships hold:

$$\begin{bmatrix} \dot{z}_{ee} \\ \dot{\varphi}_{ee} \end{bmatrix} = \frac{1}{2\pi}\begin{bmatrix} -p_{bss} & p_{bss} \\ 0 & 2\pi \end{bmatrix}\dot{\boldsymbol{q}}_{bss} = \boldsymbol{D}_{bss}\dot{\boldsymbol{q}}_{bss}\,. \tag{7}$$

The kinetic and potential energies of the ball-screw-spline can therefore be expressed with respect to the joint variables as follows:

$$T_{bss} = \frac{1}{2}\left((J_3 + \frac{m_{ee}p_{bss}^2}{4\pi^2})\dot{q}_3^2 + (J_4 + \frac{m_{ee}p_{bss}^2}{4\pi^2} + J_{ee})\dot{q}_4^2 - \frac{m_{ee}p_{bss}^2}{2\pi^2}\dot{q}_3\dot{q}_4\right) \tag{8}$$

$$U_{bss} = m_{ee}g\frac{p_{bss}}{2\pi}(q_4 - q_3), \tag{9}$$

where J_3 and J_4 are the moments of inertia of the ball screw and spline nuts (inclusive also of the inertia of the associated pulleys), m_{ee} and J_{ee} are the mass and moment of inertia of the end-effector, and g is the gravitational constant.

The mass matrix of the BSS can be obtained as the Hessian of T_{bss} with respect to $\dot{\boldsymbol{q}}_{bss}$; it results as follows:

$$\boldsymbol{M_{bss}} = \begin{bmatrix} (J_3 + \frac{m_{ee}p_{bss}^2}{4\pi^2}) & -\frac{m_{ee}p_{bss}^2}{4\pi^2} \\ -\frac{m_{ee}p_{bss}^2}{4\pi^2} & (J_4 + \frac{m_{ee}p_{bss}^2}{4\pi^2} + J_{ee}) \end{bmatrix}. \tag{10}$$

The gradient of the potential energy U_{bss} is also straightforwardly computed as follows:

$$\nabla U_{bss} = m_{ee}g\frac{p_{bss}}{2\pi}\begin{bmatrix} -1 \\ 1 \end{bmatrix}, \tag{11}$$

It should be noted that this kind of mathematical modelling neglects the elasticities of the belt transmission systems. These could be modelled using an appropriate elastic potential energy function, which, however, would require the introduction of additional non-actuated and non-measured degrees of freedom.

Considering again the entire system, the Jacobian matrix that relates the time derivatives of the joint coordinates to the end-effector velocity can be written block by block as follows:

$$\boldsymbol{D} = \begin{bmatrix} \boldsymbol{D_{5R}} & 0 \\ 0 & \boldsymbol{D_{bss}} \end{bmatrix}. \tag{12}$$

Consequently its time derivative assumes the following form:

$$\dot{\boldsymbol{D}} = \begin{bmatrix} \dot{\boldsymbol{D}}_{5R} & 0 \\ 0 & 0 \end{bmatrix}. \tag{13}$$

Equation (2) represents the classical description of the dynamics of the system expressed in the joint space. Given the goal of developing centralized control systems operating with task space variables, the equations of motion of the entire system can also be rewritten as follows:

$$\boldsymbol{M}(\boldsymbol{p})\big(\boldsymbol{D}^{-1}(\ddot{\boldsymbol{p}} - \dot{\boldsymbol{D}}\boldsymbol{D}^{-1}\dot{\boldsymbol{p}})\big) + \boldsymbol{C}(\boldsymbol{p}, \dot{\boldsymbol{p}})\big(\boldsymbol{D}^{-1}\dot{\boldsymbol{p}}\big) + \nabla U(\boldsymbol{p}) = \boldsymbol{\tau}. \tag{14}$$

Finally, through the suitable collection of the relevant terms it becomes possible to write the task space formulation of the following equations of motion:

$$\boldsymbol{M_p}(\boldsymbol{p})\ddot{\boldsymbol{p}} + \boldsymbol{n_p}(\boldsymbol{p}, \dot{\boldsymbol{p}}) = \boldsymbol{\tau}. \tag{15}$$

This last form is the one used for the development of the centralized control systems described in the following section:

2.2. Control Schemes

In this work, a modern model-based approach to the development of the robot's control system is adopted, whereby the synthesis of the regulator uses all the information commonly available from the device's overall mechatronic design. As a result, the detailed knowledge of the geometric and mass properties of the system, already included into its kinematic and dynamic models, is furthermore leveraged to improve the control performance. The proposed controllers are therefore of the inverse dynamic type, and are directly synthesized in the task space. The first one features only proportional and derivative feedback; as detailed below, the second also includes integral actions. These controllers are experimentally investigated and characterized through comparison with two joint space decentralized controllers of the PD and PID types, which, due to their widespread use, constitute useful benchmarks. The centralized control systems are based on task space variables and, in principle, allow the designer to directly specify the desired dynamics of the end-effector position error $\widetilde{\boldsymbol{p}} = \boldsymbol{p}_{sp} - \boldsymbol{p}$ (with $\boldsymbol{p}_{sp}$ being the setpoint and $\boldsymbol{p}$ the actual end-effector position). Figure 2 represents one of these regulators, the task space inverse dynamics controller (TSIDC). Its centralized nature stems from the following properties:

- Both the mass matrix $\boldsymbol{M}_p$ of the system and the feedback linearization term $\boldsymbol{n}_p$ are computed as a function of the four task space coordinates of the system according to the dynamic model expressed in Equation (15);
- Each torque setpoint is calculated according to the errors on more than one coordinate, since the mass matrix is not purely diagonal.

Considering, in particular, diagonal matrices $\boldsymbol{K}_p$ and $\boldsymbol{K}_d$, the torque setpoint is determined according to the following equation:

$$\boldsymbol{\tau}_{sp} = \boldsymbol{M}_p(\boldsymbol{p})(\ddot{\boldsymbol{p}}_{sp} + \boldsymbol{K}_d\dot{\widetilde{\boldsymbol{p}}} + \boldsymbol{K}_p\widetilde{\boldsymbol{p}}) + \boldsymbol{n}_p(\boldsymbol{p}, \dot{\boldsymbol{p}}), \tag{16}$$

in which several torque components can be singled out and defined as follows:

$$\boldsymbol{\tau}_p^{(TSID)} = \boldsymbol{M}_p\boldsymbol{K}_p\widetilde{\boldsymbol{p}} \tag{17}$$

$$\boldsymbol{\tau}_d^{(TSID)} = \boldsymbol{M}_p\boldsymbol{K}_d\dot{\widetilde{\boldsymbol{p}}} \tag{18}$$

$$\boldsymbol{\tau}_{M_p}^{(TSID)} = \boldsymbol{M}_p\ddot{\boldsymbol{p}}_{sp} \tag{19}$$

$$\boldsymbol{\tau}_{n_p}^{(TSID)} = \boldsymbol{n}_p(\boldsymbol{p}, \dot{\boldsymbol{p}}). \tag{20}$$

The torque setpoint $\boldsymbol{\tau}_{sp}$ is saturated in relation to the maximal performances achievable by the power drive systems and then fed into the driver, which generates the actual torque $\boldsymbol{\tau}$.

Assuming a perfectly matched dynamic model of the system and as a first approximation an identity between $\boldsymbol{\tau}$ and $\boldsymbol{\tau}_{sp}$, substitution of Equation (16) into Equation (15) leads to the task space error dynamics equation in the form $\ddot{\widetilde{\boldsymbol{p}}} + \boldsymbol{K}_d\dot{\widetilde{\boldsymbol{p}}} + \boldsymbol{K}_p\widetilde{\boldsymbol{p}} = \boldsymbol{0}$, in which $\boldsymbol{K}_p$ and $\boldsymbol{K}_d$ can be selected, e.g., through pole placement.

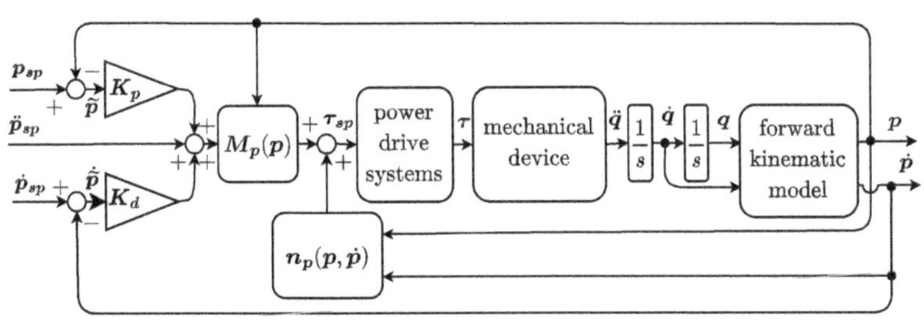

Figure 2. Task space inverse dynamics control scheme.

However, the feedback linearization of the system is imperfect due to unmodelled effects such as friction actions and non-idealities of the inner torque control loops; as a result, the actual error dynamics remains excited by a residual disturbance term $\boldsymbol{d}$ appearing on the right-hand side, and do not, in practice, converge to zero as follows:

$$\ddot{\widetilde{\boldsymbol{p}}} + \boldsymbol{K}_d\dot{\widetilde{\boldsymbol{p}}} + \boldsymbol{K}_p\widetilde{\boldsymbol{p}} = \boldsymbol{d}\,. \tag{21}$$

To counteract this effect several strategies have been proposed in the literature, ranging from adaptive control schemes [24–27], to sliding mode robust controllers [28–31]. The present paper investigates a centralized control strategy in which an integral contribution is added to the base TSID regulator, yielding the task space integral inverse dynamics controller (TSIIDC); its block diagram is represented in Figure 3. The integrator is implemented as a discrete time-forward Euler accumulator, which temporarily stops its summation whenever an upper or lower bound has been reached. This overall behaviour, which prevents the windup of the integral term, is indicated with slight abuse of notation in Figure 3 by the saturation symbol. Since the TSIIDC works in the task space, the output of the accumulator, denoted as $\widetilde{\boldsymbol{P}}$, is closely related to the integral of the task space position errors. The overall equation that yields the torque setpoint is as follows:

$$\boldsymbol{\tau}_{sp} = \boldsymbol{M}_p(\boldsymbol{p})(\ddot{\boldsymbol{p}}_{sp} + \boldsymbol{K}_d\dot{\widetilde{\boldsymbol{p}}} + \boldsymbol{K}_p\widetilde{\boldsymbol{p}} + \boldsymbol{K}_i\widetilde{\boldsymbol{P}}) + \boldsymbol{n}_p(\boldsymbol{p},\dot{\boldsymbol{p}})\,, \tag{22}$$

where again $\boldsymbol{K}_p$, $\boldsymbol{K}_d$ and also $\boldsymbol{K}_i$ are diagonal matrices. From Equation (22) several torque contributions may be isolated as follows:

$$\boldsymbol{\tau}_p^{(TSIID)} = \boldsymbol{M}_p\boldsymbol{K}_p\widetilde{\boldsymbol{p}} \tag{23}$$
$$\boldsymbol{\tau}_d^{(TSIID)} = \boldsymbol{M}_p\boldsymbol{K}_d\dot{\widetilde{\boldsymbol{p}}} \tag{24}$$
$$\boldsymbol{\tau}_i^{(TSIID)} = \boldsymbol{M}_p\boldsymbol{K}_i\widetilde{\boldsymbol{P}} \tag{25}$$
$$\boldsymbol{\tau}_{M_p}^{(TSIID)} = \boldsymbol{M}_p\ddot{\boldsymbol{p}}_{sp} \tag{26}$$
$$\boldsymbol{\tau}_{n_p}^{(TSIID)} = \boldsymbol{n}_p(\boldsymbol{p},\dot{\boldsymbol{p}})\,. \tag{27}$$

The error dynamics in this case assumes the following form:

$$\ddot{\widetilde{\boldsymbol{p}}} + \boldsymbol{K}_d\dot{\widetilde{\boldsymbol{p}}} + \boldsymbol{K}_p\widetilde{\boldsymbol{p}} + \boldsymbol{K}_i\widetilde{\boldsymbol{P}} = \boldsymbol{d}\,. \tag{28}$$

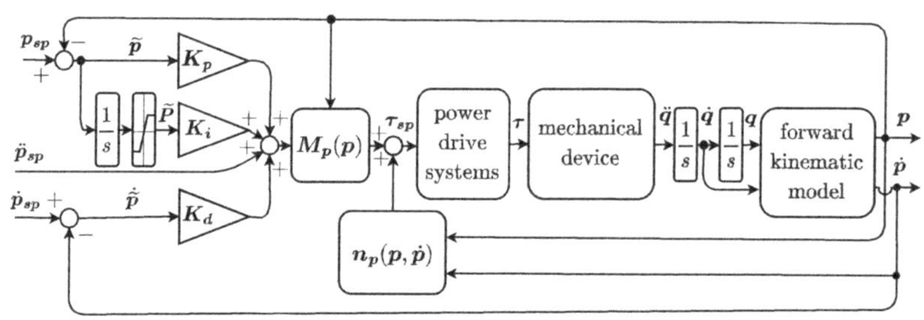

Figure 3. Task space inverse dynamics control with integral contribution scheme.

Figure 4 depicts the PD regulator. Given the joint position and velocity setpoints $\boldsymbol{q}_{sp}$, $\dot{\boldsymbol{q}}_{sp}$, the joint position and velocity errors $\widetilde{\boldsymbol{q}} = \boldsymbol{q}_{sp} - \boldsymbol{q}$ and $\dot{\widetilde{\boldsymbol{q}}} = \dot{\boldsymbol{q}}_{sp} - \dot{\boldsymbol{q}}$ are computed and used to generate a torque setpoint $\boldsymbol{\tau}_{sp}$ according to the following relationship:

$$\boldsymbol{\tau}_{sp} = \boldsymbol{K}_d\dot{\widetilde{\boldsymbol{q}}} + \boldsymbol{K}_p\widetilde{\boldsymbol{q}} = \boldsymbol{\tau}_d^{(PD)} + \boldsymbol{\tau}_p^{(PD)}\,. \tag{29}$$

Even though Equation (29) is written in vector form, it should be noted that the control loop of each joint is completely independent from the others owing to the fact that the matrices K_d and K_p are diagonal. The PID controller, illustrated in Figure 5, features an additional feedback action operating on the integral of the joint position error. The implementation of the integrator is analogous to the one already described for the TSIIDC, with its output being denoted as $\widetilde{Q}$; as can be inferred from the figure, the equation associated with the PID regulator is as follows:

$$\tau_{sp} = K_d\dot{\widetilde{q}} + K_p\widetilde{q} + K_i\widetilde{Q} = \tau_d^{(PID)} + \tau_p^{(PID)} + \tau_i^{(PID)}. \quad (30)$$

Again Equation (30) is written in vector form, but since K_d, K_p, and K_i are diagonal matrices the PID control loops operate without interactions between each other.

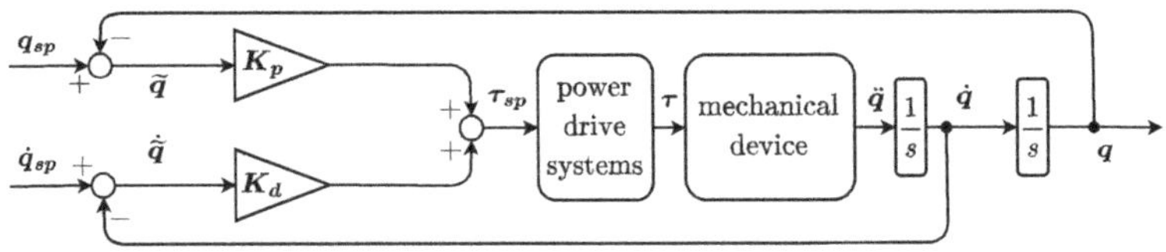

Figure 4. Joint space PD control scheme.

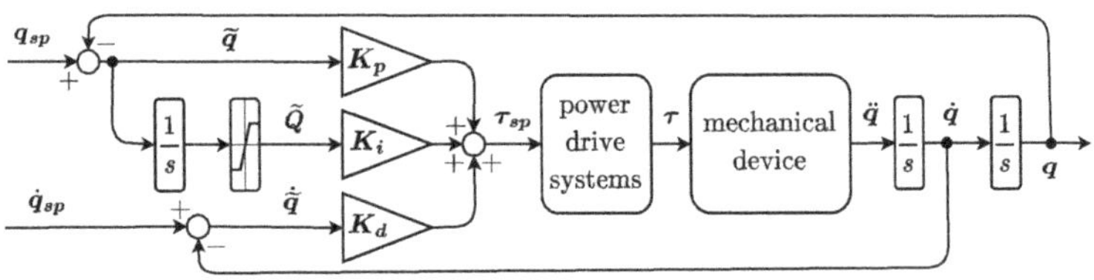

Figure 5. Joint space PID control scheme.

2.3. Workcycle Description

The regulators are tested on a workcycle generated by the motion planning system previously developed by the authors [21] specifically to define fast pick-and-place operations on moving objects such as items carried by conveyor belts. The trajectory is first defined geometrically, taking into account the shape of the singularity-free workspace of the 5R manipulator; subsequently the position, velocity, and acceleration setpoints are generated as a function of time, taking care to minimize the overall execution time of the trajectory given the constraints determined by the physical properties of the robot and of its actuation systems.

Figure 6 illustrates the workcycle used as a test case. The path, expressed in the task space coordinates, is represented in the foreground, with the color indicating the value of the end-effector rotation. In the background, the outline of the two conveyor belts that are envisioned for this type of application is represented in green, and the footprint of the useful workspace of the robot is also represented by the black line. The red markers placed on the trajectory and labelled from (a) to (i) divide the overall pick-and-place cycle into its constitutive phases. These are as follows:

- (a)–(b): Pick intercept motion that brings manipulator's end-effector directly above and aligned with its target item; at point (b) the velocity of the robot matches that of the pick conveyor;
- (b)–(c): Descent motion that brings the end-effector in contact with the item to be picked; during this phase the velocity of the conveyor belt is tracked in the longitudinal direction;

- (c)–(d): Grasping motion, during which the conveyor velocity is tracked as the gripping tool on the end-effector operates to collect the item;
- (d)–(e): Ascent motion, in which the end-effector moves away from the conveyor while still tracking its longitudinal velocity to avoid any collision with other items on the belt;
- (e)–(f): Intercept motion needed to reach the moving place position (e.g., an empty box carried by the place conveyor);
- (f)–(g): Descent motion, in which the item held by the robot is lowered to the level of the place conveyor;
- (g)–(h): Deposit motion, in which the item is released on the conveyor;
- (h)–(i): Ascent motion;
- (i)–(a): Auxiliary deceleration motion that brings the robot to a resting state.

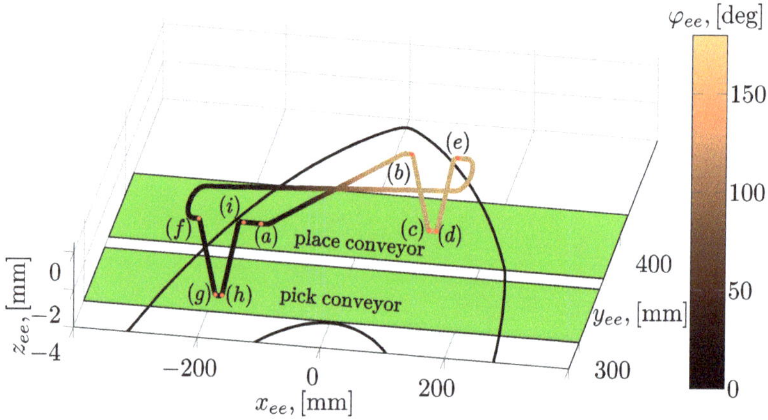

Figure 6. Reference path.

From the description above it is clear that a full pick-and-place cycle is composed of two similarly structured halves, each constituted by a target intercept motion followed by a conveyor tracking movement. The conveyor tracking phase is itself subdivided into the descent movement, the item grasping or release, and the ascent motion.

The task space setpoints are represented as a function of time in Figure 7, at the position, velocity, and acceleration levels. The gray areas highlight the conveyor tracking phases, whereas the remaining portions correspond to the fast target intercept motions.

All included, the full pick-and-place cycle lasts 2.46 s; a peak linear velocity of $1.16\,\mathrm{m\,s^{-1}}$ is reached, whereas the peak linear acceleration is equal to $11.3\,\mathrm{m\,s^{-2}}$. Concerning the workcycle geometry, it can be seen in Figure 6 that it occupies a large portion of the useful workspace. The more dynamic portions of the workcycle are constituted by the target intercept motions, during which the end-effector should quickly reach its destination; within these phases the setpoint tracking accuracy is not paramount, while the attainment of high speed is more important given the need to reduce the cycle time. As will be shown in the following section, the required torques are quite significant during these phases, and briefly reach the saturation levels, which were set to the maximal torque compatible with the power drive systems. Torque saturations were purposefully allowed in order to ensure the full exertion of the system's dynamic capabilities. On the contrary, the conveyor tracking motions are performed at a lower and constant end-effector velocity, compatible with typical industrial applications. In these portions of the workcycle, the manipulator should accurately track the conveyor and guarantee high precision, due to the finer nature of the grasping and release operations. Therefore, the proposed test cycle covers a wide range of situations, and allows the investigation both of highly dynamic working conditions, characterized by high accelerations and velocities, and of the quasi-stationary motions experienced during the conveyor tracking phases.

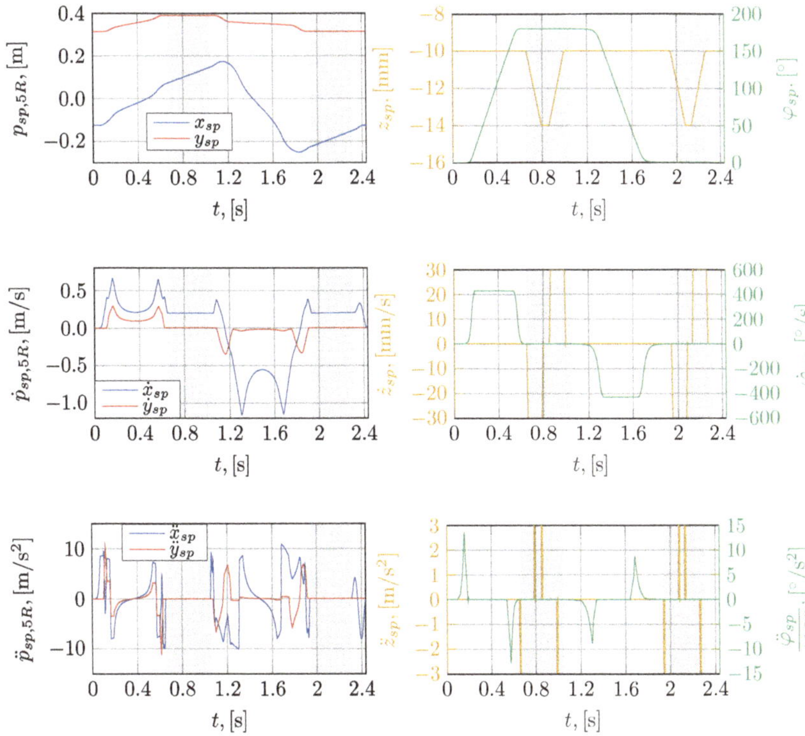

Figure 7. Task space setpoints.

2.4. Experimental Setup

In Figure 8, the experimental setup of the manipulator is shown. In the figure, the following elements are highlighted: the mechanical structure of the robot, with its two subsystems (the planar linkage and the BSS); the actuation systems; the desktop computer on which the control algorithms are executed; and the laptop used for software development. All computations were implemented in Matlab/Simulink 2017a.

In the Figure 9, a schematic of the software and communication architecture is shown. The system consists of the following two macro-components: the Simulink Real-Time (SLRT) target and the EtherCAT field bus with its connected slaves.

The SLRT target is a computer that handles all the control logic of the system, relying on the services provided by the Simulink Real-Time operating system. Figure 9 schematizes the main components of the developed Simulink program, which features the following:

- The state machine, which implements the main operating logic;
- The safety logic subsystem protecting the experimental setup from user or programming errors;
- The EtherCAT communication Master, natively included in Simulink Real-Time, which deals with communication from and to the field devices;
- The implementation of the previously described controllers.

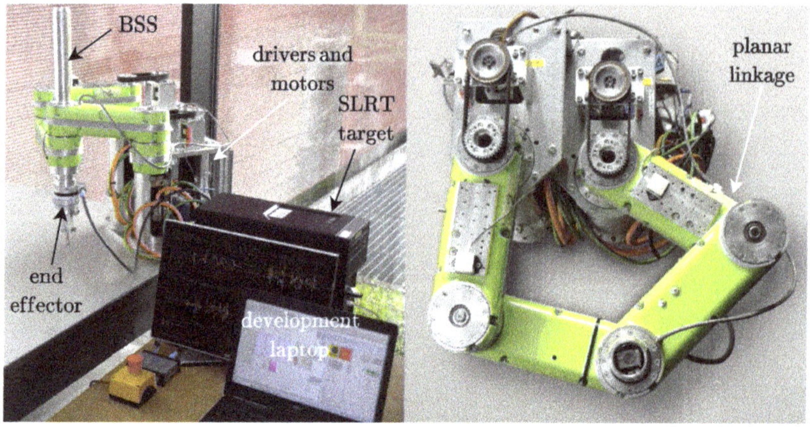

Figure 8. Experimental setup showing the main components of the robot, the SLRT target PC, and the development laptop.

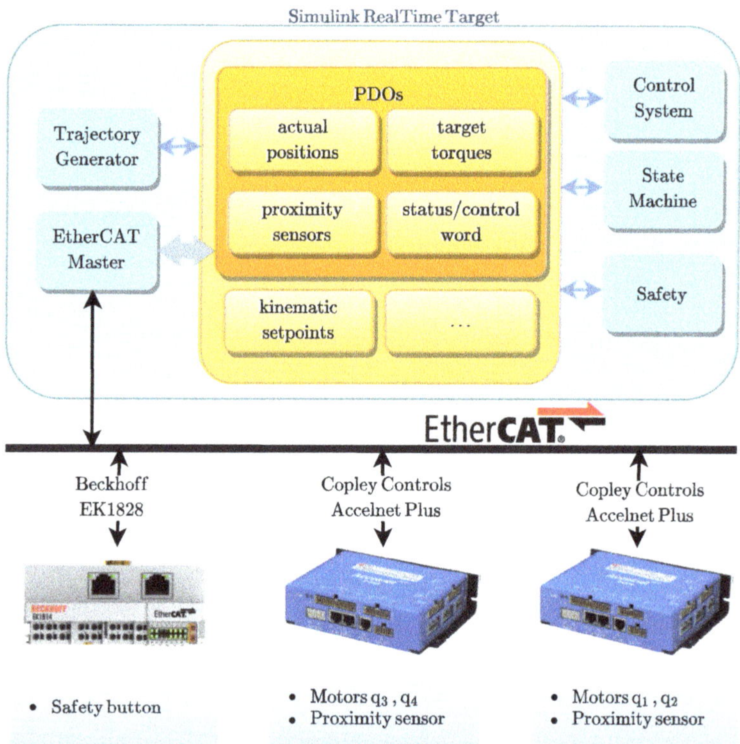

Figure 9. Control system architecture, including the main software functions and the EtherCAT fieldbus.

Figure 9 depicts the complete EtherCAT network needed to exchange data between the EtherCAT master and the field devices (drivers and digital inputs and outputs). The Simulink Real-Time target and the EtherCAT network have been set up to operate at a sampling frequency f_s equal to 4 kHz. The resulting sampling time $T_s = 0.25$ ms determines the deadline that the control systems must meet for all their computations, which, in the case of the TSIDC and TSIIDC, involve the full evaluation of the robot's kinematic and dynamic models.

Two key elements of the experimental setup are the two Accelnet BE2-090-20-R drivers, each capable of controlling two brushless motors using different operating modes and configured as EtherCAT slaves. Since the developed control schemes are all conceived to directly generate the torque setpoints, the drivers have been set up to work in torque mode.

3. Results and Discussion

In this section, the main experimental findings related to the performance of the task space centralized controllers are reported and discussed also through comparison with the joint space PD and PID controllers. Some of the reported results concern the entire workcycle, whereas others focus solely on the higher speed second half of the pick-and-place cycle in order to present an uncluttered overview of the main findings. Whenever this distinction is not immediately evident, it is made explicit in the course of the discussion. The tests were performed through the execution of the workcycle described in Section 2.3, once for each controller. The quantities of interest were logged using the functionalities offered by the Simulink Real-Time operating system, and subsequently post-processed and organized in the graphical form presented in the following discussion. The controllers were tuned using different procedures according to their type. For the TSIDC, the gains are such that the dynamics of the in-plane error components nominally have poles at 30 Hz with damping of 0.7, whereas those of the remaining ones have poles placed at 28 Hz, with a damping of 0.6. The integral gains of the TSIIDC were, on the other hand, experimentally tuned, leaving the proportional and derivative gains unaltered. Figure 10 shows the transfer functions from the torque disturbances to the task space in-plane position errors achieved with the TSID and TSIID controllers; in particular, from Figure 10a it can be observed that the introduction of the integral term improves the rejection of the disturbance's low frequency components.

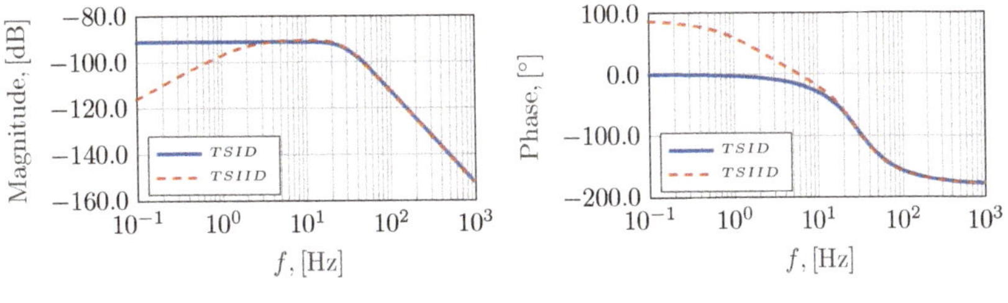

(**a**) Magnitude (**b**) Phase

Figure 10. Transfer functions between torque disturbances and task space error.

The decentralized controllers' parameters, being less amenable to clear interpretation and representation, were selected with ad hoc tuning aiming at the maximization of the regulator's performance. Figure 11 shows the norm of the errors on the horizontal plane, projected along the normal direction of the path and magnified by a factor of 20. The path, whose planar projection is shown in the figure, coincides with the one reported in Figure 6, and described in detail in Section 2.3. Here, it may be remarked that the conveyor tracking phases take place between points (b)–(e) and (f)–(i), whereas the remaining portions of the trajectory are fast intercept motions. Figure 11a in particular shows the errors achieved using the PD and TSID controllers, whereas Figure 11b depicts the errors obtained using the PID and TSIID controllers. It can be seen that the TSID controller generally performs better than the PD regulator, similarly, the TSIIDC outperforms the PID controller. It therefore appears that the introduction of model-based components is beneficial irrespectively of the presence of an integral term. Through the comparison of Figure 11a to Figure 11b it can be concluded that, in general, the introduction of an integral control action leads to significant performance improvements, especially in those phases of the workcycle that are characterized by low accelerations (i.e., the conveyor tracking phases (b)–(e) and (f)–(i), along with the central portions of the intercept motions).

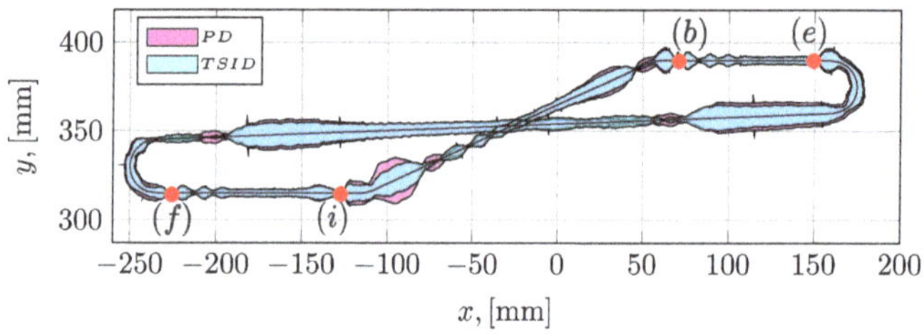

(**a**) PD and TSID controllers.

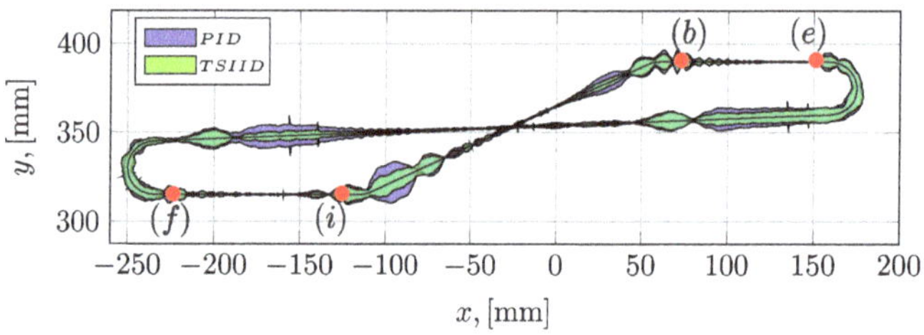

(**b**) PID and TSIID controllers.

Figure 11. Error norms in the xy plane, projected along the path normal, and magnified by a factor of 20.

Figure 12 presents the position errors in the time domain, focusing on the second half of the workcycle; the gray area highlights the tracking phase; the same convention is adopted also for subsequent figures. Figure 12a is clearly consistent with the results already presented in Figure 11, whereas Figure 12b,c confirm the analysis above also in relation to the vertical translation and rotation errors. In particular, in Figure 12b, it can be seen that during the conveyor tracking phases, where the z_{ee} setpoint is not constant, both the PD controller and the TSIDC lead to relatively large errors, with the TSIDC performing better than the PD regulator. On the contrary, the controllers having an integral contribution are better able to bring the error along the vertical direction to zero. The TSIIDC error displays, however, a more oscillating behaviour compared both to the decentralized controllers and to the TSIDC.

In Figure 12c, it can be seen that during the intercept motions, where the end-effector should perform a 180° rotation, the PD regulator and the TSIDC lag behind the setpoint; additionally, they are unable to bring the error to zero also during the ensuing tracking phase, where the end-effector does not rotate anymore. A significantly better behaviour is instead achieved through the introduction of the integral component, with the PID controller and the TSIIDC performing equally well. Figure 13 presents the achieved peak and RMS errors computed along the entire workcycle, for each end-effector coordinate and for each controller. This kind of global representation has not been commonly found in the reviewed literature, which focuses mostly on the errors' time history; however, these synthetic indices are also useful to compare at an aggregate level to the performance of the controllers. Concerning the peak $|\widetilde{x}|_{max}$ and $|\widetilde{y}|_{max}$ errors, an analogous trend can be discerned, namely, the centralized controllers are able to markedly reduce them. Along the x direction, the introduction of an integral term appears to be useful to reduce the RMS error, as the PID and TSIID controllers outperform the remaining two. On the other hand, along the y direction the best performers in terms of the RMS errors are the two centralized controllers. The peak and RMS $\widetilde{z}$ errors seem to suggest that, in aggregate terms, the best performer is the TSID regulator; however, a review of Figure 12b shows that the controllers that also feature an integral term are better able to bring the error to zero in quasi-stationary conditions, occurring, e.g., during the release or grasping phases. As can be seen in Figure 13d, the rotational error $\widetilde{\varphi}$, both in peak and RMS values progressively improves going from the PD to the TSID and then to the PID and TSIID controllers. It is possible to deduce that for this coordinate, the introduction of an integral term is the most important factor as follows: indeed, notwithstanding the fact that the TSID controller significantly outperforms the PD regulator, further non-negligible improvements are achieved in almost equal measure by the PID and TSIID controllers.

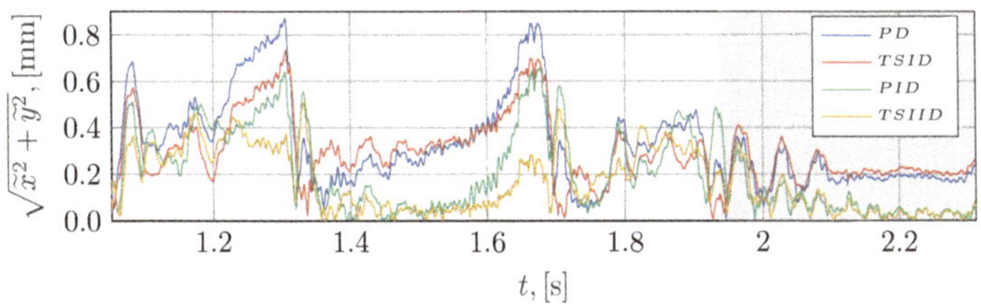

(**a**) Error in the horizontal plane.

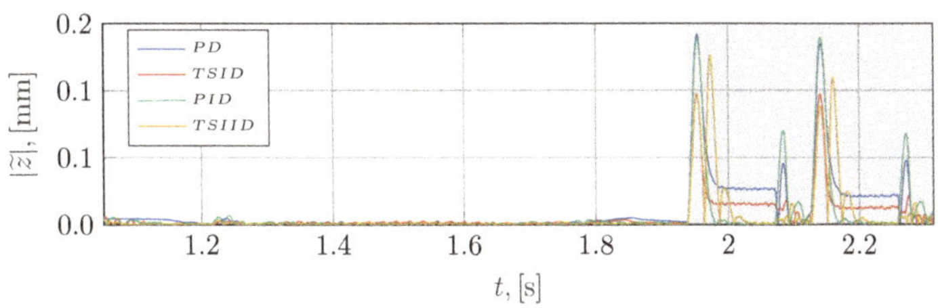

(**b**) Error in the vertical direction.

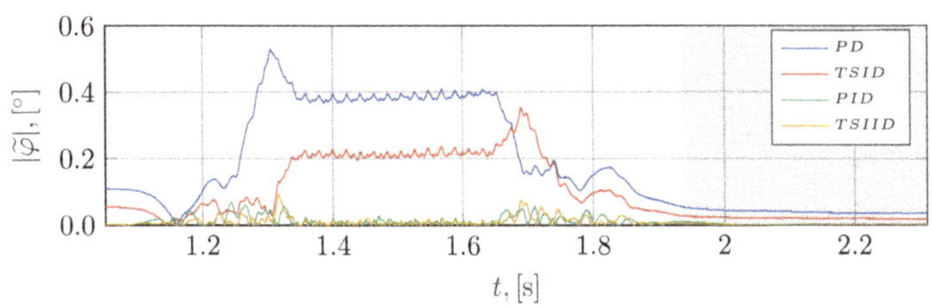

(**c**) End-effector rotation error.

Figure 12. Norm of the in-plane errors; absolute values of the vertical translation and rotation errors.

The articles reviewed in the introduction seldom report a measure of the control torques (or currents), with the greatest focus being only on the achieved errors. On the contrary, in this work, not only the overall control torques but also their constitutive components are highlighted, both instantaneously and at the aggregate level. This is done in order to clearly show how each controller operates, and which components of the control systems are actually relevant during the different phases of the test workcycle. In fact, a peculiarity of the case study presented here is the alternation of highly dynamic motions in which inertial forces dominate and of the almost constant velocity phases during which static and viscous friction is expected to act as a relevant source of disturbance. Figures 14 and 15 illustrate the torque setpoints generated by each controller, respectively, at the motors actuating the joint coordinates q_1, q_2 and q_3, q_4. As already introduced, the possibility of short-term torque saturation was purposely allowed to ensure that the behaviour of the system is investigated up to maximally challenging conditions. Concerning Figure 14, it can be seen that the overall behaviour is very similar for all controllers, with torque saturations also being reached in all cases around times 1.25 s and 1.65 s. A close inspection reveals that for the centralized controllers, the saturations have a slightly lower duration, and also the remaining torque peaks are less pronounced. The saturations tend to be correlated on the one hand with the in-plane error peaks, and on the other with higher setpoint accelerations.

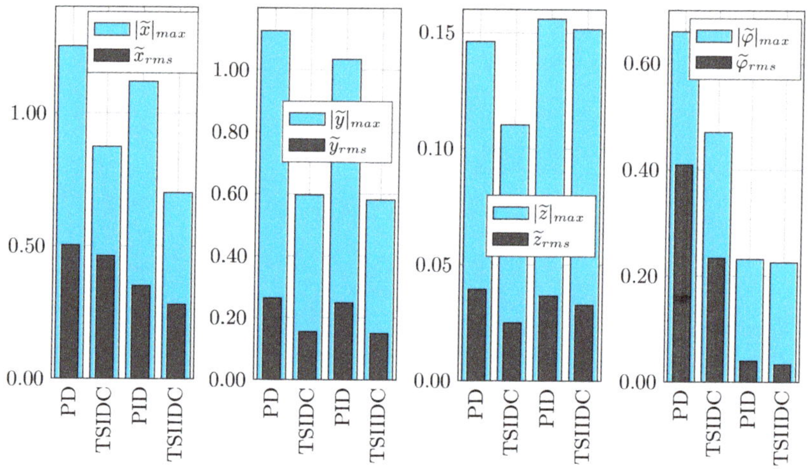

(**a**) $\widetilde{x}$ errors, [mm] (**b**) $\widetilde{y}$ errors, [mm] (**c**) $\widetilde{z}$ errors, [mm] (**d**) $\widetilde{\varphi}$ errors, [°]

Figure 13. Peak and RMS task space errors.

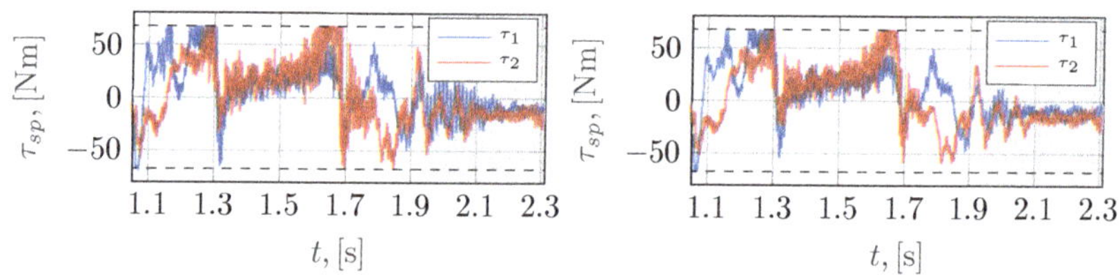

(**a**) PD controller. (**b**) TSID controller.

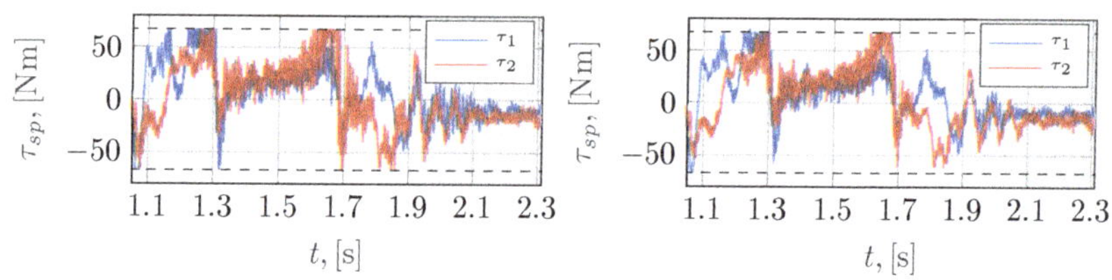

(**c**) PID controller. (**d**) TSIID controller.

Figure 14. Torque setpoints for the 5R, with saturations represented by the dashed lines.

Figure 15 shows that during the target intercept phase, the motors actuating the q_{bss} joints are not under particular strain; conversely, in the conveyor tracking phase, especially along the vertical descent and ascent motions, motor torque τ_3 has a tendency to reach its saturation value. The behaviour of the four regulators is similar in this respect; however, the torque τ_3 generated by the centralized controllers displays more marked oscillations. Figures 16–19 show the several components that contribute to the overall torque setpoint generated by each controller. These torque components are represented without considering the global saturations, since the ability to clearly tell them apart would be otherwise impaired. Comparing Figure 16 to Figure 17, it can be clearly seen that the proportional components are not distributed, in the case of the PD controller, around the zero level. Conversely, in the case of the PID controller, the proportional terms, especially in the portions of the trajectory characterized by less pronounced accelerations, decay to or oscillate around zero. This effect is due to the introduction of the integral actions, which successfully prevent, in the several sub-phases of the trajectory, the permanence of non-null position errors, while the overall torques are similar, they result as already commented in quite different error patterns. No significant differences emerge by the comparison between the derivative components of the PD and PID controllers.

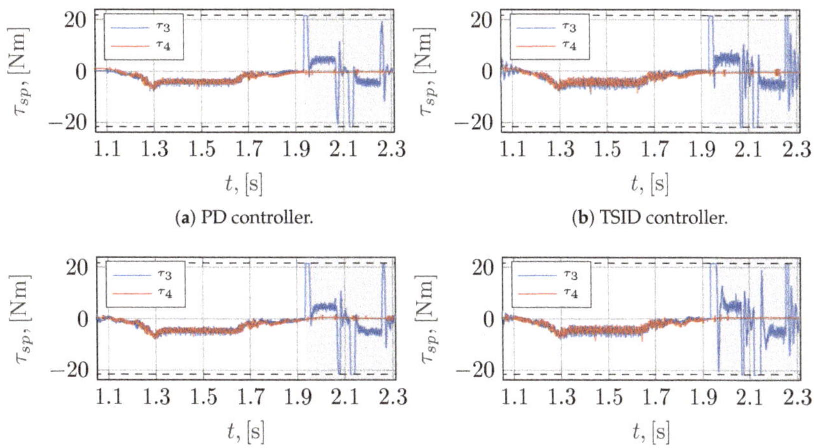

(**a**) PD controller. (**b**) TSID controller.

(**c**) PID controller. (**d**) TSIID controller.

Figure 15. Torque setpoints for the BSS, with saturations represented by the dashed lines.

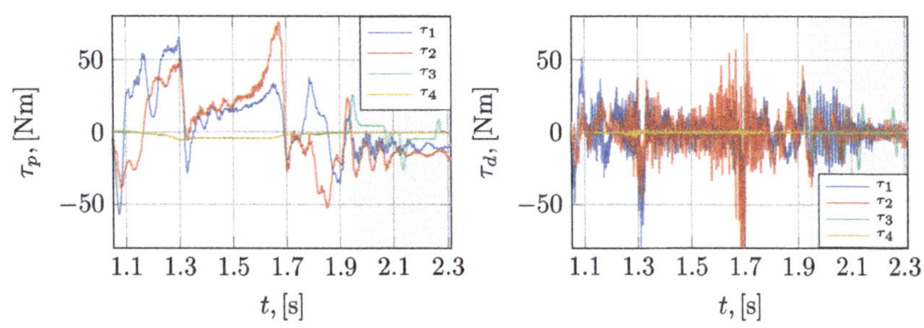

(**a**) Proportional components. (**b**) Derivative components.

Figure 16. PD torque components.

The joint analysis of Figures 18 and 19 shows that, quite obviously, the $\tau_{M_p,j}$ and $\tau_{n_p,j}$ components are, respectively, exactly or practically the same for the TSID and TSIID controllers. The feedback linearization contributions $\tau_{n_p,3}$ and $\tau_{n_p,4}$ to torques $\tau_{sp,3}$ and $\tau_{sp,4}$ are almost null, and due exclusively to the almost negligible gravitational actions. Conversely, $\tau_{n_p,1}$ and $\tau_{n_p,2}$ do contribute to torques $\tau_{sp,1}$ and $\tau_{sp,2}$, but significantly so only in the vicinity of the velocity peaks. Their entity is at any rate lower than the acceleration feedforward components $\tau_{M_p,1}$ and $\tau_{M_p,2}$.

Comparing the proportional components of the PD and TSID controllers, and also observing the acceleration feedforward components $\tau_{M_p,j}$, it is possible to ascertain that the peak values tend to occur synchronously, in the trajectory portions characterized by high accelerations. The acceleration feedforward components of the TSIDC appear to absorb part of the torque from the proportional terms, leading to the reduction in their peak values. As occurred for the PD regulator, also the proportional torque components of the TSIDC do not coalesce around zero, since position errors attributable to unmodelled friction actions persist also during quasi-stationary conditions.

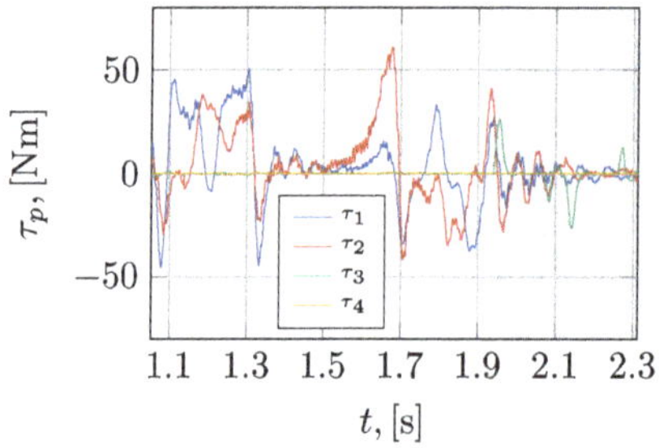

(**a**) Proportional components.

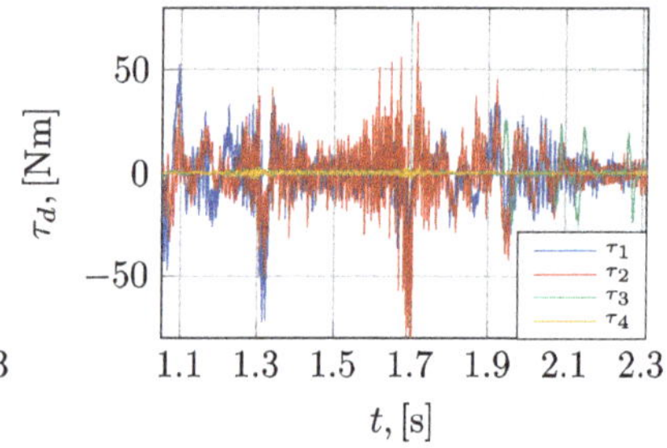

(**b**) Derivative components.

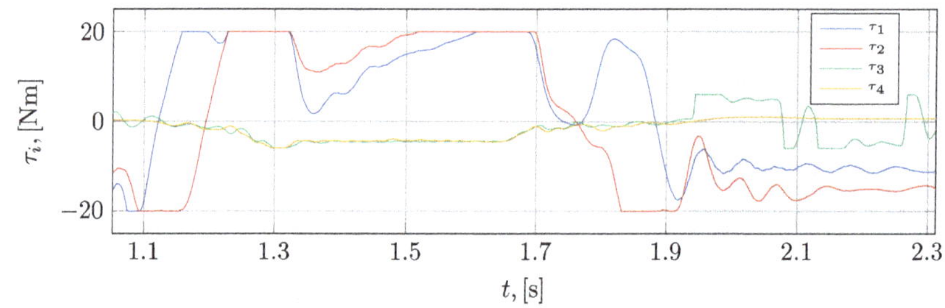

(**c**) Integral components.

Figure 17. PID torque components.

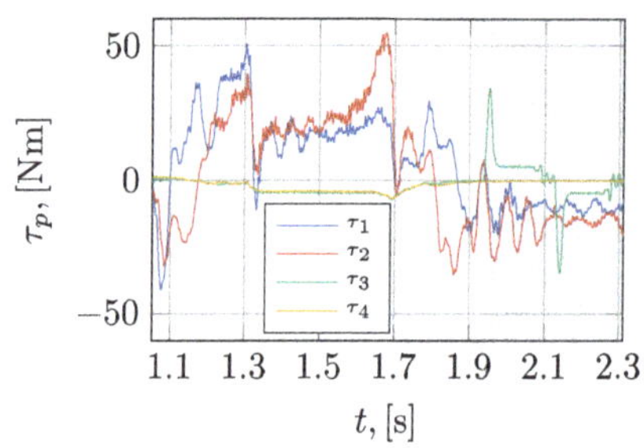

(**a**) Proportional components.

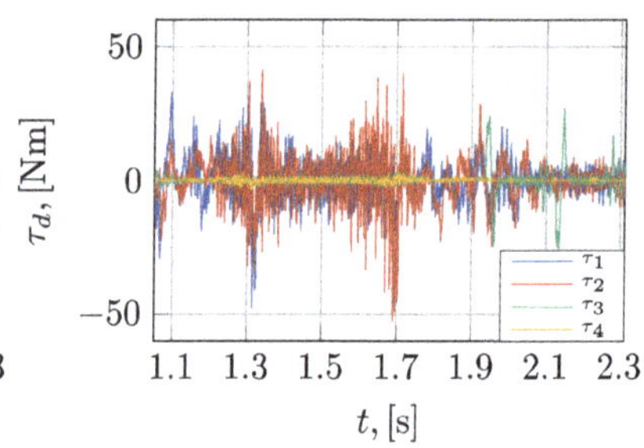

(**b**) Derivative components.

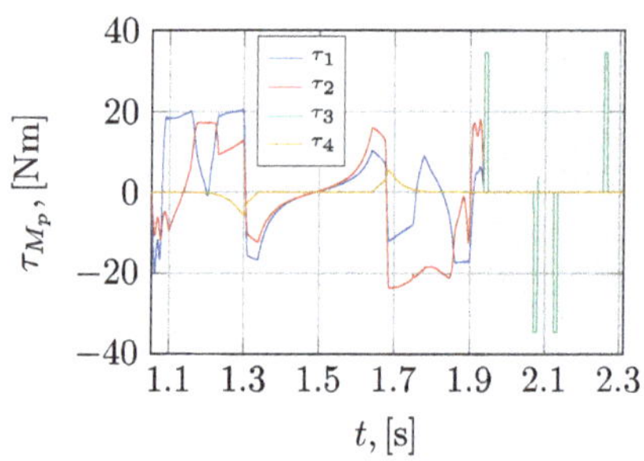

(**c**) Acceleration feedforward components.

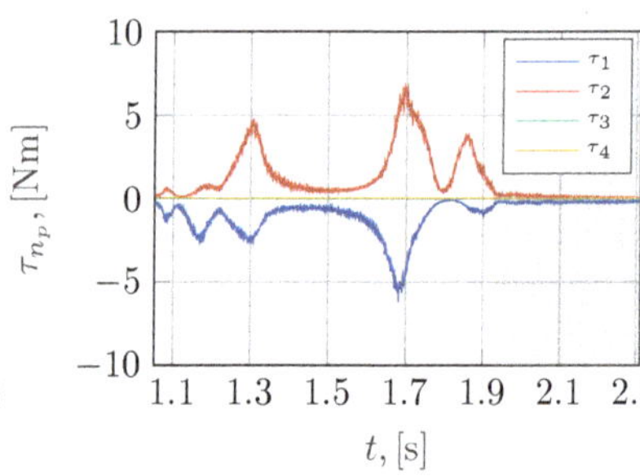

(**d**) Feedback linearization components.

Figure 18. TSIDC torque components.

The introduction of the integral terms in the TSIIDC counteracts this effect, as these contributions absorb the low-frequency components of $\tau_p^{(TSID)}$. The overall supplied torque remains largely unchanged, but the position errors induced by the unmodelled terms that characterize the actual robot's dynamics are reduced. The adoption of centralized control methods finally appears to reduce the peak values of the components based on the velocity errors. Figure 20 shows how the torque setpoints are divided into the different components at an aggregate level, over the course of the entire workcycle. In particular, the RMS values of each torque component were evaluated, and are expressed in the figure as a percentage of their overall sum. In Figure 20a,b, which concern, respectively, $\tau_{sp,1}$ and $\tau_{sp,2}$, almost identical trends can be observed. Going from the PD regulator to the TSID and then to the PID and TSIID controllers, the proportional torque contribution is progressively reduced. The PID and TSIID controllers feature an integral term of roughly equal weight. The feedback linearization terms are of low entity in both the TSIDC and the TSIIDC. Using the PD as the base case, it can be observed that the derivative contributions decrease especially when control centralization is introduced, and conversely, are less affected by additional integral components. Similar considerations can be drawn from Figure 20c. In Figure 20d, which concerns the fourth degree of freedom, it is apparent that the integral contribution is especially important whenever it is introduced. In particular, for both the PID and the TSIID controllers, the proportional and derivative terms have a similarly low relevance.

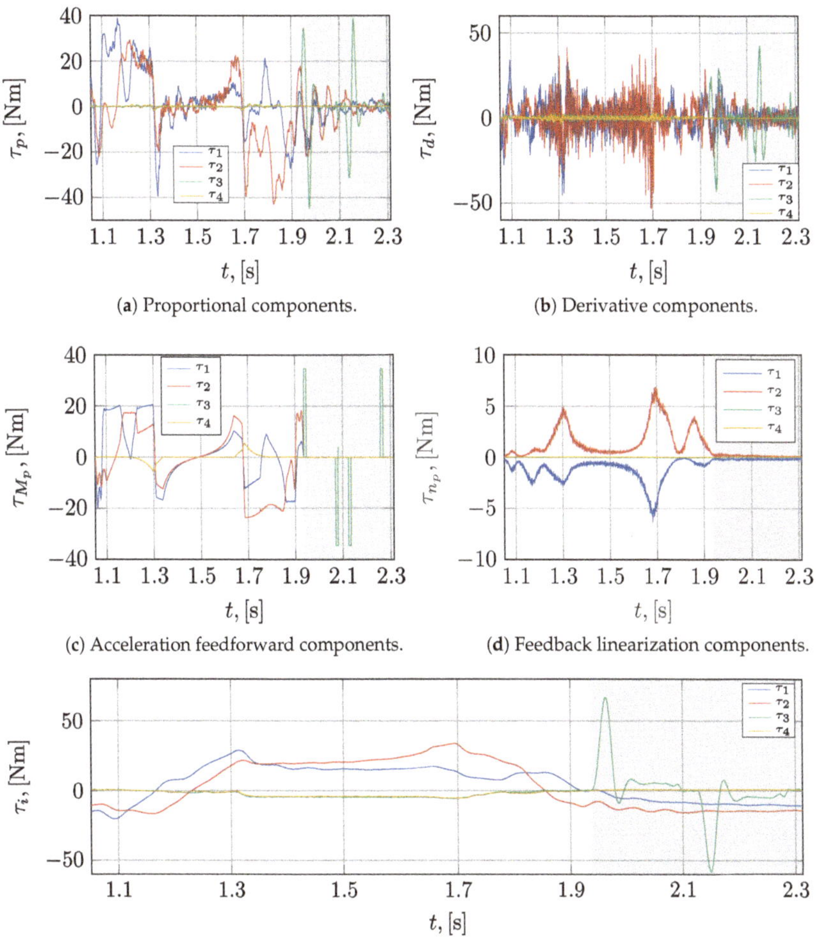

(**a**) Proportional components. (**b**) Derivative components.

(**c**) Acceleration feedforward components. (**d**) Feedback linearization components.

(**e**) Integral components.

Figure 19. TSIIDC torque components.

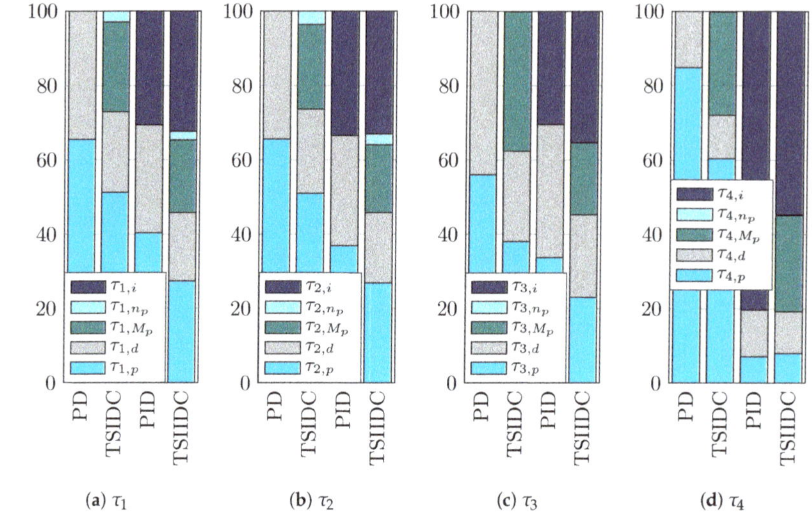

Figure 20. RMS torque percentage contributions.

4. Conclusions

In this work, centralized model-based controllers for a 4-DOF parallel kinematic robot are synthesized using the information derived from the mechatronic design of the entire system. Given the availability of an accurate characterization of the geometric and mass properties of the robot, which are established at design time, it is possible to build kinematic and dynamic models useful for control system development and for dynamic performance maximization. A task space formulation of the inverse dynamics control was first adopted and further refined through the introduction of an integral action term. The computational load stemming from the evaluation of the mathematical model of the robot was observed to be compatible with modern commodity hardware, despite the relatively high sampling frequency of 4 kHz. The behaviour of the centralized controllers was experimentally characterized through the execution of a high-speed pick-and-place cycle representative of the robot's intended use. One peculiarity of this application lies in the wide range of experienced working conditions, which alternate between highly dynamic motions and slower movements in which the accuracy requirements are stringent. Classic PD and PID regulators were also considered as benchmarks.

The TSIDC performed well during the high speed and high acceleration phases, but was unable to bring the errors to zero during the quasi-stationary motions due to disturbances arising from unmodelled friction actions. These were, therefore, counteracted through the introduction of an integral term, yielding the TSIID controller, which was found to be able to operate satisfactorily in both kinds of situation. The PD and PID decentralized regulators were reliably outperformed by their centralized counterparts. Moreover, despite their simple structure, their experimental tuning was found to be time consuming. On the contrary, the synthesis of the TSID and TSIID controllers required an undeniably more complex implementation, which, however, finds a natural collocation within the overall design of the mechatronic system. Once developed, moreover, their free parameters were tuned with lower effort thanks to their clear interpretability.

Based on these considerations, it can be stated that, for applications similar to the one presented in this paper, the use of a centralized controller and, in particular, of a TSIID controller, allows the system to achieve high positioning performances both in high dynamics operations and in semi-stationary conditions. The adoption of centralized controllers in general appears particularly advantageous whenever a high-confidence estimate of the main system parameters is available as a by-product of the device design process.

Moreover, the proposed investigation highlighted the several components that contribute to the control torques. The results show that, in general, the centralized control actions account for a sizable part of the overall generated torque, and are especially beneficial for the reduction in the peak errors; moreover, the integral action introduced within the TSIIDC also has a relevant weight and leads to significantly lower RMS errors. These findings suggest that while the inertial effects have been successfully compensated thanks to the dynamic model of the system, the friction actions are non-negligible terms within the actual mechanical dynamics. The authors therefore plan to specifically investigate this issue, with the overall goal of providing an alternative to the integral torque components in the form of a suitably derived friction model.

Author Contributions: Conceptualization, P.R. and F.C.; methodology, P.R., F.C., J.S. and F.T.; software, J.S., F.T. and F.C.; validation, F.C. and F.T.; investigation, P.R., R.S., F.C. and J.S.; resources, A.G.; writing—original draft preparation, P.R., R.S., F.C., J.S. and F.T.; supervision, P.R. All authors have read and agreed to the published version of the manuscript.

Funding: The research was funded by the University of Bergamo. Fund code: 60RIGH21.

Data Availability Statement: Data are contained within the article.

Acknowledgments: Authors thank the company Mechatronics and Dynamic Devices s.r.l. for the design and the realization of the robot.

Conflicts of Interest: Author Andrea Ginammi was employed by the company Mechatronics and Dynamic Devices s.r.l, Italy. The remaining authors declare that the research was conducted in the absence of any commercial or financial relationships that could be construed as a potential conflict of interest.

References

1. Saied, H.; Chemori, A.; El Rafei, M.; Francis, C.; Pierrot, F. From Non-model-Based to Model-Based Control of PKMs: A Comparative Study. *Mech. Mach. Sci.* **2019**, *58*, 153–169. [CrossRef]
2. Sancak, K.V.; Bayraktaroglu, Z.Y. Nonlinear Computed Torque Control of 6-Dof Parallel Manipulators. *Int. J. Control. Autom. Syst.* **2022**, *20*, 2297–2311. [CrossRef]
3. Shang, W.; Cong, S. Nonlinear computed torque control for a high-speed planar parallel manipulator. *Mechatronics* **2009**, *19*, 987–992. [CrossRef]
4. Vu, M.T.; Alattas, K.A.; Bouteraa, Y.; Rahmani, R.; Fekih, A.; Mobayen, S.; Assawinchaichote, W. Optimized fuzzy enhanced robust control design for a Stewart parallel robot. *Mathematics* **2022**, *10*, 1917. [CrossRef]
5. Tajdari, F.; Ebrahimi Toulkani, N. Implementation and intelligent gain tuning feedback–based optimal torque control of a rotary parallel robot. *J. Vib. Control* **2022**, *28*, 2678–2695. [CrossRef]
6. Hu, H.; Chen, S.; Zhao, J.; Luo, J.; Jia, S.; Zhou, J.; Zhang, J.; Xiong, C.; Zhang, C.; Yang, G. Robust Adaptive Control of a Bimanual 3T1R Parallel Robot with Gray-Box-Model and Prescribed Performance Function. *IEEE/ASME Trans. Mechatron.* **2024**, *29*, 466–475. [CrossRef]
7. Wu, G.; Lin, Z. On the Dynamics and Computed Torque Control of an Asymmetric Four-Limb Parallel Schonflies Motion Generator. *J. Robot. Autom.* **2020**, *4*, 168–178. [CrossRef]
8. Lin, Z.; Cui, C.; Wu, G. Dynamic modeling and torque feedforward based optimal fuzzy pd control of a high-speed parallel manipulator. *J. Robot. Control* **2021**, *2*, 527–538. [CrossRef] [PubMed]
9. Fontes, J.V.D.C.; Colombo, F.T.; Silva, N.B.F.D.; Silva, M.M.D. Model-based joint and task space control strategies for a kinematically redundant parallel manipulator. *Robotica* **2022**, *40*, 1570–1586. [CrossRef]
10. Li, P.; Shu, T.; Xie, W.F.; Tian, W. Dynamic visual servoing of a 6-RSS parallel robot based on optical CMM. *J. Intell. Robot. Syst.* **2021**, *102*, 40. [CrossRef]
11. Yen, P.L.; Lai, C.C. Dynamic modeling and control of a 3-DOF Cartesian parallel manipulator. *Mechatronics* **2009**, *19*, 390–398. [CrossRef]
12. Rodríguez, E.Y.; Mckinley, J.R.; Ramírez, J.V. Calculated Torque Control of a Planar Parallel Robot 2-RR. *J. Auton. Intell.* **2022**, *5*, 72. [CrossRef]
13. Coutinho, A.G.; Hess-Coelho, T.A. Improving the performance of parallel robots by applying distinct hybrid control techniques. *Robotica* **2022**, *40*, 951–975. [CrossRef]
14. Yu, H. Modeling and Control of Hybrid Machine Systems-a Five-bar Mechanism Case. *Int. J. Autom. Comput.* **2006**, *3*, 235–243. [CrossRef]
15. Krishan, G.; Singh, V. Motion control of five bar linkage manipulator using conventional controllers under uncertain conditions. *Int. J. Intell. Syst. Appl.* **2016**, *8*, 34. [CrossRef]

16. Le, T.D.; Kang, H.J.; Suh, Y.S.; Ro, Y.S. An online self-gain tuning method using neural networks for nonlinear PD computed torque controller of a 2-dof parallel manipulator. *Neurocomputing* **2013**, *116*, 53–61. [CrossRef]
17. Salas, F.; Soto, I.; Juarez, R.; Ponce, I.U. A finite-time nonlinear PID set-point controller for a parallel manipulator. In *Advanced Topics on Computer Vision, Control and Robotics in Mechatronics*; Springer: Cham, Switzerland, 2018; pp. 241–264. [CrossRef]
18. Righettini, P.; Strada, R.; Cortinovis, F. Modal kinematic analysis of a parallel kinematic robot with low-stiffness transmissions. *Robotics* **2021**, *10*, 132. [CrossRef]
19. MDQUADRO s.r.l. | Advanced Mechatronic Solutions. Available online: https://www.mdquadro.com (accessed on 14 March 2024).
20. Righettini, P.; Strada, R.; Cortinovis, F. General Procedure for Servo-Axis Design in Multi-Degree-of-Freedom Machinery Subject to Mixed Loads. *Machines* **2022**, *10*, 454. [CrossRef]
21. Righettini, P.; Strada, R.; Cortinovis, F. Neural Network Mapping of Industrial Robots' Task Times for Real-Time Process Optimization. *Robotics* **2023**, *12*, 143. [CrossRef]
22. Bourbonnais, F.; Bigras, P.; Bonev, I.A. Minimum-Time Trajectory Planning and Control of a Pick-and-Place Five-Bar Parallel Robot. *IEEE/ASME Trans. Mechatron.* **2015**, *20*, 740–749. [CrossRef]
23. Liu, X.J.; Wang, J.; Pritschow, G. Kinematics, singularity and workspace of planar 5R symmetrical parallel mechanisms. *Mech. Mach. Theory* **2006**, *41*, 145–169. [CrossRef]
24. Visioli, A.; Adamini, R.; Legnani, G. Adaptive friction compensation for industrial robot control. In Proceedings of the IEEE/ASME International Conference on Advanced Intelligent Mechatronics, AIM, Como, Italy, 8–12 July 2001; Volume 1, pp. 577–582. [CrossRef]
25. Burkan, R.; Mutlu, A. Robust control of robot manipulators with an adaptive fuzzy unmodelled parameter estimation law. *Robotica* **2022**, *40*, 2365–2380. [CrossRef]
26. Shen, X.; Zhou, K.; Yu, R.; Wang, B. Design of Adaptive RBFNN and Computed-torque Control for Manipulator Joint Considering Friction Modeling. *Int. J. Control. Autom. Syst.* **2022**, *20*, 2340–2352. [CrossRef]
27. Wang, Q.; Zhuang, H.; Duan, Z.; Wang, Q. Robust control of uncertain robotic systems: An adaptive friction compensation approach. *Sci. China Technol. Sci.* **2021**, *64*, 1228–1237. [CrossRef]
28. Xu, H.; Li, M.; Lu, C.; Wang, W. Nonlinear sliding mode control of manipulator based on iterative learning algorithm. *J. Electr. Syst.* **2021**, *17*, 421–437.
29. Chen, J.; Chen, L.; Zou, Q. Sliding mode control for manipulator based on fuzzy switching gain adjustment. In Proceedings of the IEEE International Conference on Mechatronics and Automation, ICMA 2019, Tianjin, China, 4–7 August 2019; pp. 822–826. [CrossRef]
30. Vyas, D.R.; Markana, A. Setpoint tracking control using modified Higher Order Sliding Mode Control: Application to robotic manipulator. In Proceedings of the International Conference on Thermal Engineering, Gandhinagar, India, 23–26 February 2019; Volume 2019.
31. Saied, H.; Chemori, A.; El Rafei, M.; Francis, C. A Novel Model-Based Robust Super-Twisting Sliding Mode Control of PKMs: Design and Real-Time Experiments. In Proceedings of the 2021 IEEE/RSJ International Conference on Intelligent Robots and Systems (IROS), Prague, Czech Republic, 27 September–1 October 2021; pp. 8029–8035. [CrossRef]

Author Index

MDPI
St. Alban-Anlage 66
4052 Basel
Switzerland
www.mdpi.com

Robotics Editorial Office
E-mail: robotics@mdpi.com
www.mdpi.com/journal/robotics

www.ingramcontent.com/pod-product-compliance
Lightning Source LLC
Chambersburg PA
CBHW061558210926
51850CB00009B/2140
* 9 7 8 3 7 2 5 8 1 2 0 7 3 *